Springer Tracts in Civil Engineering

Springer Tracts in Civil Engineering (STCE) publishes the latest developments in Civil Engineering - quickly, informally and in top quality. The series scope includes monographs, professional books, graduate textbooks and edited volumes, as well as outstanding PhD theses. Its goal is to cover all the main branches of civil engineering, both theoretical and applied, including:

- Construction and Structural Mechanics
- Building Materials
- Concrete, Steel and Timber Structures
- Geotechnical Engineering
- Earthquake Engineering
- Coastal Engineering; Ocean and Offshore Engineering
- Hydraulics, Hydrology and Water Resources Engineering
- Environmental Engineering and Sustainability
- Structural Health and Monitoring
- Surveying and Geographical Information Systems
- Heating, Ventilation and Air Conditioning (HVAC)
- Transportation and Traffic
- Risk Analysis
- Safety and Security

Indexed by Scopus

To submit a proposal or request further information, please contact:
Pierpaolo Riva at Pierpaolo.Riva@springer.com (Europe and Americas) Wayne Hu at wayne.hu@springer.com (China)

Wei Huang · Jian Xu

Optimized Engineering Vibration Isolation, Absorption and Control

Springer

Wei Huang
SINOMACH (China)
Beijing, China

Jian Xu
SINOMACH (China)
Beijing, China

ISSN 2366-259X ISSN 2366-2603 (electronic)
Springer Tracts in Civil Engineering
ISBN 978-981-99-2212-3 ISBN 978-981-99-2213-0 (eBook)
https://doi.org/10.1007/978-981-99-2213-0

This Springer imprint is published by the registered company Springer Nature Singapore Pte Ltd.
The registered company address is: 152 Beach Road, #21-01/04 Gateway East, Singapore 189721, Singapore

Preface

Nowadays, the engineering vibration control technology has been applied widely with the rapid development of China's industry and great progress in science and technology. This technology has played an important role in the vibration control of large power equipment, ultra-precision equipment, and building structures. Moreover, this is a key support technology in the fields of machinery, electronics, electricity, metallurgy, weapons, and aviation.

Improper vibration control will negatively affect the normal operation and service life of the equipment, the normal measurement of instruments and meters, the health of staff and residents nearby, and even the service life and safety of nearby industrial buildings.

With the continuous improvement of equipment, the vibration load, the frequency range, and the demand for low-frequency micro-vibration control of the equipment increased. However, traditional passive vibration isolation cannot fully adapt to the increased working frequency band. This is due to the fact that the damping and stiffness characteristics are not adjustable once the vibration isolation parameters are determined.

Active vibration control has the advantages of large output, good control effect, and the ability to continuously adjust the control output according to the change of excitation. Notably, the design and parameter optimization of the control system has an important impact on the control effect. The emergence of smart materials and smart dampers has promoted the rapid development of semi-active control technology, such as magnetorheological (MR) and electrorheological (ER). The control algorithm and the target design based on active control effect are important research concerns.

This book focuses on the system and parameter optimization of vibration isolation, absorption, active and semi-active control of engineering vibration, as well as the optimal arrangement of sensors. The research in this book is an attempt to develop the advanced technology of engineering vibration control and, hopefully, plays a certain guiding role for practical engineering.

We expect that the researchers and practitioners can explore engineering vibration control technology and discipline development from different perspectives and viewpoints.

Beijing, China

Jian Xu
A Member of Chinese Academy
of Engineering; The chief scientist
of SINOMACH

Acknowledgements This book was completely supported by the Youth Fund of SINOMACH Academy of Science and Technology Co. Ltd., *Research and application of key technologies for multi-source vibration control, operation and maintenance monitoring of laboratory cluster in high-rise building*, and it was also fully supported by the Key Youth Fund of SINOMACH, *Research and application of key technologies for micro-nano environmental vibration control of chinese major science and technology infrastructures.*

Contents

About the Authors

Wei Huang, Ph.D. is a registered senior structural engineer. He was born in Hanshan County, Anhui Province, China, in 1988. He graduated from Hefei University of Technology with a bachelor's degree in civil engineering in 2009. From 2009 to 2013, he served as a lecturer at Hefei University of Technology and began his postgraduate and doctoral studies at the same university in 2011. In 2016, he obtained his doctoral degree in structural engineering. He joined China National Machinery Industry Group Co., Ltd. in 2016, mainly engaged in engineering projects and scientific research on engineering vibration control. Besides, he worked as a postdoctoral researcher at the mobile station of Tsinghua University in 2018 to conduct studies related to numerical simulation of practical micro-vibration control and left the station in 2020. His research interests include vibration control and performance optimization of industrial equipment, engineering structures, etc. He has contributed to many projects funded by various agencies, such as the National Natural Science Foundation of China, the National Key R&D Program, and the major projects of the China Machinery Industry Group. He is the "youth high potential" of SINOMACH and the deputy secretary general of the Construction Vibration Professional Committee of China Engineering Construction Standardization Association.

Jian Xu is a member of Chinese Academy of Engineering. He received his postgraduate degree from Hunan University in 1988. He was promoted to a professor-level senior engineer by the Design and Research Institute of the Ministry of Machinery Industry in 1994, and he served as the dean there from 1998 to 2001. He formerly served as the general manager and the chief engineer of China National Machinery Industry Corporation (SINOMACH). He is a young and middle-aged expert with outstanding contributions in China, and he is a directly contacted expert by the Organization Department of the Central Committee of the CPC. He was elected as the first and second levels of the national BaiQianWan Talents Program and enjoyed the Special Government Allowances of the State Council. He is a national first-class registered structural engineer, a national registered consulting engineer, and a national first-class construction project manager.

He has been engaged in engineering vibration control and seismic vibration control of industrial buildings for a long time. He has prepared 25 national and professional standards as a chief editor or a participant and has published 28 professional books and more than 60 academic papers.

He has won 2 Second-Class Prizes for STP of China and 4 First-Class Prizes for STP at the provincial and ministerial levels.

He serves as a part-time vice chairman of the China Association for Engineering Construction Standardization, a part-time vice chairman of the China Exploration and Design Association, and a part-time vice chairman of the China National Association of Engineering Consultants.

Chapter 1
Introduction

Abstract In this chapter, the background, purpose, and significance of the studies presented in this book are carried out, and the importance of vibration control for two types of industrial equipment, i.e., power equipment and sensitive equipment is indicated. Besides, single-stage and double-stage isolation system with passive, active and semi-active control strategies are studied by reviewing lots of literature, and some advanced vibration control techniques and artificial intelligence methods are studied, such as magnetorheological damper, dynamic vibration absorber, particle swarm optimization etc.

1.1 Background, Purpose, and Significance

With the rapid development of modern industrial engineering, vibration control technology plays a crucial role in engineering construction. For instance, if the vibration risks are not adequately eliminated, they will affect the equipment's regular operation and service life, the standard measurement of instruments and meters, the physical health of workers and nearby residents, and even the safety of the nearby industrial buildings.

The equipment commonly used in industrial engineering can be divided into power equipment and vibration-sensitive equipment. The power equipment includes large-scale slewing, reciprocating, impact, and random vibration devices. These equipment have played an essential role in the national economy and defense construction. However, the vibration generated during their use causes damage to the equipment itself and harms the operators, industrial buildings, and the surrounding environment (Fig. 1.1).

On the other hand, vibration-sensitive equipment includes high-precision microscopy, optical interference detection, biochemical analyzers, precision grinding, and processing machine tools. These equipment are distributed in astronomical optics, military engineering, rapid detection, nanotechnology, laser devices, ultra-thin metals, grating scribing, and other fields. During their use, deviations and incorrect results are caused by minimal environmental disturbances (Fig. 1.2).

W. Huang and J. Xu, *Optimized Engineering Vibration Isolation, Absorption and Control*, Springer Tracts in Civil Engineering, https://doi.org/10.1007/978-981-99-2213-0_1

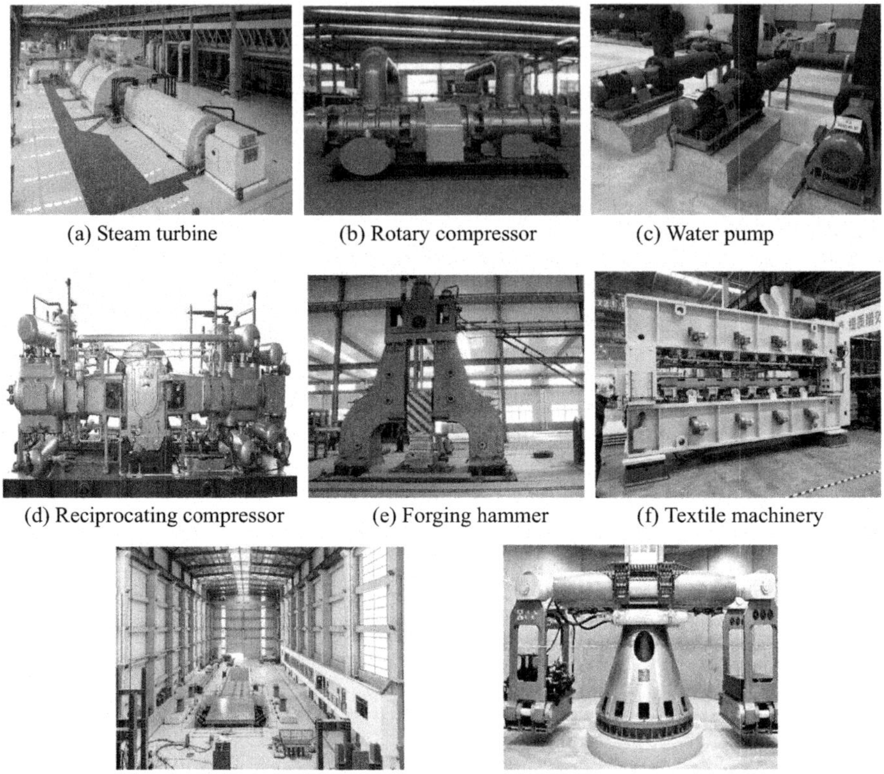

(a) Steam turbine (b) Rotary compressor (c) Water pump

(d) Reciprocating compressor (e) Forging hammer (f) Textile machinery

(g) Shaking table (h) Centrifuge

Fig. 1.1 Typical modern power equipment

There is still a particular gap between China's vibration control technology and the advanced international level, which restricts the rapid and high-quality development of China's industry to a certain extent. The painful lesson of the 'ZTE incident' is still remembered. The media has made sharp judgments about this event, 'Powerful nations still need to be hard on their own, and wars without smoke will eventually require scientific and technological researchers to develop high-end core technologies… today's core technology has been blocked by foreign swords … ZTE's tragedy will be staged if we still do not recognize the support of advanced technology …'. In fact, high-quality chips are closely related to micro-vibration control technologies, which are crucial links made from the cultivation of crystals. If the environmental vibration exceeds the standard, crystal damage is typically caused, significantly reducing the chip's quality. This is the epitome of the critical role of vibration protection technologies in modern industrial engineering.

Effective vibration isolation and control measures for power equipment can reduce its vibrations and decrease their adverse impact on the surrounding environment. For

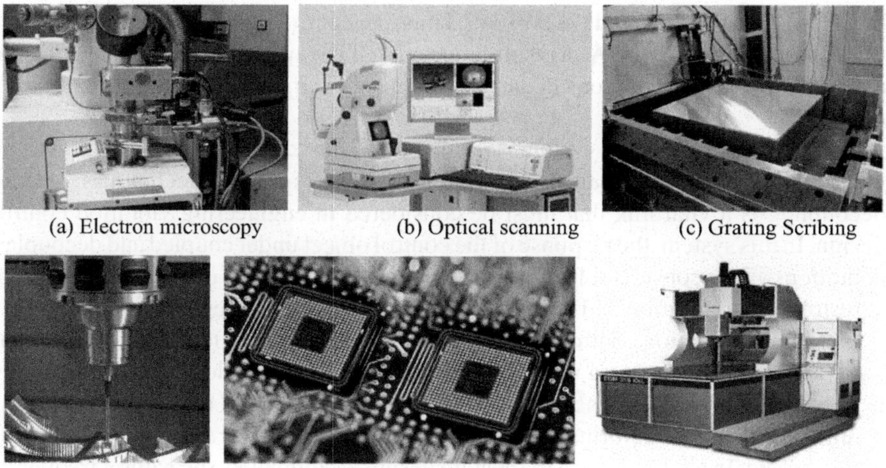

(a) Electron microscopy (b) Optical scanning (c) Grating Scribing

(d) Probe (e) IC manufacturing (f) Precision machine tools

Fig. 1.2 Typical modern sensitive equipment

vibration-sensitive equipment, vibration suppression can effectively lower the impact of ambient vibration on the equipment and ensure the regular use of the equipment.

Passive isolation is simple, easily implemented, and widely used system in vibration control engineering. However, its parameters are fixed, and it cannot actively adjust to changes in external interference once the system is designed (such as changes in frequency and excitation form). As a result, active control systems are typically considered. In such devices, the response of the controlled object is obtained in real-time and sent to the active controller. Thereafter, the instructions are sent simultaneously to drive the actuator by which active control force is generated to counteract severe vibrations in real-time. Hence, its control effect can be adjusted autonomously according to the changes in external interference.

In recent years, scholars have investigated system improvement, vibration isolation parameters optimization, and advanced algorithms development for hybrid passive isolation and active control systems. However, there is still extra room for theoretical and applied studies focusing on advanced control, especially those aiming to improve the effectiveness of active control technology. Besides, there are still gaps in the implementation and integration of such systems in actual engineering applications. Currently, the finite element method (FEM) is considered as a vital technique in modern engineering vibration control. A key issue herein is to improve the computational accuracy of engineering vibration control and reliably approximate the theoretical calculation as much as possible. If the theoretical analysis and numerical calculation results significantly differ, it is difficult for research and development personnel and engineers to choose reliable results. These problems cause critical issues in using passive and active control systems. The current research shows that most of the problems are based on simplified calculations using

MATLAB/SIMULINK and LABVIEW. Thus, the control object's actual characteristics cannot be considered. As a result, improving the way to perform detailed passive and active control calculations in the finite element environment is an urgent research direction.

Indeed, passive control system optimization is an effective way to improve the control effect and performance. Among the currently available techniques, system decoupling is a vital link that must be considered in engineering vibration control design. In this system, the response of the control object under coupled and decoupled conditions varies considerably. The most obvious disadvantage of the coupled system is restraining translation and torsion, which significantly causes adverse effects.

For micro-vibration control, the vibration source hazards often have low frequencies. Therefore, designing a vibration isolation system with a low natural frequency is necessary. A low frequency for a steel spring damper with quasi-linear stiffness results in substantial deformation, and the actual engineering application is a challenging issue. In addition, once the system design is completed, the natural frequency changes with the change in the load mass. On the other hand, introducing air springs into the micro-vibration control has gradually overcome the problems involved in low-frequency design and large static deformation. Alternatively, these issues can be overcome by ensuring a constant chamber's effective height. A dual-chamber air spring is typically added with a specific additional air chamber volume for a single chamber. The upper and lower air chambers are usually connected through orifices, which effectively address the load-dependent natural frequency changes. Accordingly, the air spring floating system is suitable for medium and large loads. In contrast, a quasi-zero stiffness system can be considered for low-frequency micro-vibration control of small loads, which is a more forward-looking research direction.

1.2 Literature Research and Review

Figure 1.3a and b describe passive vibration isolation for power and sensitive equipment, which are two different single-stage vibration isolation systems. This passive isolation strategy is an early researched and applied method that has a simple structure and convenient design [1–4]. However, single-stage systems are often not ideal in the frequency domain, and two-stage systems are required [5]. Figure 1.4 shows the corresponding model. The dynamic characteristics of the two-stage vibration isolation system based on the minimum mean method have been previously analyzed [6]. Wei et al. [5] proposed a hybrid method for parameter optimization design of a two-stage vibration isolation system based on a genetic algorithm (GA) [7] using the principle of maximum entropy optimization. Farshidianfar et al. [8] concluded that calculating the vibration isolation of equipment using a two-stage vibration isolation system is necessary and effective.

The above passive vibration isolation system does not consider external energy input and does not rely on other automatic control systems. Once the system is designed, its structural parameters and damping and stiffness characteristics are

Fig. 1.3 Uncontrolled vibration isolation systems of two devices (single-stage)

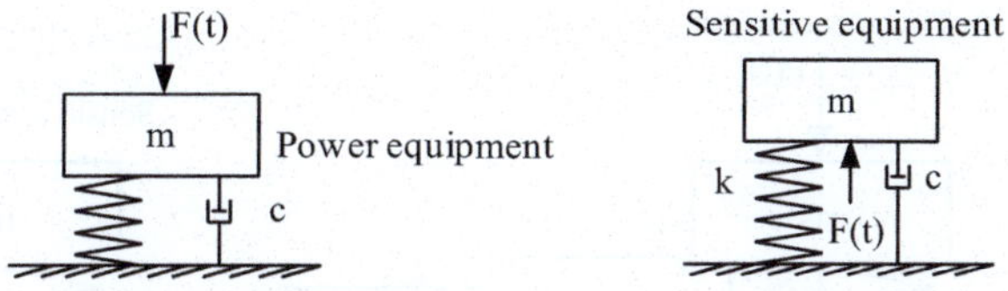

(a) power equipment (b) sensitive equipment

Fig. 1.4 Uncontrolled vibration isolation systems of two devices (double-stage)

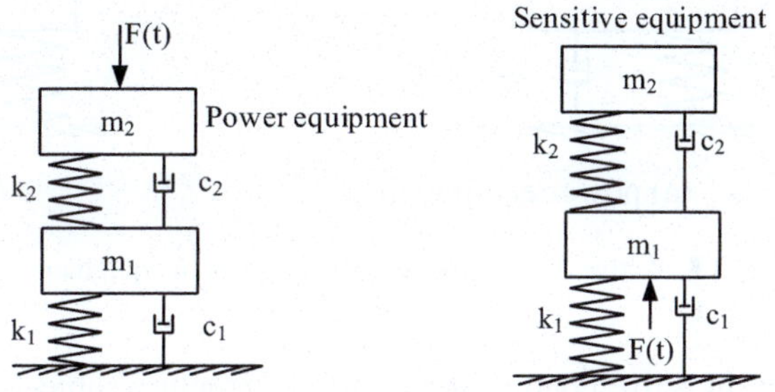

(a) power equipment (b) sensitive equipment

fixed. Besides, it cannot fully adapt to a wider operating frequency band. Certain limitations exist in passive control systems, such as not being conducive to vibration isolation design under low-frequency excitation and often failing to reach the expected vibration isolation effect. It cannot adaptively adjust to external interference, such as changes in amplitude, frequency, or excitation [9–12]. At this time, active control with energy input must be considered. Figure 1.5 depicts the active control for power and vibration-sensitive equipment with external controllers. This model was proposed by Farshidianfar [8] in 2012. Modern control methods, including proportion-integral-differential (PID) control [13], linear quadratic optimal control (such as the linear quadratic regulator, LQR [14], linear quadratic gauss, and LQG [15]), and H_2/H_∞ [16] control. The above control methods require establishing accurate computational models (such as transfer function or state space models). However, this method has some limitations when there is uncertainty in the mass, stiffness, or damping of the vibration system or when the system has strong nonlinearity [17]. For this reason, many scholars have carried out research on intelligent control methods, such as fuzzy logic control (FLC) [18], neural network control (NNC) [19], and fuzzy neural network control (FNNC) [20]. The above control methods have been widely used in vibration control fields such as vehicle vibration reduction, structural earthquake resistance, and structural wind resistance. Taghirad [21] carried out active control on the vehicle vibration reduction system based on the LQR/LQG control method. Kar et al. [22] conducted active control for thin plate structures using the H_∞ control method and developed a feedback controller to stabilize the control system. Pourzeynali et al. [23] studied the performance of the FLC control method

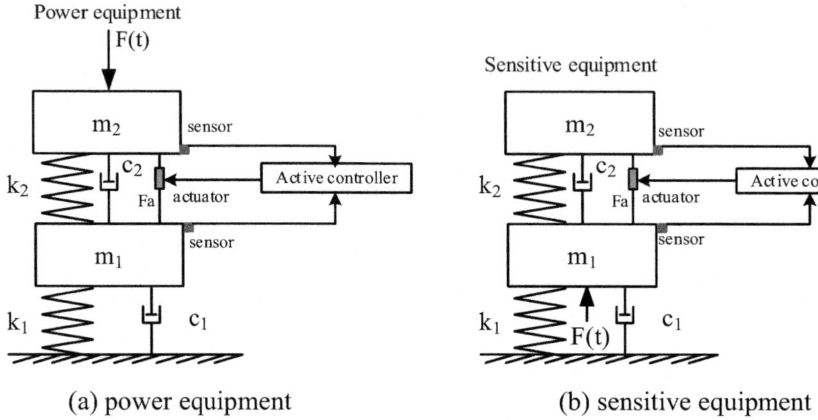

Fig. 1.5 Active control systems of two types of equipment

as an active control of a high-rise structure subjected to wind vibration. Nevertheless, these methods are rarely used for power and vibration-sensitive equipment in the active control field. Moreover, the literature still lacks research that forms a complete active control system for both types of equipment. Therefore, this method is mainly applied to the active control of typical equipment and is significant for engineering needs.

In the active control strategy, the actuator output is crucial to achieving a good control effect. However, it still has some disadvantages, such as complicated sensor/actuator system design, troublesome vibration data collection and processing approaches, large control energy consumption, and adverse economic effects. In addition, active control systems often have a time lag phenomenon. When the time lag is large, it may reduce the vibration control effect and diverges the system response [17, 24, 25]. For this reason, scholars have proposed a method between uncontrolled vibration isolation and active control known as semi-active control. This method requires only a small amount of energy to maintain the regular operation of the relevant electronic and electrical components. In these systems, the external power provides direct control, and the need for devices that induce the control forces and energy to support active control is eliminated [26]. From the international researchers' perspective, semi-active control mainly includes semi-active variable stiffness control and semi-active variable damping control. The semi-active variable stiffness control performs the calculations according to the preset control law and output control instructions and sends them to the mechanical device to finally apply the control strategy to the controlled object, as shown in Fig. 1.6 [27, 28]. On the other hand, the semi-active variable damping control is generally based on a hydraulic damper or a viscous fluid damper and a servo to form a damper with an adjustable fluid flow. It can continuously change the damping force and control a wide range of exciting vibration capabilities, as shown in Fig. 1.7 [29–31].

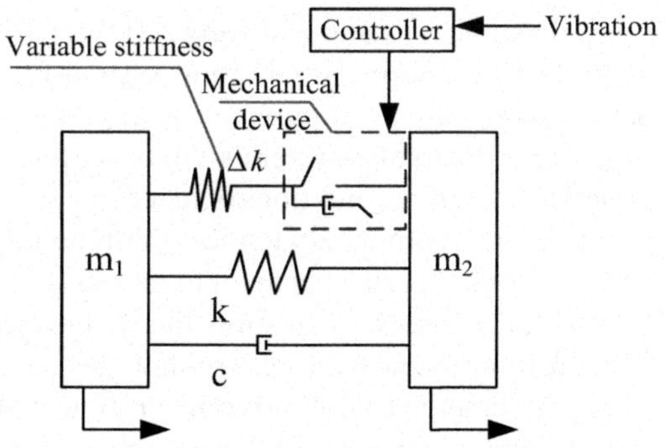

Fig. 1.6 Semi-active variable stiffness control

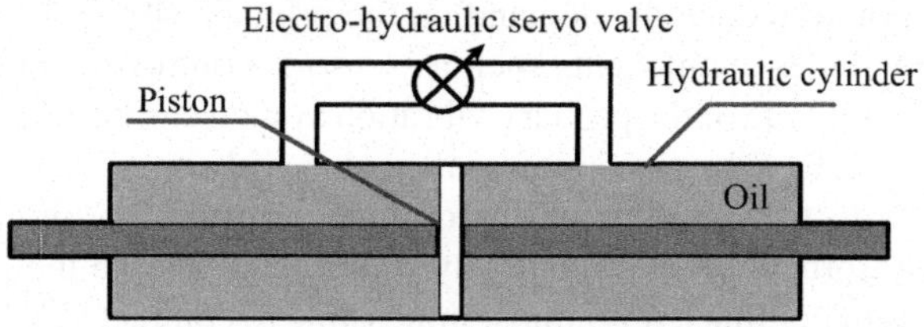

Fig. 1.7 Semi-active variable damping control

Recently, the traditional semi-active control technology has been greatly improved and promoted with the advancement of smart materials and dampers. Electro-rheological damper (ERD) is a new damper type that uses electro-rheological fluid (ERF) smart materials. As a result, its damping viscosity can change with the applied electric field strength. When lacking an electric field, the ER fluid flows freely. Once the applied electric field reaches a certain value, the ER fluid instantly gets into a gel state, with the response changing in milliseconds and reversible [32, 33]. Wang et al. [34] utilized the ERD device in building structures. Furthermore, Choi [35] applied the ERD system to study the semi-active control of fixed beam structures. Shortly after the invention of ERD, scientists discovered the magnetorheological fluid (MRF) and invented the magnetorheological damper (MRD). Compared to the ERF, MRF offers significant advantages in driving the ERF voltage substantially, up to several thousand volts, while MRF achieves a few volts to tens of volts [17]. Additionally, the MRF shear strength is much greater than ERF. Therefore, the volume of MRF in the damper is generally 100–1000 times smaller than ERF. Besides, the MRF is not sensitive to impurities in the body, and the temperature adaptation range is wider than the ERF. Therefore, in recent years, many scholars have utilized the MRD in semi-active control tasks. Yao et al. [36] applied MRD to semi-active control of the vehicle vibration reduction system and analyzed it with the Bouc-Wen model. Dyke et al. [37] established an MRD semi-active control system for structural seismic control based on sliding mode control. However, the current application of MRD in the semi-active control of power and sensitive equipment is scarce, which requires intensive research and exploration.

A dynamic vibration absorber (DVA) is a mass-spring-damping mechanism attached to the primary vibration system [38, 39]. In practical engineering applications, the passive vibration absorber exerts a better energy dissipation effect when the primary vibration system's vibration frequency remains unchanged. Due to the complex operating conditions of mechanical equipment and the changeable operating environment, it is not easy to meet this requirement in practical applications [40, 41]. In order to simulate the behavior of vibration absorbers, many scholars have improved traditional devices by adjusting and optimizing their essential parameters. With the introduction of new theories, technologies, and materials, some new research directions have emerged in vibration absorption, mainly including active vibration absorption technology, adaptive vibration absorption technology, and nonlinear vibration absorption technology [42–44]. An active vibration absorber adopts active intervention to reduce the demand on the primary vibration system, and its interior mainly includes components such as sensors, controllers, and actuators. Theoretically, the amplitude of the primary vibration system can be zero when the actuator force on the primary vibration system is equal to and opposite to the excitation force it receives. With the development of optimal control algorithms, studies on active vibration absorbers have dramatically progressed. Due to the lag between the feedback and actuator links, it is almost impossible to completely offset the force generated by the actuator and the exciting force in an ideal state. The primary vibration system may be unstable, especially for complex vibration systems. Therefore, active dynamic vibration absorber is mainly used in theoretical analysis, numerical simulation, and laboratory research stages, while it has relatively few actual engineering applications (Fig. 1.8).

Industrial equipment and civil structures are inseparable. Taking into account the vibration of equipment and structures as a composite system has important practical significance Yang et al. [45] studied the composite vibration system of structure and equipment under seismic excitation, Igusa et al. [46] modeled the composite vibration system of equipment and structure as a two-degree-of-freedom system to

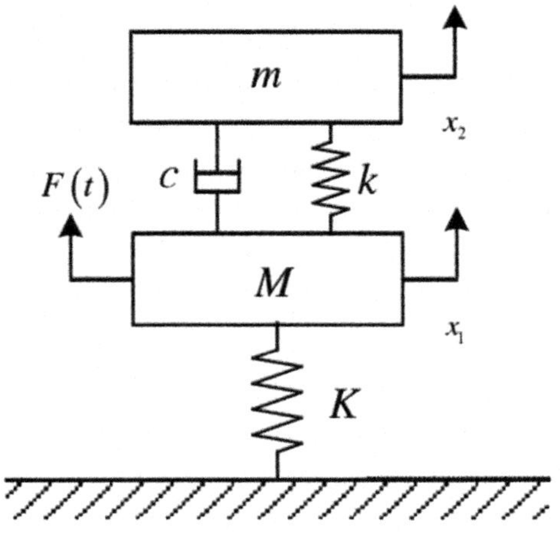

Fig. 1.8 Mechanical model of the dynamic absorber

perform a seismic control investigation. Xu et al. [47] assessed the vibration control effect of precision equipment attached to frame structure under seismic vibration. Ismail et al. [48] proposed a new vibration isolator to study the disturbance of equipment in the structure under seismic load. Murnal et al. [49] introduced a variable frequency pendulum vibration isolator to study the damping effect of equipment in structures under seismic loads. Currently, the literature lacks extensive research on the engineering vibration of structures caused by power equipment. Besides, seismic input shaking intensity is primarily considered in the vibration input, whereas other interference forms affecting sensitive equipment and vibration control methods are rarely taken into account. Generally, the equipment and structure are considered a two-degree-of-freedom system for calculation. Therefore, research on the composite system of equipment and building structures has great practical significance for the vibration control of modern industrial engineering.

High-tech equipment used in producing semiconductors and optical microscopes is very expensive. In order to ensure the high quality of ultra-precision products, high-tech equipment used to manufacture these products requires a normal working environment with minimal vibration intensities. Accordingly, the top priority herein is to find an effective way to ensure that the functions of high-tech equipment are not affected by the micro-vibration of the building structure. The main sources of micro-vibration that affect the normal operation of high-tech equipment are ground motion caused by traffic, ground vibration caused by machinery, and direct interference [50]. Previously, many researchers have studied the measurement and prediction of ground motion caused by traffic and vibration caused by machinery [50–52]. The main frequency range of building structure floor vibration caused by machinery mainly depends on the machinery's rotation speed and the characteristics of the beams and slabs in the building.

Power equipment placed upstairs in structure is inevitable for modern industrial and social development. Power equipment station rooms, air conditioners, range hoods, and power pipes cause excessive building vibration, severely reduce the comfort of personnel, induce operational failure in the building's precision equipment, and even cause structural damage (Fig. 1.9).

Sensor deployment is widely used in modern industrial engineering structures. The main concept herein is to arrange a certain number of sensors in a limited two-dimensional plane or three-dimensional space structure to adequately cover the entire monitored area. This problem in engineering is called optimal sensors deployment (OSD). The OSD technique is widely used in large civil engineering structures' health monitoring and data collection [53–58]. There are many types of sensor deployment strategies, including modal kinetic energy (MKE) [59], MinMAC [60], QR decomposition [61], and probabilistic sensing models [62]. For a structure in which the equipment is installed in industrial engineering, there are two types of situations in which sensors are arranged on planar and three-dimensional space structures. Solving the OSD problems of these two conditions is of great significance for modern industrial engineering.

This book mainly focuses on passive vibration isolation and optimized active and semi-active control systems for power and precision equipment. In this regard,

Fig. 1.9 Vibration and isolation for building structure with equipment

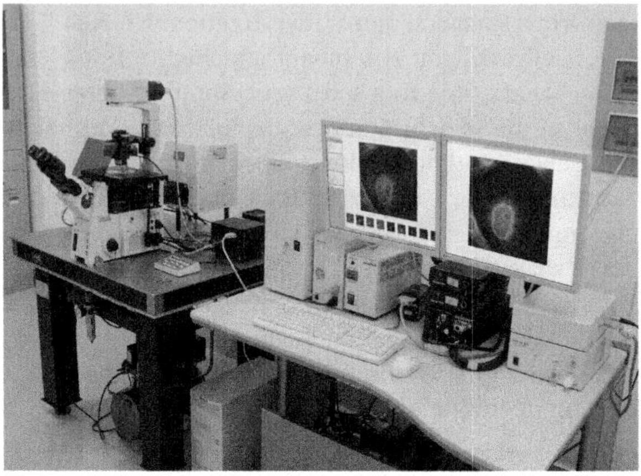

(a) A building with sensitive equipment

(b) A building with power equipment

Fig. 1.9 (continued)

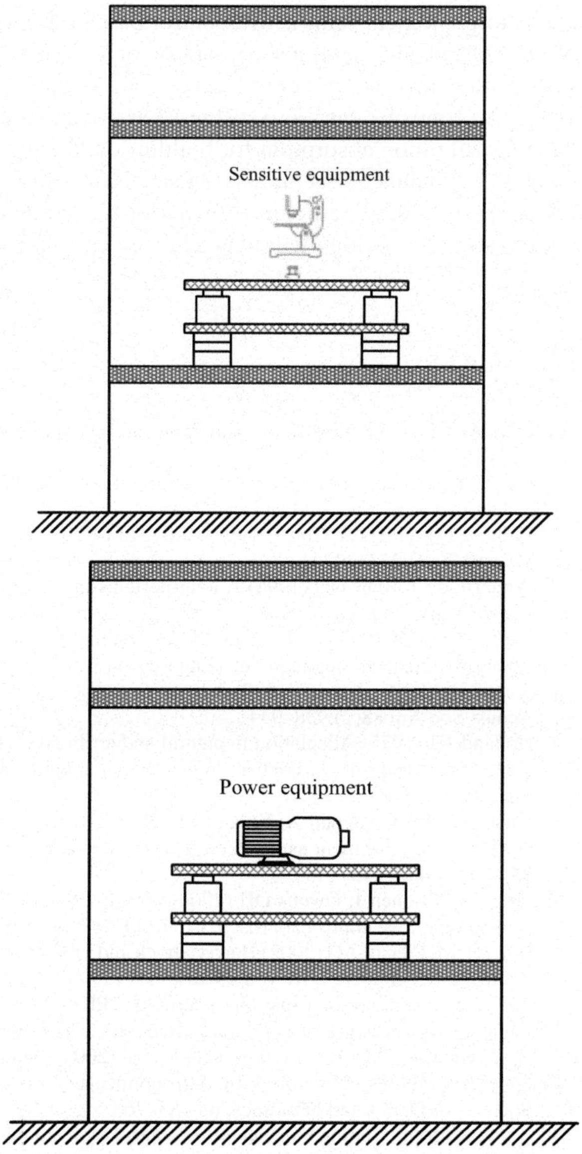

(c) Schematic diagram of vibration and isolation model for building with equipment magnetorheological semi-active control based on tracking the active control effect, active control and semi-active control of coupled building equipment vibration system, and fine active control using finite element analysis, decoupling passive and active control, quasi-zero stiffness passive and active control, passive and active dynamic vibration absorption for buildings and equipment, optimal sensor deployment on 2D planner structure and in special 3D structure will be discussed. Within this series of studies, particle swarm optimization and multi-objective particle swarm optimization technology will play an essential role in the optimization strategy.

References

1. Snowdon JC (1979) Vibration isolation: use and characterization. J Acoust Soc Am 66(5):1245–1274
2. Liu J, Winterflood J, Blair DG (1995) Transfer function of an ultralow frequency vibration isolation system. Rev Sci Instrum 66(5):3216–3218
3. Hensley JM, Peters A, Chu S (1999) Active low frequency vertical vibration isolation. Rev Sci Instrum 70(6):2735–2741
4. Winterflood J, Blair DG (1996) A long-period conical pendulum for vibration isolation. Phys Lett A 222(3):141–147
5. Wei YD, Lai XB, Chen DZ et al (2006) Optimal parameters design of two stage vibration isolation system. J Zhejiang Univ (Eng Ed) 40(5):893–896
6. Sommerfeldt SD, Tichy J (1990) Adaptive control of a two-stage vibration isolation mount. J Acoust Soc Am 88(2):938–944
7. Holland JH (1975) Adaptation in natural and artificial systems: an introductory analysis with applications to biology, control, and artificial intelligence. University of Michigan Press, Michigan
8. Farshidianfar A, Saghafi A, Kalami SM et al (2012) Active vibration isolation of machinery and sensitive equipment using H∞ control criterion and particle swarm optimization method. Meccanica 47(2):437–453
9. Daley S, Hätönen J, Owens DH (2006) Active vibration isolation in a "smart spring" mount using a repetitive control approach. Control Eng Pract 14(9):991–997
10. Harris CM, Piersol AG (2002) Harris' shock and vibration handbook. McGraw-Hill, New York
11. Beard AM, Schubert DW, von Flotow AH (1994) Practical product implementation of an active/passive vibration isolation system. In: SPIE's 1994 international symposium on optics, imaging, and instrumentation. International Society for Optics and Photonics, pp 38–49
12. Bronowicki AJ, MacDonald R, Gursel Y et al (2003) Dual stage passive vibration isolation for optical interferometer missions. In: Astronomical telescopes and instrumentation. International Society for Optics and Photonics, pp 753–763
13. Guclu R (2006) Sliding mode and PID control of a structural system against earthquake. Math Comput Model 44(1):210–217
14. Ang KK, Wang SY, Quek ST (2002) Weighted energy linear quadratic regulator vibration control of piezoelectric composite plates. Smart Mater Struct 11(1):98
15. Bai MR, Lin GM (1996) The development of a DSP-based active small amplitude vibration control system for flexible beams by using the LQG algorithms and intelligent materials. J Sound Vib 198(4):411–427
16. Karimi HR, Zapateiro M, Luo N (2009) Vibration control of base-isolated structures using mixed H_2/H_∞ output-feedback control. Proc Inst Mech Eng Part I: J Syst Control Eng 223(6):809–820
17. Ou JP (2003) Structural vibration control, active, semi-active and intelligent control. Science Press, Beijing

18. Guclu R, Yazici H (2008) Vibration control of a structure with ATMD against earthquake using fuzzy logic controllers. J Sound Vib 318(1):36–49
19. Snyder SD, Tanaka N (1995) Active control of vibration using a neural network. IEEE Trans Neural Netw 6(4):819–828
20. Lin FJ, Hwang WJ, Wai RJ (1999) A supervisory fuzzy neural network control system for tracking periodic inputs. IEEE Trans Fuzzy Syst 7(1):41–52
21. Taghirad HD, Esmailzadeh E (1998) Automobile passenger comfort assured through LQG/LQR active suspension. J Vib Control 4(5):603–618
22. Kar IN, Miyakura T, Seto K (2000) Bending and torsional vibration control of a flexible plate structure using H_∞-based robust control law. IEEE Trans Control Syst Technol 8(3):545–553
23. Pourzeynali S, Lavasani HH, Modarayi AH (2007) Active control of high rise building structures using fuzzy logic and genetic algorithms. Eng Struct 29(3):346–357
24. Dong X, Yu M, Liao C et al (2010) Comparative research on semi-active control strategies for magneto-rheological suspension. Nonlinear Dyn 59(3):433–453
25. Jalili N (2002) A comparative study and analysis of semi-active vibration-control systems. J Vib Acoust 124(4):593–605
26. Li HN, Li ZX, Qi A, Jia Y (2005) Structural vibration and control. China Construction Industry Press, Beijing
27. Liu Y, Matsuhisa H, Utsuno H (2008) Semi-active vibration isolation system with variable stiffness and damping control. J Sound Vib 313(1):16–28
28. Du H, Li W, Zhang N (2011) Semi-active variable stiffness vibration control of vehicle seat suspension using an MR elastomer isolator. Smart Mater Struct 20(10):105003
29. Liu Y, Waters TP, Brennan MJ (2005) A comparison of semi-active damping control strategies for vibration isolation of harmonic disturbances. J Sound Vib 280(1):21–39
30. Iba D (2011) Vibration control using harmonically-varying damping. J Syst Des Dyn 5(5):727–736
31. Hirohata S, Iba D (2012) Frequency response analysis of multi-degree-of-freedom system with harmonically varying damping. In: SPIE smart structures and materials+ nondestructive evaluation and health monitoring. International Society for Optics and Photonics, 83453R-15
32. Kamath GM, Werely NM (1997) Nonlinear viscoelastic-plastic mechanisms-based model of an electrorheological damper. J Guid Control Dyn 20(6):1125–1132
33. Sims ND, Peel DJ, Stanway R et al (2000) The electrorheological long-stroke damper: a new modelling technique with experimental validation. J Sound Vib 229(2):207–227
34. Wang KW, Kim YS, Shea DB (1994) Structural vibration control via electrorheological-fluid-based actuators with adaptive viscous and frictional damping. J Sound Vib 177(2):227–237
35. Choi SB (1999) Vibration control of a flexible structure using ER dampers. J Dyn Syst Meas Contr 121(1):134–138
36. Yao GZ, Yap FF, Chen G et al (2002) MR damper and its application for semi-active control of vehicle suspension system. Mechatronics 12(7):963–973
37. Dyke SJ, Spencer BF Jr, Sain MK et al (1996) Modeling and control of magnetorheological dampers for seismic response reduction. Smart Mater Struct 5(5):565
38. Wei W (2014) Study of control strategy of variable mass dynamic vibration absorber. Changan University
39. Ren MZ (2011) Analysis, control and calculation method of mechanical vibration. Machinery Industry Press
40. Liu YZ, Yu DL, Zhao HG et al (2007) Review of passive dynamic vibration absorbers. Chin J Mech Eng 43(3):14–21
41. Li CX, Liu YX, Wang ZM (2003) A review of mass dampers. Adv Mech 33(2)
42. Liu K, Liu J (2005) The damped dynamic vibration absorbers: revisited and new result. J Sound Vib 284(3):1181–1189
43. Arnold FR (1955) Steady-state behavior of systems provided with nonlinear dynamic vibration absorbers. J Appl Mech 22(4)
44. Hunt JB, Nissen JC (1982) The broadband dynamic vibration absorber. J Sound Vib 83(4):573–578

45. Yang YB, Huang WH (1998) Equipment–structure interaction considering the effect of torsion and base isolation. Earthquake Eng Struct Dynam 27(2):155–171
46. Igusa T, Der Kiureghian A (1985) Dynamic characterization of two-degree-of-freedom equipment-structure systems. J Eng Mech 111(1):1–19
47. Xu YL, Li B (2006) Hybrid platform for high-tech equipment protection against earthquake and micro vibration. Earthquake Eng Struct Dynam 35(8):943–967
48. Ismail M, Rodellar J, Ikhouane F (2009) Performance of structure–equipment systems with a novel roll-n-cage isolation bearing. Comput Struct 87(23):1631–1646
49. Murnal P, Sinha R (2004) Aseismic design of structure–equipment systems using variable frequency pendulum isolator. Nucl Eng Des 231(2):129–139
50. Ungar EE, Sturz DH, Amick CH (1990) Vibration control design of high technology facilities. Sound Vib 24(7):20–27
51. Xu YL, Liu HJ, Yang ZC (2003) Hybrid platform for vibration control of high-tech equipment in buildings subject to ground motion. Part 1: experiment. Earthquake Eng Struct Dynam 32(8):1185–1200
52. Yang ZC, Xu YL, Chen J et al (2003) Hybrid platform for vibration control of high-tech equipment in buildings subject to ground motion. Part 2: analysis. Earthquake Eng Struct Dynam 32(8):1201–1215
53. Soua R, Saidane L, Minet P (2010) Sensors deployment enhancement by a mobile robot in wireless sensor networks. In: 2010 ninth international conference on networks (ICN). IEEE, pp 121–126
54. Yingying Z, Xia L, Shiliang F (2010) A research on the deployment of sensors in underwater acoustic wireless sensor networks. In: 2010 2nd international conference on information science and engineering (ICISE). IEEE, pp 4312–4315
55. Zou Y, Chakrabarty K (2003) Sensor deployment and target localization based on virtual forces. In: Twenty-second annual joint conference of the IEEE computer and communications (INFOCOM 2003), vol 2. IEEE Societies. IEEE, pp 1293–1303
56. Wang YC, Hu CC, Tseng YC (2005) Efficient deployment algorithms for ensuring coverage and connectivity of wireless sensor networks. In: Proceedings. First international conference on wireless internet. IEEE, pp 114–121
57. Yi TH, Li HN, Gu M (2011) Optimal sensor placement for structural health monitoring based on multiple optimization strategies. Struct Design Tall Spec Build 20(7):881–900
58. Yi TH, Li HN, Gu M (2011) Optimal sensor placement for health monitoring of high-rise structure based on genetic algorithm. Math Probl Eng 2011
59. Papadopoulos M, Garcia E (1998) Sensor placement methodologies for dynamic testing. AIAA J 36(2):256–263
60. Carne TG, Dohrmann CR (1995) A modal test design strategy for model correlation. In: Proceedings-SPIE the international society for optical engineering. SPIE International Society for Optical, pp 927–927
61. Schedlinski C, Link M (1996) An approach to optimal pick-up and exciter placement. In: Proceedings—SPIE the international society for optical engineering. SPIE International Society for Optical, pp 376–382
62. Wu Q, Rao NSV, Du X et al (2007) On efficient deployment of sensors on planar grid. Comput Commun 30(14):2721–2734

Chapter 2
Particle Swarm Optimization

Abstract In this chapter, a new artificial intelligence optimization tool, a detailed introduction for particle swarm optimization (PSO) is presented here. The basic structure and the main characteristics of PSO algorithm and the multi-objective PSO (MOPSO) algorithm is described and elaborated, and some standard numerical examples for PSO and MOPSO are tested.

Natural systems, such as flocks of birds and schools of fish, often have some impressive, collision-free, synchronized interactions. This behavior is based on the inherent reaction of everyone in the group, although the reason is quite complicated from a macro perspective. For instance, by keeping an appropriate distance between each bird in the flock and its neighbors, the flock's migration behavior can be simulated accurately. This distance depends on the bird's size and behavior. On the other hand, when a school of fish swims freely, the individuals maintain a large mutual distance, whereas when there is a predator, the school of fish will gather into a very close group.

A similar phenomenon also exists in physical systems. A typical example is particle aggregation due to Brownian motion or fluid shear force. Human beings also have homogenous behavior characteristics, especially in forming social organization hierarchies and beliefs. However, unlike physical systems, people can hold the same idea or viewpoint without disagreement. These simplified aggregation behaviors in natural, physical, and human social systems enable researchers to conduct more in-depth experiments and simulation studies, thereby laying a foundation for developing swarm intelligence. Although the material structures of these systems are different, they have common properties with the following five basic swarm intelligence principles:

(1) Distance: the ability to perform space and time calculations.
(2) Quality: the ability to respond to environmental quality factors.
(3) Diverse reactions: the ability to make various reactions.
(4) Stability: the ability to maintain stable behavior under slight environmental changes.
(5) Adaptability: the ability to change behavior under the decision of external factors.

W. Huang and J. Xu, *Optimized Engineering Vibration Isolation, Absorption and Control*, Springer Tracts in Civil Engineering, https://doi.org/10.1007/978-981-99-2213-0_2

Furthermore, sharing social information between individuals in the system provides evolutionary advantages. Based on the studies mentioned above, in 1995, J. Kennedy and R. C. Eberhan officially published an article titled Particle swarm optimization at the IEEE International Neural Network Academic Conference, which marked the birth of the PSO algorithm. Since then, this algorithm has been extensively used, promoted, and generalized in the literature [1–7].

The research on vibration control of industrial engineering equipment involves many disciplines, such as civil engineering, machinery, automation, and computer engineering. Simple and effective optimization tools are significant for solving such complex problems. Traditional gradient-based optimization [8] requires continuously calculating sensitivity factors and eigenvectors during the iteration process, which greatly increases the computational cost, reduces the convergence speed, and makes determining the optimal solution challenging [9]. The proposal of GA can effectively solve this issue. However, when there are situations where the target is a highly recognized optimization object, the parameters to be optimized have a high intercorrelation. Besides, the GA optimization ability is insufficient when the parameter's dimension is large. Eberhart and Kennedy [10] proposed a new swarm intelligence optimization algorithm, the particle swarm optimization (PSO) algorithm, in 1995 to overcome these shortcomings. The main idea herein is to find the optimal solution (particle) based on the interparticles' cooperation and competition. This algorithm has the advantages of simplicity, easy implementation, fast convergence, and few adjustable parameters. As a result, it has been widely used for handling optimization tasks in the engineering field [11, 12]. Coello et al. [13] proposed a multi-objective particle swarm optimization algorithm (MOPSO), whose main idea is to determine the particle flight directly through the Pareto optimal solution set and obtain the previously found non-domination in the global knowledge base vector to guide other particles in flying. Generally, PSO and MOPSO can deal with single-objective and multi-objective optimization problems, respectively. Their intersection and cooperation constitute a new chapter in modern engineering optimization [14].

The PSO algorithm starts by initializing a group of particles without volume and mass, where each particle is considered a potential solution to the optimization problem. Thereafter, a pre-defined fitness function is used to determine the particles' quality. In general, all particles move in the problem's search space, and the speed variable limits the particles' direction and distance. Usually, the particle seeks the current optimal position in each generation, where each particle follows the individual and neighbor optimal position. The particle swarm optimization algorithm is a new intelligent optimization algorithm that comes from the simple social simulation of birds and fish schools. Therefore, particle swarm optimization algorithms can be used. Simulating this interactive process provides a new way to solve decision-making issues in complex environments.

The PSO is a random optimization method that can be considered an artificial intelligence method. As mentioned above, this algorithm is inspired by the social behavior of animals and insects, such as bird flocks and fish schools. The PSO generally employs a swarm of multiple particles, each with its position and velocity. All particles share information with each other, and efficient searching is obtained through interaction between the particles. Figure 2.1 depicts the PSO's basic structure. It can be seen that the direction of the particle motion is a linear combination of individual particles with random coefficients for each vector, which ultimately results in an improved particle position. In this way, a good search within the space is performed, and appropriate solutions are obtained for the optimization problem. Figure 2.2 illustrates a schematic diagram of the PSO.

The state of each particle in the PSO algorithm can be described by a group of position and velocity vectors representing all possible solutions and movement directions within the search space. The desired solution can be determined by constantly

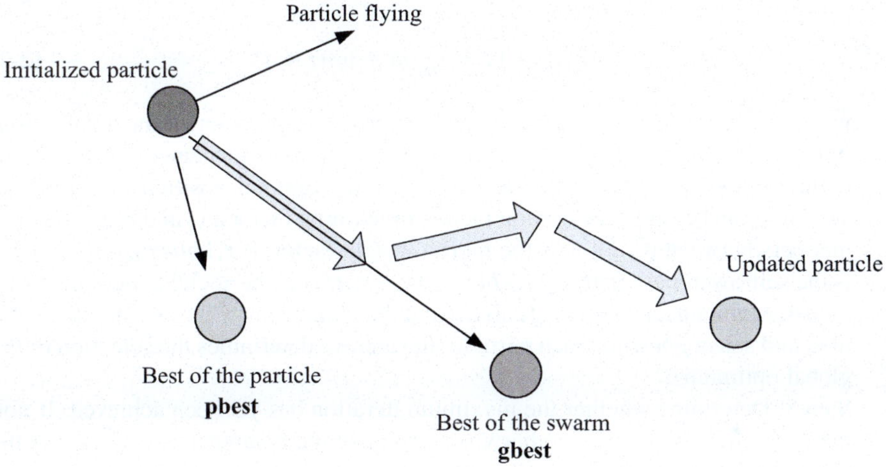

Fig. 2.1 Basic structure of the PSO algorithm

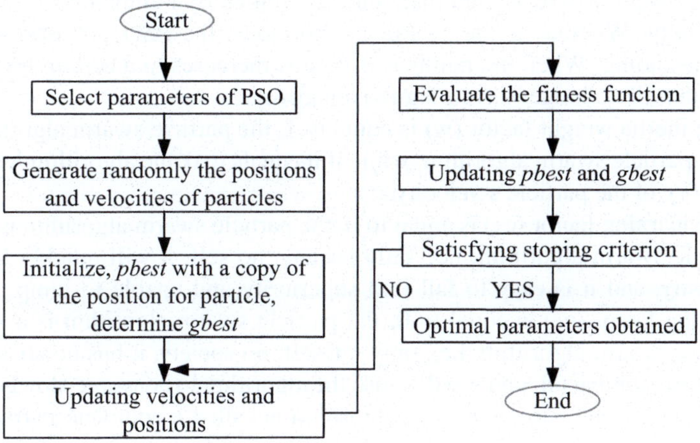

Fig. 2.2 Schematic diagram of the PSO algorithm

updating the optimal neighbor solutions each time new global and optimal solutions are discovered. The main steps of the MOPSO can be summarized as follows:

Step 1: Initialize the population, compute the particles' corresponding objective vectors, and add the non-dominated solutions to the external archive.
Step 2: Initialize the local optimum *pbest* of particles and the global optimum *gbest*.
Step 3: Adjust the velocities and positions of the particles by evaluating the following equations to generate a new *pbest*.

$$v_{ij} = wv_{ij}(t) + c_1 r_1 (pbest_{ij}(t) - x_{ij}(t)) + c_2 r_2 (gbest_{ij}(t) - x_{ij}(t)) \quad (2.1)$$

$$x_{ij}(t+1) = x_{ij}(t) + v_{ij}(t+1) \quad (2.2)$$

where i is the ith particle; j is the jth dimension of each particle; t is the tth iteration; $v_{ij}(t)$ is the flight velocity vector; $x_{ij}(t)$ is the flight displacement vector; *pbest* is the optimal location component; *gbest* is the optimal position of the whole swarm; c_1 and c_2 are acceleration factors or learning factors; r_1 and r_2 are random numbers between (0, 1); w is the inertia weight factor; $c_1 r_1 (pbest_{ij}(t) - x_{ij}(t))$ is the self-awareness part; $c_2 r_2 (gbest_{ij}(t) - x_{ij}(t))$ is the social experience part.
Step 4: Maintain an external archive using the obtained new non-dominated solution, and select *gbest* for each particle (the archive determines the selection of the global optimum).
Step 5: Determine whether the maximum iteration has yet been achieved. If not, the program will continue, otherwise, terminate the computation and output the optimal *Pareto* solution set and *gbest*.

The constituent elements of the particle swarm algorithm are as follows:

1. If the swarm's size is minimal, the possibility of falling into a situation is very high. Whereas if the swarm is enormous, the PSO optimization ability is outstanding. When the number of groups increases to a certain level, further growth will no longer have a significant effect.
2. If the inertia weight factor (w) is equal to 1, the particle swarm algorithm is the basic particle swarm algorithm, while if it equals to 0, the algorithm will lose the memory of the particle's velocity.
3. If the learning factor (c_1) is equal to 0, the particle swarm algorithm is a selfless particle swarm algorithm, i.e., only society, no self, it will quickly lose group diversity, and it is easy to fall into superiority and unable to jump out. If the learning factor (c_2) is equal to 0, the particle swarm algorithm is a self-aware particle swarm algorithm, i.e., only oneself, no society, it has no social sharing of information at all, and it will cause the algorithm to converge slowly. If c_1 and c_2 are not 0, the particle swarm algorithm is called a complete particle swarm algorithm, which can easily maintain the convergence speed, and the balanced search effect is a better choice.

4. The maximum velocity ($v_{\max}$) maintains the balance between the exploration and development abilities of the algorithm. When $v_{\max}$ is large, the exploration ability is enhanced, but the particles can easily fly through the optimal solution. When $v_{\max}$ is small, the development ability is enhanced, but the particle swarm algorithm easily falls into the local optimum. The $v_{\max}$ is typically set to 10–20% of the variable range in each dimension.

It is essential to point out that when a traditional multi-objective optimization is utilized, a set of equivalent solutions will be generated, resulting in difficulty in determining the desired solution. Pareto dominating is the most direct method of solving this problem since it considers all non-dominated solutions in the archive and then determines the leader. The density measuring technique is commonly employed to determine the global optimum. The nearest neighbor density estimation (NNDE) method [15] based on the nearest neighbor congestion evaluation of particles is adopted in this chapter. In the NNDE technique, the crowding distance can be measured by a cuboid that consists of neighboring particles, with the long perimeter indicating a low-density distribution of individuals and good adaptability. Figure 2.3 depicts a schematic diagram of the NNDE method. Other similar methods can be alternatively used, such as the kernel density estimation method [16].

The main difference between the MOPSO and PSO is selecting the optimal global solution and setting and updating external files. Among them, the update of the external file will directly determine the optimal global solution, as shown in Fig. 2.4. In this figure, the NDs represent non-dominated solution sets, and s_1–s_5 represents the solution sets.

Regarding the selection of the global optimal solution *gbest* in the multi-objective optimization, there is a set of equivalent optimal solutions for direct calculation, where it is difficult to determine an optimal solution from each iteration. Parsopoulos and Vrahatis [17] proposed a MOPSO algorithm using the goal aggregation concept. The algorithm's core is to use an adaptive weight to convert a multi-objective optimization problem into a single-objective optimization one. Fieldsend and Singh [18]

Fig. 2.3 Schematic diagram of the nearest neighbor density estimation method

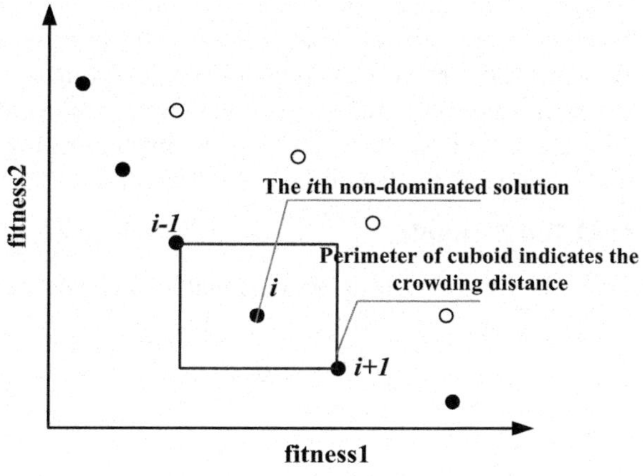

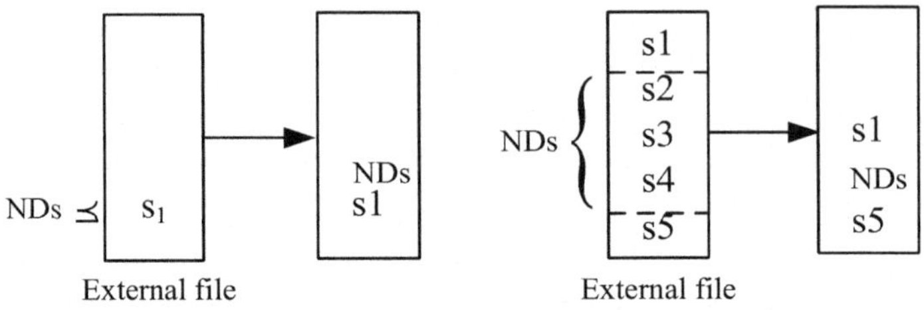

Fig. 2.4 Updating of external archive

Fig. 2.5 Pareto frontier

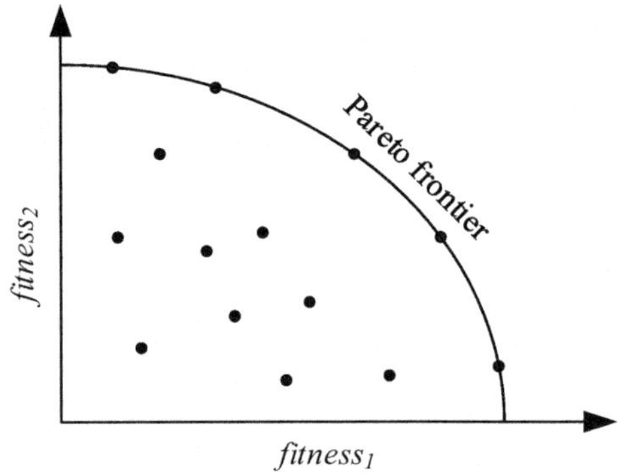

developed a dominating tree's data structure that can obtain the particle swarm's elite solution. Li et al. [19] introduced the multi-objective optimization idea in the NSGA-2 algorithm to the particle swarm algorithm so that the PSO can achieve or approach the multi-objective optimization effect of the former to the greatest extent possible. In multi-objective optimization, the most direct method is to use the Pareto domination concept by considering all non-inferior solutions in the archives and determining a leader from them to form a set of Pareto frontiers, as shown in Fig. 2.5. In this study, the nearest neighbor density estimation method based on the particle nearest neighbor crowding degree is utilized to select the optimal solution from the Pareto front [15, 20]. The nearest neighbor density estimation method is typically adopted to determine the congestion degree by measuring the rectangular perimeter of adjacent particles as vertices. The longer the perimeter, the lower the density of the individual distribution and the better the adaptability [20].

PSO Test Example

Solve the following four-dimensional Rosenbrock function optimization problem:

$$\min f(\boldsymbol{x}) \sum_{i=1}^{3} [100(x_{i+1} - x_i^2)^2 + (x_i - 1)^2], \quad x_i \in [-30, 30]\ (i = 1, 2, 3, 4).$$

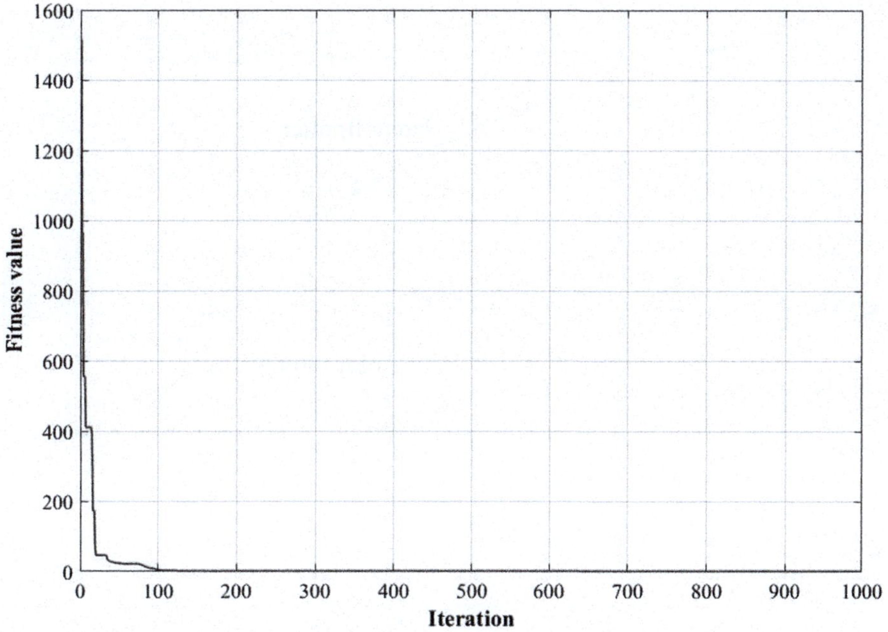

Fig. 2.6 Fitness convergence of Rosenbrock test function using PSO

The PSO algorithm parameters: the swarm size is 200, $c_1 = 2$, $c_2 = 1$, the maximum number of iterations is 1000, the inertia weight factor $w = 0.99^t$, t is the t-th iteration, the parameter optimization range of x_i is $[-30, 30]$, the velocity range of the particle is $[-60, 60]$. Figure 2.6 shows the fitness value convergence curve after calculation. It can be seen that the optimal solution is 1.2043, 1.4241, 1.9918, 3.9665, and the optimized function's minimum value is 1.4074.

MOPSO Test Example 1

Solve the following multi-objective optimization problem:

$$\begin{cases} \min \ f_1(x) = 1 - \exp\left(-\sum\limits_{i=1}^{3}\left(x_i - 1/\sqrt{3}\right)^2\right) \\ \min \ f_2(x) = 1 - \exp\left(-\sum\limits_{i=1}^{3}\left(x_i + 1/\sqrt{3}\right)^2\right) \\ \text{s.t.} \quad x_i \in [-4, 4], \quad i = 1, 2, 3 \end{cases}$$

The MOPSO algorithm parameters are: the swarm size is 100, $c_1 = 2$, $c_2 = 1$, the maximum number of iterations is 100, the inertia weight factor $w = 0.99^t$ [21], the parameter optimization range of x_i is $[-4, 4]$. Figure 2.7 illustrates the Pareto frontier and the optimal solution after calculation. It can be seen that the Pareto frontier is characterized by *convex*, and the optimal solution is –0.1952, –0.2028, –0.1937.

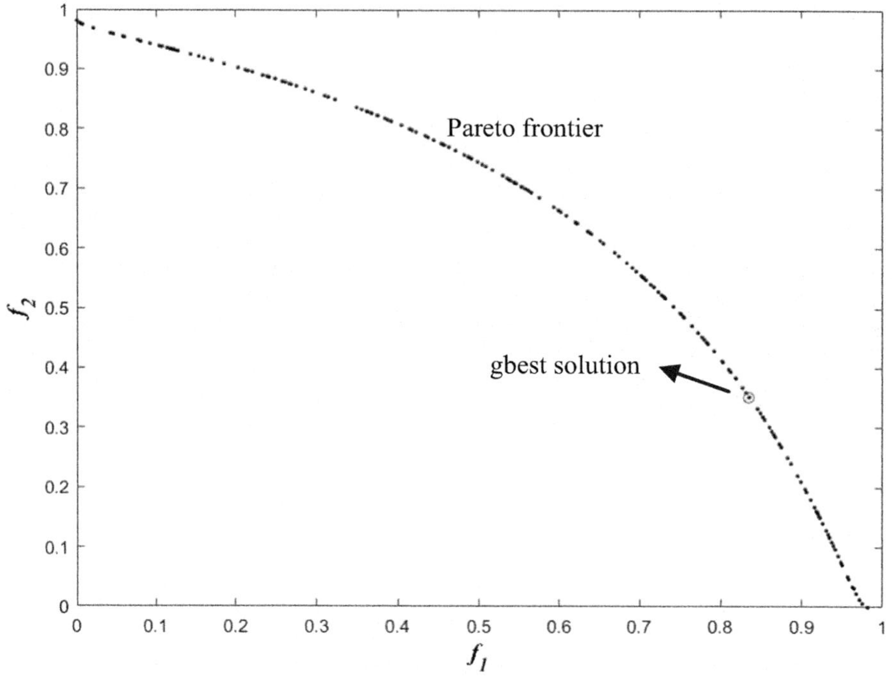

Fig. 2.7 Pareto frontier and optimal solution of test function 1 using MOPSO

MOPSO Test Example 2

Solve the following multi-objective optimization problem:

$$\begin{cases} \min \ f_1(x) = 1 - \exp(-4x_1)\sin^6(6\pi x_1) \\ \min \ f_2(x) = g(x)\left[1 - (x_1/g(x))^2\right] \\ \quad g(x) = 1 + 9\left[\sum_{i=2}^{n}(x_i/(n-1))\right]^{0.25} \\ \text{s.t.} \ x_i \in [0, 1], \quad i = 1, 2, \ldots, 10 \end{cases}$$

The MOPSO algorithm parameters are: the swarm size is 100, $c_1 = 2$, $c_2 = 1$, the maximum number of iterations is 100, the inertia weight factor $w = 0.99^t$, and the parameter optimization range of x_i is $[-1, 1]$. Figure 2.8 depicts the Pareto frontier and the optimal solution. It indicates that the Pareto frontier is characterized by *concave*, and the optimal solution is −0.9182, −0.0032, −0.0005, −0.1481, 0.0662, 0.0262, 0.0614, −0.0156, −0.1306, −0.0175.

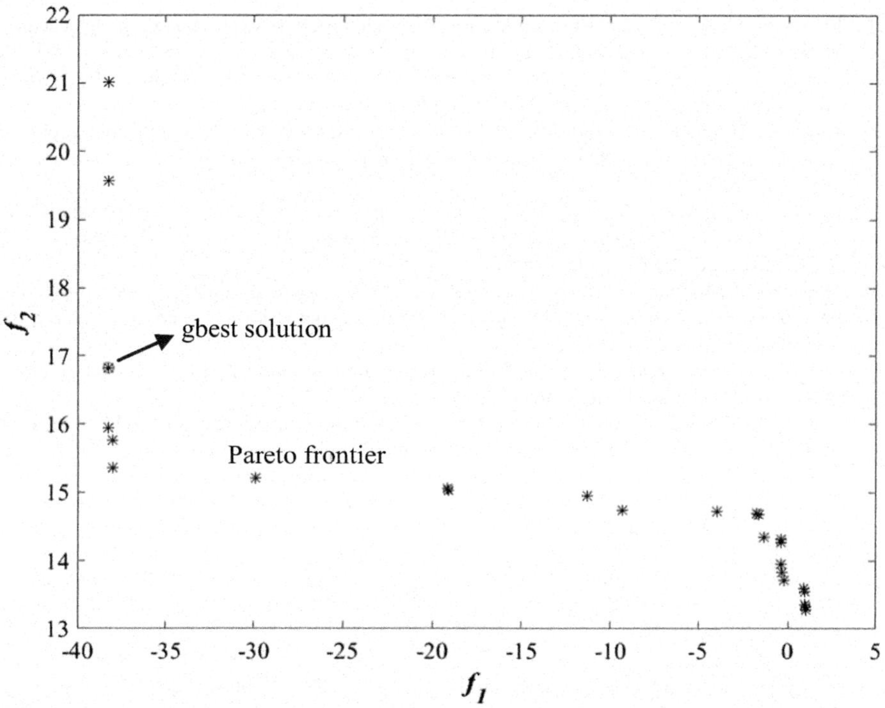

Fig. 2.8 Pareto frontier and optimal solution of test function 2 using MOPSO

References

1. Yoshida H, Kawata K, Fukuyama Y et al (2000) A particle swarm optimization for reactive power and voltage control considering voltage security assessment. IEEE Trans Power Syst 15(4):1232–1239
2. Gaing ZL (2003) Particle swarm optimization to solving the economic dispatch considering the generator constraints. IEEE Trans Power Syst 18(3):1187–1195
3. Salman A, Ahmad I, Al-Madani S (2002) Particle swarm optimization for task assignment problem. Microprocess Microsyst 26(8):363–371
4. Nouiri M, Bekrar A, Jemai A et al (2018) An effective and distributed particle swarm optimization algorithm for flexible job-shop scheduling problem. J Intell Manuf 29(3):603–615
5. Eberhart RC, Shi Y (2000) Comparing inertia weights and constriction factors in particle swarm optimization. In: Proceedings of the 2000 congress on evolutionary computation. CEC00 (Cat. No. 00TH8512), vol 1. IEEE, pp 84–88
6. Huang T, Mohan AS (2005) A hybrid boundary condition for robust particle swarm optimization. IEEE Antennas Wirel Propag Lett 4:112–117
7. Robinson J, Rahmat-Samii Y (2004) Particle swarm optimization in electromagnetics. IEEE Trans Antennas Propag 52(2):397–407
8. Conn AR, Elfadel IM, Molzen Jr WW et al (1999) Gradient-based optimization of custom circuits using a static-timing formulation. In: Proceedings of the 36th annual ACM/IEEE design automation conference. ACM, pp 452–459
9. Shayeghi A, Shayeghi H, Kalasar HE (2009) Application of PSO technique for seismic control of tall building. World Acad Sci Eng Technol 3(4):1116–1123

10. Eberhart RC, Kennedy J (1995) A new optimizer using particle swarm theory. In: Proceedings of the sixth international symposium on micro machine and human science, vol 1, pp 39–43
11. Marinaki M, Marinakis Y, Stavroulakis GE (2010) Fuzzy control optimized by PSO for vibration suppression of beams. Control Eng Pract 18(6):618–629
12. Amini F, Hazaveh NK, Rad AA (2013) Wavelet PSO-based LQR algorithm for optimal structural control using active tuned mass dampers. Comput-Aided Civil Infrastruct Eng 28(7):542–557
13. Coello CAC, Lechuga MS (2002) MOPSO: a proposal for multiple objective particle swarm optimization. In: Proceedings of the 2002 congress on evolutionary computation, 2002 (CEC'02), vol 2. IEEE, pp 1051–1056
14. Marinaki M, Marinakis Y, Stavroulakis GE (2011) Fuzzy control optimized by a multi-objective particle swarm optimization algorithm for vibration suppression of smart structures. Struct Multidiscip Optim 43(1):29–42
15. Deb K, Pratap A, Agarwal S et al (2002) A fast and elitist multi-objective genetic algorithm: NSGA-II. IEEE Trans Evol Comput 6(2):182–197
16. Goldberg DE, Richardson J (1987) Genetic algorithms with sharing for multimodal function optimization. In: Genetic algorithms and their applications: proceedings of the second international conference on genetic algorithms. Lawrence Erlbaum, Hillsdale, NJ, pp 41–49
17. Parsopoulos KE, Vrahatis MN (2002) Particle swarm optimization method in multiobjective problems. In: Proceedings of the 2002 ACM symposium on applied computing. ACM, pp 603–607
18. Fieldsend JE, Singh S (2002) A multi-objective algorithm based upon particle swarm optimisation, an efficient data structure and turbulence. In: U.K. workshop on computational intelligence, pp 37–44
19. Li X (2003) A non-dominated sorting particle swarm optimizer for multiobjective optimization. In: Genetic and evolutionary computation—GECCO 2003. Springer, Berlin, Heidelberg, pp 37–48
20. Li H, Zhang Q (2009) Multiobjective optimization problems with complicated Pareto sets, MOEA/D and NSGA-II. IEEE Trans Evol Comput 13(2):284–302
21. Farshidianfar A, Saghafi A, Kalami SM et al (2012) Active vibration isolation of machinery and sensitive equipment using H_∞ control criterion and particle swarm optimization method. Meccanica 47(2):437–453

Chapter 3
Optimization of Passive Isolation Systems

Abstract In this chapter, the transmissibility characteristics of single-stage and two-stage uncontrolled vibration isolation systems for power and sensitive equipment are studied by using MOPSO algorithm, and the damping ratio, mass ratio, frequency ratio and other parameters are investigated and the obtained *gbest* solution can perform an optimal design for single-stage and two-stage uncontrolled vibration isolation.

3.1 Uncontrolled Vibration Isolation for Power Equipment

3.1.1 Single-Stage Vibration Isolation System

Consider the single-stage vibration isolation system of the power equipment shown in Fig. 1.3a [1–3] and assume that the equipment's dynamic load is a simple harmonic form represented by $F(t) = \mathrm{F}_0 \sin(\omega_{ref} t)$, where F_0 is the amplitude of the load generated by the power equipment, ω_{ref} is the vibration circle frequency defined as $\omega_{ref} = 2\pi f$, f is the frequency, and t is the time. Let the power equipment's vibration displacement, velocity, and acceleration be expressed as x, $\dot{x}$, $\ddot{x}$. Thus, the system's dynamic equation can be defined as follows:

$$\mathrm{m}\ddot{x} + \mathrm{c}\dot{x} + \mathrm{k}x = F(t) \tag{3.1}$$

Assume that the initial state is zero, perform Laplace transformation on Eq. (3.1), and obtain its transmissibility can be expressed using Eq. (3.2)

$$T_g = \frac{F_g(s)}{F(s)} = \frac{\mathrm{c}s + \mathrm{k}}{\mathrm{m}s^2 + \mathrm{c}s + \mathrm{k}} \tag{3.2}$$

where s is the complex frequency defined as $\mathrm{s} = \mathrm{j}\omega_{ref}$ with j being the imaginary unit, k is the stiffness coefficient, c is the damping coefficient, m is the mass, and

W. Huang and J. Xu, *Optimized Engineering Vibration Isolation, Absorption and Control*, Springer Tracts in Civil Engineering, https://doi.org/10.1007/978-981-99-2213-0_3

F_g is the force transferred from the power equipment to the foundation defined as $F_g = c\dot{x} + kx$ with $X(s)$ being the Laplace transform form of displacement.

Equation (3.2) can be rearranged as follows:

$$|T_g| = \left|\frac{F_g(s)}{F(s)}\right| = \sqrt{\frac{\left(\frac{\omega}{\omega_{ref}}\right)^4 + \left(2\frac{\omega}{\omega_{ref}}\eta\right)^2}{\left[1 - \left(\frac{\omega}{\omega_{ref}}\right)^2\right]^2 + \left(2\frac{\omega}{\omega_{ref}}\eta\right)^2}} \tag{3.3}$$

where $\omega = \sqrt{k/m}$ is the natural frequency, $\eta = c/(2\omega m)$ is the damping ratio, $|T_g|$ is the magnitude of the force transmitted from the power equipment to the foundation, which is an evaluation index of the external environment disturbed by the power equipment vibration.

The power equipment's vibration inertia force can be defined as $F_I = m\ddot{x}$. The inertia force is typically used to evaluate the equipment's safe use and working life. The transmissibility expression is defined as follows:

$$|T_I| = \left|\frac{F_I(s)}{F(s)}\right| = \frac{1}{\sqrt{\left[1 - \left(\frac{\omega}{\omega_{ref}}\right)^2\right]^2 + \left(2\frac{\omega}{\omega_{ref}}\eta\right)^2}} \tag{3.4}$$

Equations (3.3) and (3.4) can be illustrated in Fig. 3.1. It can be seen that the increase in the damping ratio can effectively reduce the resonance zone's peak value, but it reduces the ideal vibration isolation zone's effect. Accordingly, the main issue herein is that the ideal vibration isolation areas $|T_g|$ and $|T_I|$ are distributed on both sides of the resonance zone, and it is not easy to consider and optimize them simultaneously.

In order to explain this point more intuitively, this study utilizes the MOPSO algorithm with 200 iterations, an inertia weight factor of $\omega = 0.99^t$ [4], $c_1 = 2$, $c_2 = 1$ acceleration factors, and η and ω/ω_{ref} parameter search range of [0.01, 0.01] ~ [100, 100].

The results of the four tests are shown in Fig. 3.2. Generally, it is difficult to achieve a small transmissibility value using both $|T_g|$ and $|T_I|$. Moreover, the gbest solutions obtained from all experiments have $|T_g| = 1$, $|T_I| = 0$. The calculation rules of the gbest solution obtained from the Pareto frontier in this study are described in Sect. 1.3.2 and will not be repeated in the following section.

To this end, the performance of the two-stage vibration isolation system will be investigated.

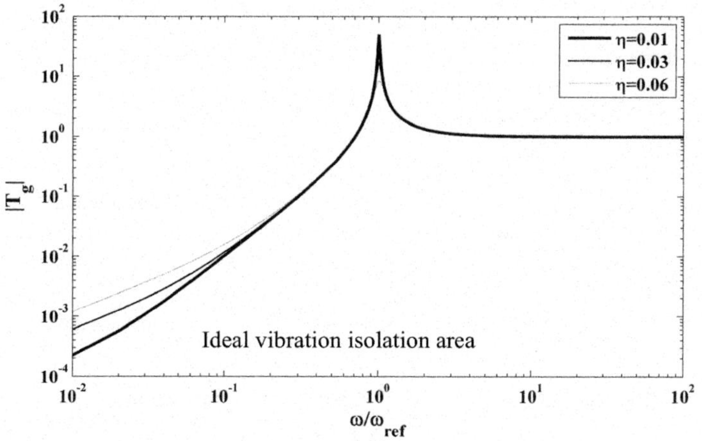

(a) $|T_g|$

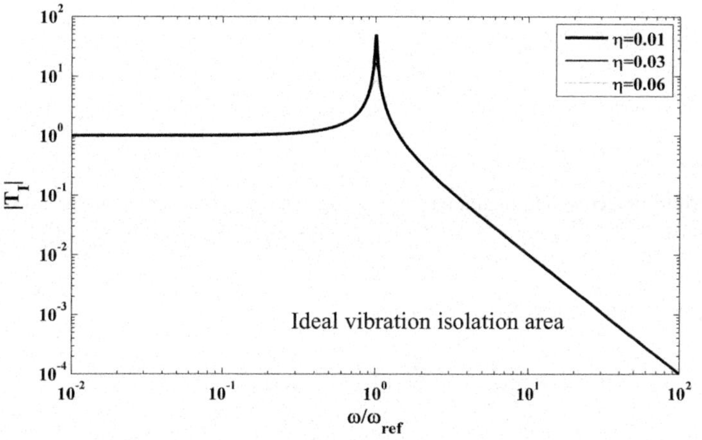

(b) $|T_I|$

Fig. 3.1 Schematic diagram of $|T_g|$ and $|T_I|$ with respect to the η (single-stage) value

3.1.2 Two-Stage Vibration Isolation System

In the two-stage vibration isolation system shown in Fig. 3.4a [5–8], m_2 is the mass of power equipment, k_2 and c_2 are the stiffness and damping of the vibration isolation system, respectively, m_1, k_1, and c_1 are the mass, stiffness, and damping of the foundation or supporting structure, respectively, and the vibration load form is the same as the single-stage vibration isolation system. In this case, the device's displacement, velocity, and acceleration are denoted as x_2, $\dot{x}_2$, $\ddot{x}_2$, respectively. Besides, foundation's displacement, velocity, and acceleration are denoted as x_1, $\dot{x}_1$, $\ddot{x}_1$, respectively.

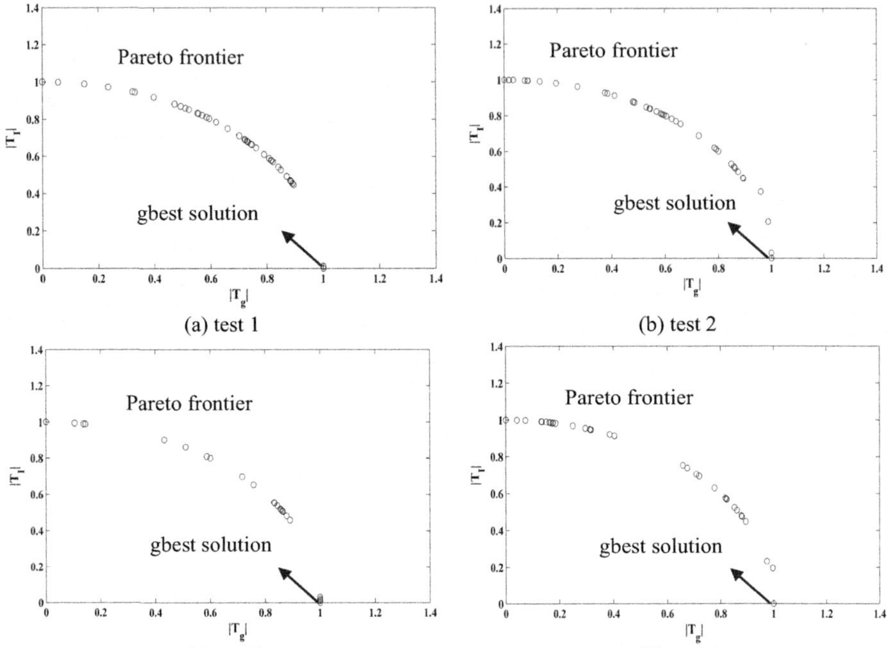

(a) test 1 (b) test 2 (c) test 3 (d) test 4

Fig. 3.2 Multi-objective optimization results of $|T_g|$ and $|T_I|$ with respect to the MOPSO (single-stage) value

The dynamic equation at this time is:

$$\begin{cases} \mathrm{m}_2 \ddot{x}_2 + \mathrm{c}_2 \dot{x}_2 - \mathrm{c}_2 \dot{x}_1 + \mathrm{k}_2 x_2 - \mathrm{k}_2 x_1 = F(t) \\ \mathrm{m}_1 \ddot{x}_1 - \mathrm{c}_2 \dot{x}_2 + (\mathrm{c}_2 + \mathrm{c}_1) \dot{x}_1 - \mathrm{k}_2 x_2 + (\mathrm{k}_2 + \mathrm{k}_1) x_1 = 0 \end{cases} \tag{3.5}$$

The transmissibility of the foundation's force and that of the equipment's inertial force can be obtained by performing Laplace transformation on Eq. (3.5) as follows:

$$\begin{aligned} T_g &= \frac{F_g(s)}{F(s)} \\ &= \frac{\mathrm{c}_1\mathrm{c}_2 s^2 + \mathrm{c}_1\mathrm{k}_2 s + \mathrm{c}_2\mathrm{k}_1 s + \mathrm{k}_1\mathrm{k}_2}{\mathrm{m}_1\mathrm{m}_2 s^4 + (\mathrm{m}_2\mathrm{c}_1 + \mathrm{m}_2\mathrm{c}_2 + \mathrm{m}_1\mathrm{c}_2)s^3 + (\mathrm{m}_2\mathrm{k}_1 + \mathrm{m}_2\mathrm{k}_2 + \mathrm{m}_1\mathrm{k}_2 + \mathrm{c}_1\mathrm{c}_2)s^2 + (\mathrm{c}_2\mathrm{k}_1 + \mathrm{c}_1\mathrm{k}_2)s + \mathrm{k}_1\mathrm{k}_2} \end{aligned} \tag{3.6a}$$

$$\begin{aligned} T_I &= \frac{F_I(s)}{F(s)} \\ &= \frac{\mathrm{m}_1\mathrm{m}_2 s^4 + \mathrm{c}_1\mathrm{m}_2 s^3 + \mathrm{c}_2\mathrm{m}_2 s^3 + \mathrm{k}_1\mathrm{m}_2 s^2 + \mathrm{k}_2\mathrm{m}_2 s^2}{\mathrm{m}_1\mathrm{m}_2 s^4 + (\mathrm{m}_2\mathrm{c}_1 + \mathrm{m}_2\mathrm{c}_2 + \mathrm{m}_1\mathrm{c}_2)s^3 + (\mathrm{m}_2\mathrm{k}_1 + \mathrm{m}_2\mathrm{k}_2 + \mathrm{m}_1\mathrm{k}_2 + \mathrm{c}_1\mathrm{c}_2)s^2 + (\mathrm{c}_2\mathrm{k}_1 + \mathrm{c}_1\mathrm{k}_2)s + \mathrm{k}_1\mathrm{k}_2} \end{aligned} \tag{3.6b}$$

where $u = m_1/m_2$, $\omega_1 = \sqrt{k_1/m_1}$, $\omega_2 = \sqrt{k_2/m_2}$, $\eta_1 = c_1/(2m_1\omega_1)$, $\eta_2 = c_2/(2m_2\omega_2)$, $\varpi = \omega_2/\omega_{ref}$, and $\gamma = \omega_1/\omega_2$.

The two absolute transmissibility expressions are defined as follows:

$$|T_g| = \left|\frac{F_g(s)}{F(s)}\right| = \sqrt{\frac{(u\gamma^2\varpi^4 - 4\eta_1\eta_2 u\gamma\varpi^2)^2 + (2\eta_1 u\gamma\varpi^3 + 2\eta_2 u\gamma^2\varpi^3)^2}{[u - (u + 4\eta_1\eta_2 u\gamma + 1 + u\gamma^2)\varpi^2 + u\gamma^2\varpi^4]^2 + [-(2\eta_2 u + 2\eta_2 + 2\eta_1\gamma u)\varpi + (2\eta_1\gamma u + 2\eta_2 u\gamma^2)\varpi^3]^2}} \tag{3.7a}$$

$$|T_I| = \left|\frac{F_I(s)}{F(s)}\right| = \sqrt{\frac{(u - u\gamma^2\varpi^2 - \varpi^2)^2 + (2\eta_1 u\gamma\varpi + 2\eta_2\varpi)^2}{[u - (u + 4\eta_1\eta_2 u\gamma + 1 + u\gamma^2)\varpi^2 + u\gamma^2\varpi^4]^2 + [-(2\eta_2 u + 2\eta_2 + 2\eta_1\gamma u)\varpi + (2\eta_1\gamma u + 2\eta_2 u\gamma^2)\varpi^3]^2}} \tag{3.7b}$$

Equations (3.7a) and (3.7b) are illustrated in Fig. 3.3. It can be seen that there are two resonance peaks in the two-stage vibration isolation system. As long as the parameters are reasonably set to move in the direction shown in Fig. 3.3, $|T_g|$ and $|T_I|$ can enter the ideal vibration isolation area at the same time. The variation law of $|T_g|$ with respect to the vibration isolation parameters is shown in Fig. 3.4. It can be seen that increasing η_2 raises the peak of $|T_g|$ and impacts the ideal vibration isolation area. Additionally, increasing γ reduce the peak value of $|T_g|$ and affects the two peaks' positions. The change rule of $|T_I|$ with respect to the vibration isolation parameters is shown in Fig. 3.5. It can be seen that increasing η_1 reduces the peak value of $|T_I|$ and slightly affects the ideal vibration isolation area. Moreover, increasing η_2 increases the peak value of $|T_I|$ and slightly influences the ideal vibration isolation zone. In contrast, raising u significantly changes the two peaks' positions, which benefits the ideal vibration isolation area. Furthermore, increasing γ reduces the peak value of $|T_I|$, which benefits the ideal vibration isolation area and affects the two peaks' positions (Fig. 3.6).

This study carries out an investigation based on the MOPSO algorithm to more intuitively present the characteristics of the two-stage vibration isolation system. The basic algorithm parameters are similar to those mentioned above, with the exception that the ranges of η_1, η_2, u, γ and ϖ were arbitrarily set to [0.01, 0.01, 0.01, 0.01, 0.01] ~ [1, 1, 10, 10, 100]. The results of the four tests are shown in Fig. 2.6. It can be seen that both $|T_g|$ and $|T_I|$ can achieve a low transmission in the two-stage vibration isolation system to develop a combination of both parameters. Additionally, the obtained gbest solution can design the optimal vibration isolation parameters.

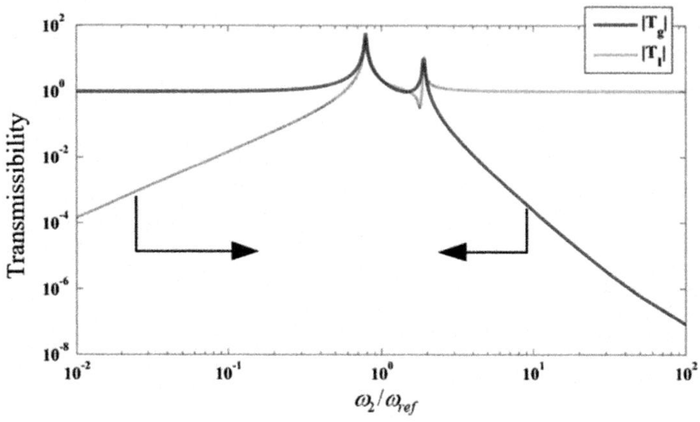

Fig. 3.3 Schematic diagram of $|T_g|$ and $|T_I|$ for the double-stage vibration isolation system

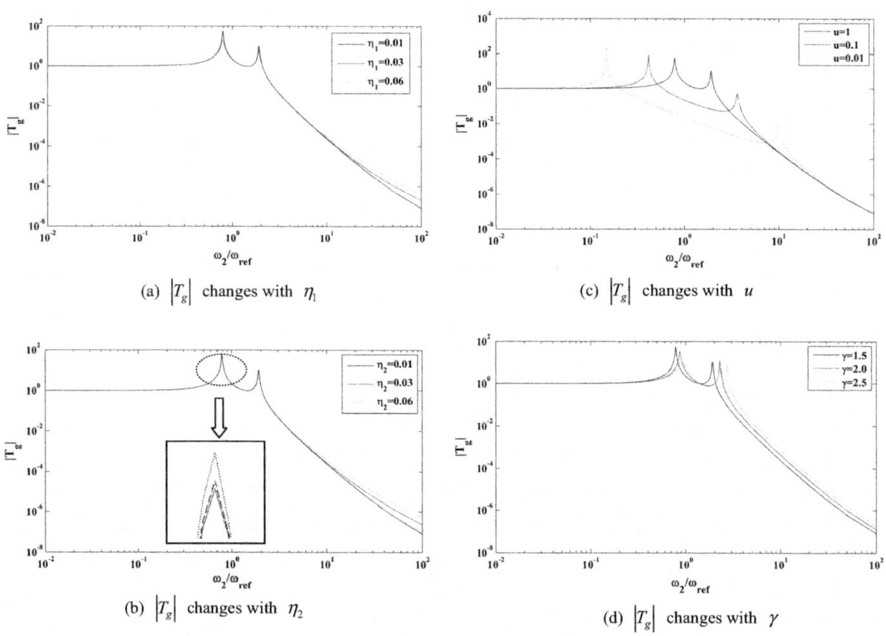

(a) $|T_g|$ changes with η_1

(c) $|T_g|$ changes with u

(b) $|T_g|$ changes with η_2

(d) $|T_g|$ changes with γ

Fig. 3.4 Schematic diagram of $|T_g|$ with respect to the isolation parameters (double-stage)

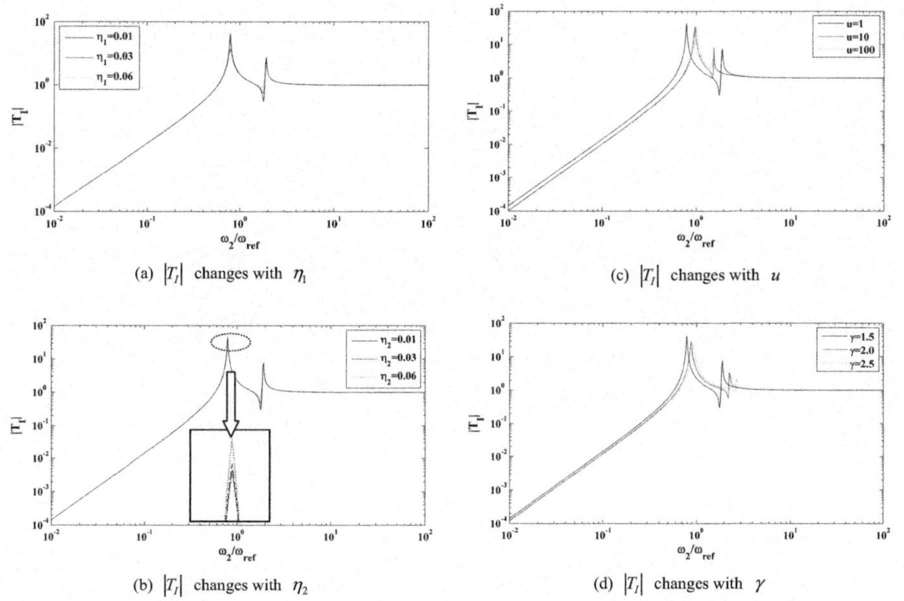

(a) $|T_I|$ changes with η_1

(c) $|T_I|$ changes with u

(b) $|T_I|$ changes with η_2

(d) $|T_I|$ changes with γ

Fig. 3.5 Schematic diagram of $|T_I|$ with respect to the isolation parameters (double-stage)

3.1.3 Uncontrolled Vibration Isolation of Vibration-Sensitive Equipment

3.1.3.1 Single-Stage Vibration Isolation System

Consider the single-stage isolation system of the vibration-sensitive equipment given in Fig. 1.3b. The interference load is a simple harmonic input vibration x_g, $\dot{x}_g$, with a vibration circle frequency of $\omega_{ref} = 2\pi f$. The displacement, velocity, and acceleration that cause the device's sensitive vibrations are denoted as x, $\dot{x}$, $\ddot{x}$. The dynamic equation of the system is defined as follows:

$$\mathrm{m}\ddot{x} + \mathrm{c}\dot{x} + \mathrm{k}x = \mathrm{k}x_g + \mathrm{c}\dot{x}_g \tag{3.8}$$

The displacement transmissibility of the equipment can be obtained through Laplace transformation as follows:

$$T_d = \frac{X(s)}{X_g(s)} = \frac{\mathrm{c}s + \mathrm{k}}{\mathrm{m}s^2 + \mathrm{c}s + \mathrm{k}} \tag{3.9}$$

From the Eq. (3.9):

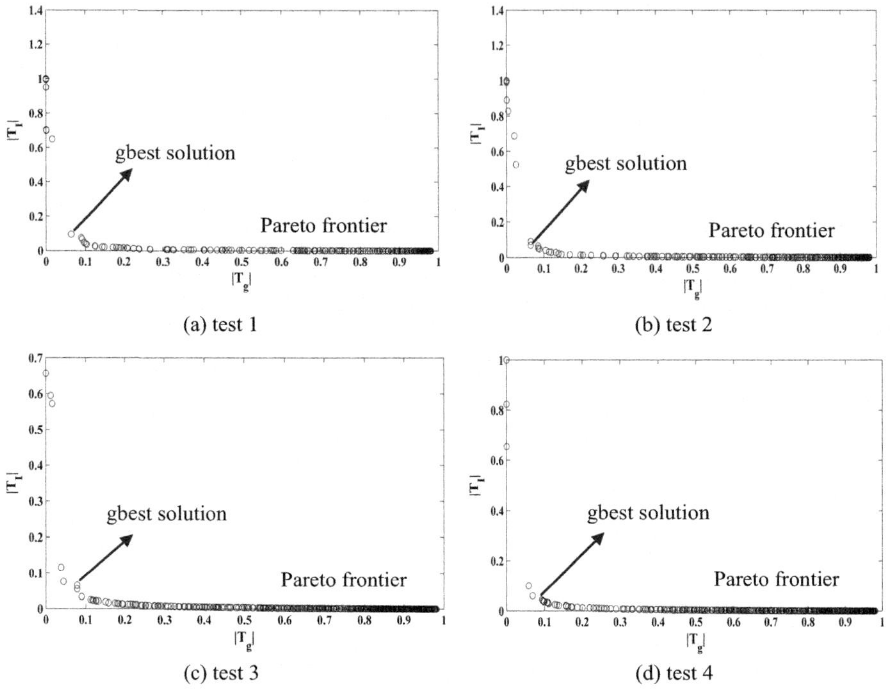

Fig. 3.6 Multi-objective optimization results of $\left|T_g\right|$ and $|T_I|$ based on the MOPSO algorithm (double-stage)

$$|T_{d1}| = \left|\frac{X(s)}{X_g(s)}\right| = \sqrt{\frac{\left(\frac{\omega}{\omega_{ref}}\right)^4 + \left(2\eta\frac{\omega}{\omega_{ref}}\right)^2}{\left[1-\left(\frac{\omega}{\omega_{ref}}\right)^2\right]^2 + \left(2\eta\frac{\omega}{\omega_{ref}}\right)^2}} \tag{3.10}$$

where $X(s)$ and $X_g(s)$ are Laplace transforms for the equipment displacement and input excitation, $\omega = \sqrt{\text{k/m}}$, and $\eta = \text{c}/(2\text{m}\omega)$.

The relative displacement transmissibility or drift transmissibility is an index that reflects the deformation level of the vibration isolator after the installation of vibration-sensitive equipment (especially large precision equipment), which directly affects the regular use of the equipment [9]. The transmissibility expression is defined as follows:

$$|T_{d2}| = \left|\frac{X(s) - X_g(s)}{X_g(s)}\right| = \frac{1}{\sqrt{\left[1-\left(\frac{\omega}{\omega_{ref}}\right)^2\right]^2 + (2\eta\frac{\omega}{\omega_{ref}})^2}} \tag{3.11}$$

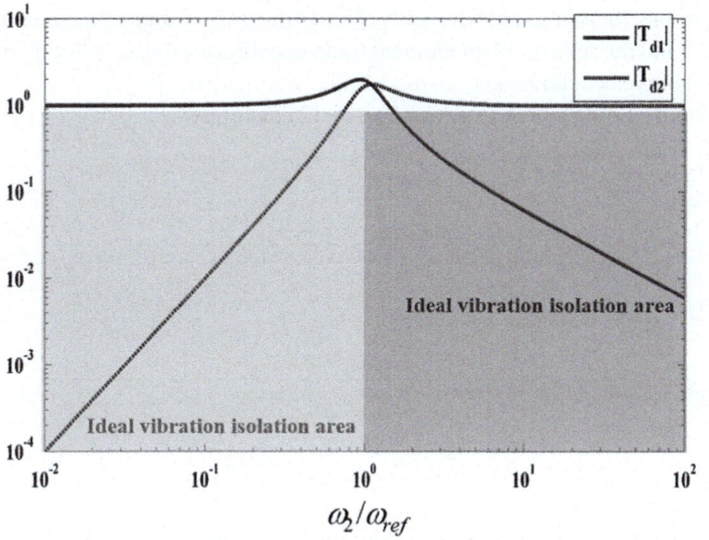

Fig. 3.7 Schematic diagram of the $|T_{d1}|$ and $|T_{d2}|$(single-stage)

Similarly, $|T_{d1}|$ and $|T_{d2}|$ will also face a problem where the vibration isolation system cannot simultaneously enter the ideal vibration isolation area, as shown in Fig. 3.7. Since Eqs. (3.10) and (3.11) are completely consistent with the expressions of the $\left|T_g\right|$ and $|T_I|$ transmissibility for power equipment with a single-stage vibration isolation system, the MOPSO investigation will not be repeated here. Accordingly, the two-stage vibration isolation system of vibration-sensitive equipment will be studied.

3.1.4 Two-Stage Vibration Isolation System

In the two-stage vibration isolation system shown in Fig. 1.4b, m_2 is the mass of sensitive equipment, k_2 and c_2 are the stiffness and damping of the vibration isolation system, m_1, k_1, and c_1 are the mass, stiffness, and damping of the foundation or placement platform, respectively, the vibration load form is consistent with the single-stage vibration isolation system, the equipment's displacement, velocity, and acceleration are denoted $x_2, \dot{x}_2, \ddot{x}_2$, respectively, the displacement, velocity, and acceleration of the installation platform or foundation are denoted $x_1, \dot{x}_1, \ddot{x}_1$, respectively. The dynamic equation at this time is:

$$\begin{cases} m_2\ddot{x}_2 + c_2\dot{x}_2 - c_2\dot{x}_1 + k_2x_2 - k_2x_1 = 0 \\ m_1\ddot{x}_1 - c_2\dot{x}_2 + (c_2 + c_1)\dot{x}_1 - k_2x_2 + (k_2 + k_1)x_1 = k_1x_g + c_1\dot{x}_g \end{cases} \tag{3.12}$$

The displacement transmissibility, velocity transmissibility, or acceleration transmissibility and the relative displacement transmissibility, relative velocity transmissibility, or relative acceleration transmissibility of sensitive equipment can be obtained by performing Laplace transform on Eq. (3.12) as follows:

$$T_{d1} = \frac{X_2(s)}{X_g(s)} = \frac{c_1c_2s^2 + c_1k_2s + c_2k_1s + k_1k_2}{m_1m_2s^4 + (m_1c_2 + m_2c_2 + m_2c_1)s^3 + (m_1k_2 + c_1c_2 + k_2m_2 + m_2k_1)s^2 + (c_1k_2 + c_2k_1)s + k_1k_2} \tag{3.13a}$$

$$T_{d2} = \frac{X_2(s) - X_1(s)}{X_g(s)} = \frac{-(m_2c_1s^3 + m_2k_1s^2)}{m_1m_2s^4 + (m_1c_2 + m_2c_2 + m_2c_1)s^3 + (m_1k_2 + c_1c_2 + k_2m_2 + m_2k_1)s^2 + (c_1k_2 + c_2k_1)s + k_1k_2} \tag{3.13b}$$

where $u = m_1/m_2$, $\omega_1 = \sqrt{k_1/m_1}$, $\omega_2 = \sqrt{k_2/m_2}$, $\eta_1 = c_1/(2m_1\omega_1)$, $\eta_2 = c_2/(2m_2\omega_2)$, $\overline{\omega} = \omega_2/\omega_{ref}$, $\gamma = \omega_1/\omega_2$.

Equation (3.13) can be rearranged into the following form:

$$|T_{d1}| = \left|\frac{X_2(s)}{X_g(s)}\right| = \sqrt{\frac{(u\gamma^2\overline{\omega}^4 - 4\eta_1\eta_2u\gamma\overline{\omega}^2)^2 + (2\eta_1u\gamma\overline{\omega}^3 + 2\eta_2u\gamma^2\overline{\omega}^3)^2}{[u - (u + 4\eta_2\eta_1u\gamma + 1 + u\gamma^2)\overline{\omega}^2 + u\gamma^2\overline{\omega}^4]^2 + [-(2\eta_2u + 2\eta_2 + 2\eta_1\gamma u)\overline{\omega} + (2\eta_1\gamma u + 2\eta_2u\gamma^2)\overline{\omega}^3]^2}} \tag{3.14a}$$

$$|T_{d2}| = \left|\frac{X_2(s) - X_1(s)}{X_g(s)}\right| = \sqrt{\frac{(u\gamma^2\varpi^2)^2 + (2\eta_1u\varpi\gamma)^2}{[u - (u + 4\eta_2\eta_1u\gamma + 1 + u\gamma^2)\overline{\omega}^2 + u\gamma^2\overline{\omega}^4]^2 + [-(2\eta_2u + 2\eta_2 + 2\eta_1\gamma u)\overline{\omega} + (2\eta_1\gamma u + 2\eta_2u\gamma^2)\overline{\omega}^3]^2}} \tag{3.14b}$$

Equation (3.14) is illustrated in Fig. 3.8. It can be seen that $|T_{d1}|$ and $|T_{d2}|$ enter the ideal vibration isolation area simultaneously and reach a small transmittance value, effectively overcoming the shortcomings of the single-stage vibration isolation system. The changes in $|T_{d1}|$ and $|T_{d2}|$ with respect to the parameters $\eta_1, \eta_2, u, \gamma$ are shown in Figs. 3.9 and 3.10. It can be seen that increasing η_1 reduces the peak value of $|T_{d1}|$, but it negatively affects the ideal vibration isolation area. The change in η_2 is similar to that of η_1. The increase in u rises the peak value of $|T_{d1}|$ and impacts the two peaks' position. The increase in γ negatively influences the ideal vibration isolation area. Moreover, it can be noticed that the increase in η_1 reduces the peak value of $|T_{d2}|$, which is not good for the ideal vibration isolation area. The increase in η_2 decreases the peak value of $|T_{d2}|$, while not significantly impacting the ideal vibration isolation area. The increase in u rises the peak value of $|T_{d2}|$, which affects the two peaks' positions but does not significantly impacts the influence on the ideal vibration isolation area. The increase in γ is not good for the ideal vibration isolation area.

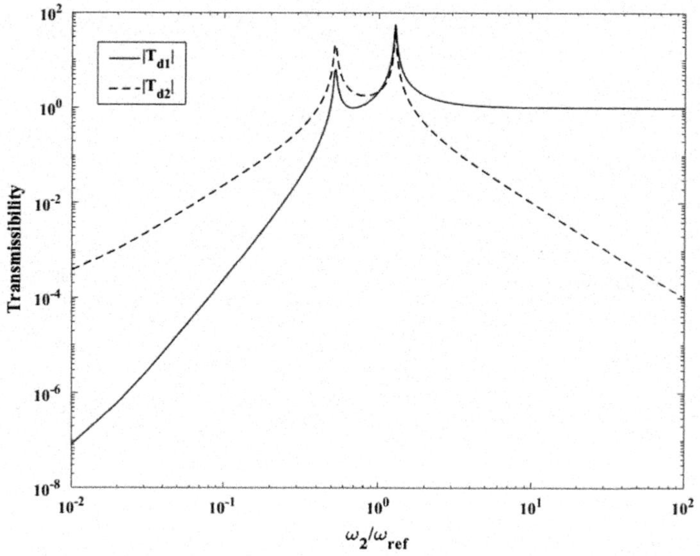

Fig. 3.8 Schematic diagram of $|T_{d1}|$ and $|T_{d2}|$ (double-stage)

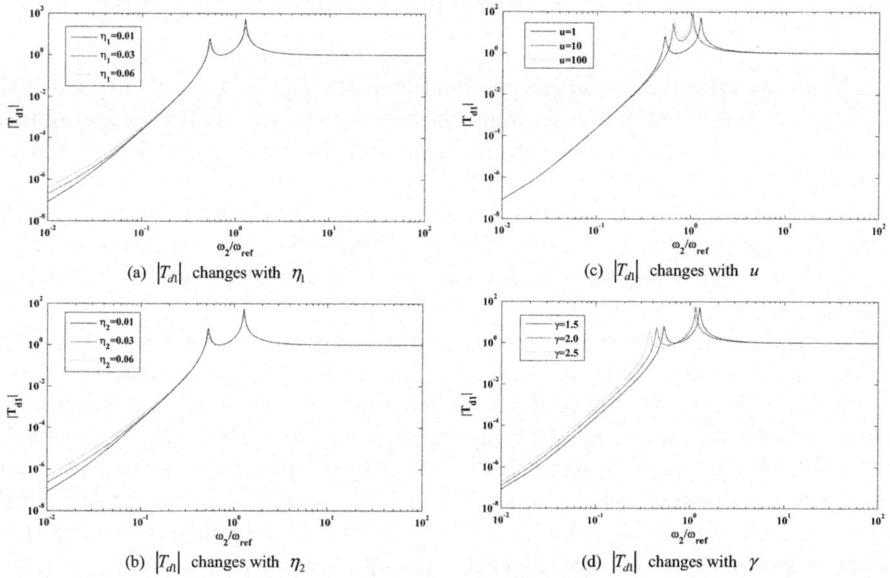

Fig. 3.9 Schematic diagram of $|T_{d1}|$ varied with isolation parameters (double-stage)

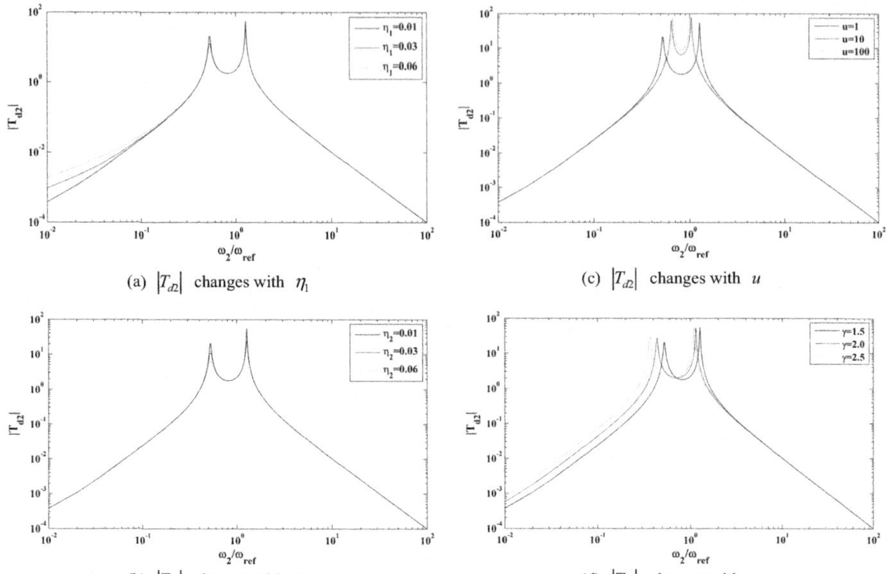

(a) $|T_{d2}|$ changes with η_1

(b) $|T_{d2}|$ changes with η_2

(c) $|T_{d2}|$ changes with u

(d) $|T_{d2}|$ changes with γ

Fig. 3.10 Schematic diagram of $|T_{d2}|$ varied with isolation parameters (double-stage)

While analyzing the two-stage vibration isolation system of the vibration-sensitive equipment using the MOPSO algorithm, the basic parameter settings were set similar to those mentioned before, with the exception that η_1, η_2, u, γ and $\overline{\omega}$ ranges were arbitrarily set to $[0.01, 0.01, 0.01, 0.01, 0.01] \sim [1, 1, 10, 10, 100]$. The results of the four tests are shown in Fig. 3.11. Under the two-stage vibration isolation system, both $|T_{d1}|$ and $|T_{d2}|$ can reach a small transmissibility value, achieving a combination of them. The obtained gbest solution can optimally design the vibration isolation parameters.

In this chapter, the dynamic transmissibility of single-stage and two-stage uncontrolled vibration isolation systems in power and sensitive equipment is derivated and studied. Within this context, MOPSO was introduced to compare the performance of these two systems. In general, the study results show that the two-stage uncontrolled vibration isolation system effectively overcomes the shortcomings of the single-stage uncontrolled vibration isolation system and is more suitable for uncontrolled vibration isolation of two equipment types. It is worth mentioning that multi-objective optimization using the MOPSO algorithm can optimize the design of the parameters for the uncontrolled vibration isolation system, hence, the obtained gbest solution.

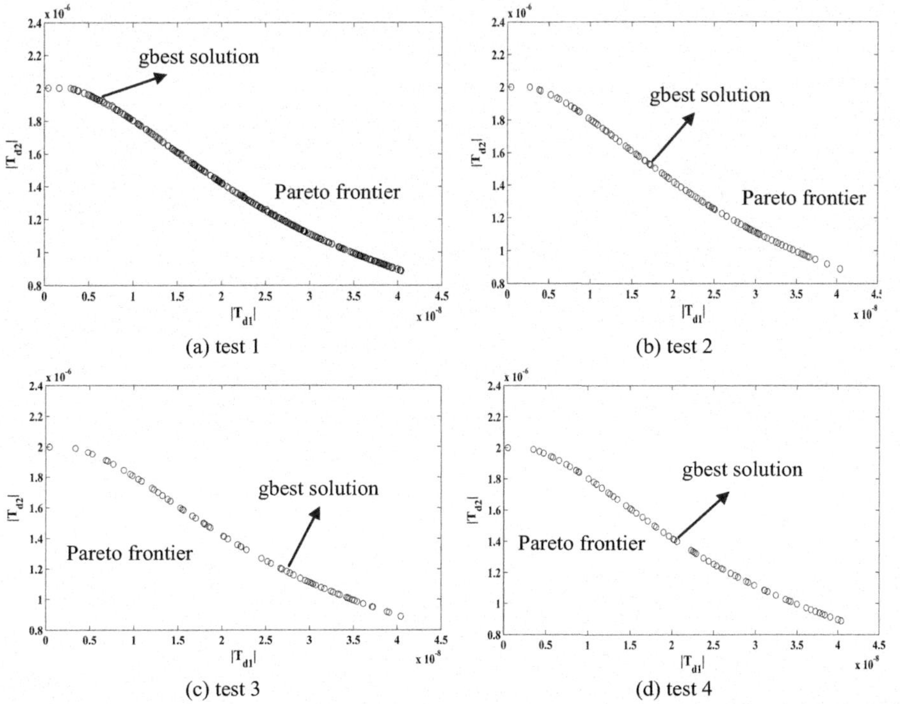

Fig. 3.11 Multi-objective optimization results of the $|T_{d1}|$ and $|T_{d2}|$ based on MOPSO (double-stage)

References

1. Voigtländer B, Coenen P, Cherepanov V et al (2017) Low vibration laboratory with a single-stage vibration isolation for microscopy applications. Rev Sci Instrum 88(2):023703
2. Preumont A, Horodinca M, Romanescu I et al (2007) A six-axis single-stage active vibration isolator based on Stewart platform. J Sound Vib 300(3–5):644–661
3. Mousavi SH (2020) Modeling and controlling a semi-active nonlinear single-stage vibration isolator using intelligent inverse model of an MR damper. J Mech Sci Technol 34(9):3525–3532
4. Farshidianfar A, Saghafi A, Kalami SM et al (2012) Active vibration isolation of machinery and sensitive equipment using H∞ control criterion and particle swarm optimization method. Meccanica 47(2):437–453
5. Wang X, Zhou J, Xu D et al (2017) Force transmissibility of a two-stage vibration isolation system with quasi-zero stiffness. Nonlinear Dyn 87(1):633–646
6. Lu Z, Yang T, Brennan MJ et al (2014) On the performance of a two-stage vibration isolation system which has geometrically nonlinear stiffness. J Vib Acoust 136(6)
7. Lu Z, Brennan MJ, Yang T et al (2013) An investigation of a two-stage nonlinear vibration isolation system. J Sound Vib 332(6):1456–1464
8. Sommerfeldt SD, Tichy J (1990) Adaptive control of a two-stage vibration isolation mount. J Acoust Soc Am 88(2):938–944
9. Xu YL, Li B (2006) Hybrid platform for high-tech equipment protection against earthquake and micro vibration. Earthquake Eng Struct Dynam 35(8):943–967

Chapter 4
Optimized Active Control for Equipment

Abstract In this chapter, active control for power equipment and sensitive equipment are performed out by using proportional-integral-differential (PID) control, Linear Quadratic Regulator (LQR) control, Linear Quadratic Gaussian (LQG) control, H_2/H_∞ control, fuzzy logic control, and the controllers are optimized by PSO technique. Besides, active control based on multi-objective control output and multi-objective H_2/H_∞ control for the two types of equipment is also performed out by MOPSO technique.

4.1 Optimized PID Active Control

4.1.1 PID Control Algorithm

PID (proportional-integral–differential) control [1] is a linear control algorithm that performs the proportional P, integral I, and differential D operations on the deviation $e(t)$, $e(t) = r(t) - y(t)$, between the input $r(t)$ and the output $y(t)$. Besides, it adds the three types of operation results to obtain the actual control output, as shown in Fig. 4.1. The expression of the PID control algorithm in the continuous time domain is as follows:

$$u(t) = k_p\left[e(t) + \frac{1}{T_i}\int_0^t e(t)\mathrm{d}t + T_d\frac{\mathrm{d}(t)}{\mathrm{d}t}\right] \tag{4.1}$$

where k_p is the proportionality coefficient, T_i is the integration time constant, and T_d is the differential time constant.

The role of each link in the PID controller is as follows:

(1) Proportional link (P): proportional adjustment of the deviation $e(t)$ in the control process, where as long as the deviation occurs, the link will immediately produce a control effect to reduce the error.

W. Huang and J. Xu, *Optimized Engineering Vibration Isolation, Absorption and Control*, Springer Tracts in Civil Engineering, https://doi.org/10.1007/978-981-99-2213-0_4

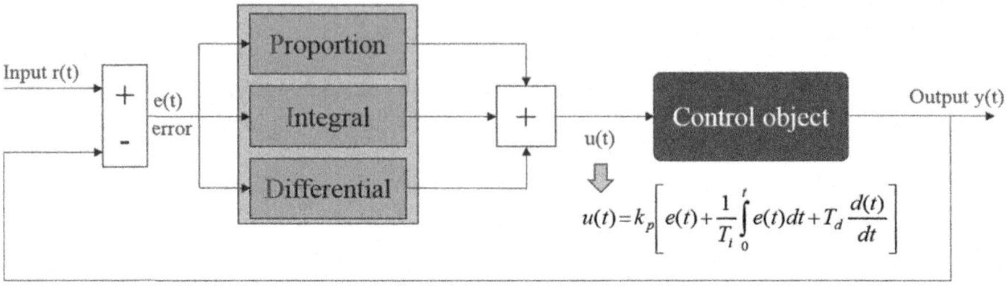

Fig. 4.1 PID control

(2) Integral link (I): eliminate the static difference to improve the invariance of the control system. The size of the integral effect depends on T_i, where the smaller the T_i, the more substantial the integral effect, and the weaker the contrary.
(3) Differential link (D): reflect the rate of change of reaction deviation used to adjust the error's differential output. It controls the error in real-time when it suddenly changes. Before the deviation signal becomes large, an early signal can be introduced into the control system to correct the signal, accelerating the control system's movement and reducing the adjustment time.

The above three links in the PID control process can obtain good control performance by combining their respective advantages.

Digital control systems are mostly sampling controls. Generally, the control performance is calculated based on the sampling time's deviation. Therefore, the differential and integral terms in Eq. (4.1) must be discretized. This can be achieved by using a series of discrete sampling moments kT to represent continuous time t, replacing the integrals b sums, and differentiating in increments, as follows:

$$t \approx kT(k = 0, 1, 2, \ldots) \tag{4.2a}$$

$$\int_0^t e(t)dt \approx T\sum_{j=0}^{k} e(jT) = T\sum_{j=0}^{k} e(j) \tag{4.2b}$$

$$\frac{\mathrm{d}e(t)}{\mathrm{d}t} \approx \frac{e(kT) - e[(k-1)T]}{T} = \frac{e(k) - e(k-1)}{T} \tag{4.2c}$$

For simplicity, $e(kT)$ can be expressed as $e(k)$, which results in the following discrete PID expression:

$$u(k) = k_p\left\{e(k) + \frac{T}{T_i}\sum_{j=0}^{k} e(j) + \frac{T_d}{T}[e(k) - e(k-1)]\right\}$$

$$= k_p e(k) + k_i \sum_{j=0}^{k} e(j)T + k_d \frac{e(k) - e(k-1)}{T} \tag{4.3}$$

where k_p is the scale factor, k_i is the integration factor, and k_d is the differential factor, $u(k)$ is the output of the controller at the kth sampling time, $e(k)$ is the deviation of the control system at the kth sampling time, $e(k-1)$ is the deviation of the control system at the $k-1$ th sampling time, and T is the sampling period.

4.1.2 Optimized PID Active Control for Power Equipment

Consider the active control system for the power equipment shown in Fig. 1.5a. Let $F_a(t)$ be the active controller that drives the control force generated by the actuator, then Eq. (4.4) describes the system's equation of motion.

$$\begin{cases} \mathrm{m}_1\ddot{x}_1 + \mathrm{k}_1 x_1 + \mathrm{c}_1\dot{x}_1 - \mathrm{c}_2(\dot{x}_2 - \dot{x}_1) - \mathrm{k}_2(x_2 - x_1) = F_a(t) \\ \mathrm{m}_2\ddot{x}_2 + \mathrm{c}_2(\dot{x}_2 - \dot{x}_1) + \mathrm{k}_2(x_2 - x_1) = F(t) - F_a(t) \end{cases} \tag{4.4}$$

Equation (4.5) can be obtained by assuming the initial state to be zero and performing Laplace transform on (4.4).

$$\begin{cases} (\mathrm{m}_1 s^2 + \mathrm{k}_1 + \mathrm{k}_2 + \mathrm{c}_1 s + \mathrm{c}_2 s)X_1(s) - (\mathrm{c}_2 s + \mathrm{k}_2)X_2(s) = F_a(s) \\ (\mathrm{m}_2 s^2 + \mathrm{c}_2 s + \mathrm{k}_2)X_2(s) - (\mathrm{c}_2 s + \mathrm{k}_2)X_1(s) = F(s) - F_a(s) \end{cases} \tag{4.5}$$

where s is the complex frequency.

Assume that the force transmitted to the foundation by the power equipment is $F_{out} = \mathrm{k}_1 x_1 + \mathrm{c}_1 \dot{x}_1$ and the Laplace transform form is $(\mathrm{k}_1 + \mathrm{c}_1 s)X_1(s)$, the following intermediate variables can be derived:

$$\begin{cases} \Delta_1 = \mathrm{m}_1 s^2 + \mathrm{c}_1 s + \mathrm{c}_2 s + \mathrm{k}_1 + \mathrm{k}_2 \\ \Delta_2 = \mathrm{c}_2 s + \mathrm{k}_2 \\ \Delta_3 = \mathrm{m}_2 s^2 + \mathrm{c}_2 s + \mathrm{k}_2 \\ \Delta_4 = \mathrm{c}_1 s + \mathrm{k}_1 \end{cases} \tag{4.6}$$

(1) The transfer function can be written, given $F_a(s) = 0$, as follows:

$$G_1(s) = \frac{F_{out}(s)}{F(s)} = \frac{\Delta_2 \Delta_4}{\Delta_1 \Delta_3 - \Delta_2^2} \tag{4.7}$$

(2) Let $F(s) = 0$, then the second transfer function can be written as follows:

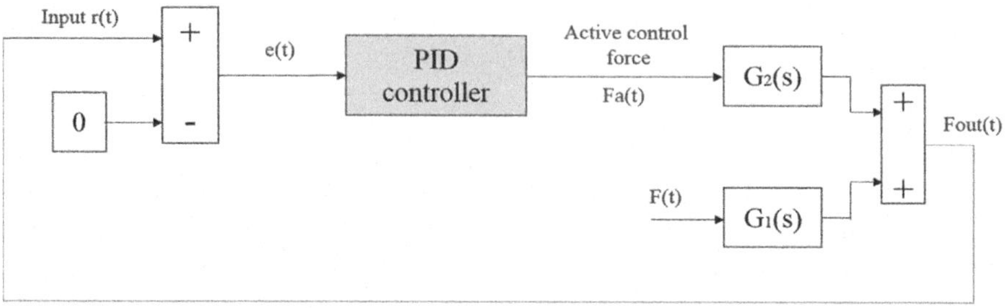

Fig. 4.2 PID active control for power equipment

$$G_2(s) = \frac{F_{out}(s)}{F_a(s)} = \frac{(\Delta_3 - \Delta_2)\Delta_4}{\Delta_1\Delta_3 - \Delta_2^2} \tag{4.8}$$

The active PID control system can be developed using Eqs. (4.7) and (4.8), as shown in Fig. 4.2.

The PID control determines the proportional, integral, and differential coefficients upon obtaining an exact mathematical model of the controlled object based on a specific parameter tuning algorithm [1]. As mentioned above, each of the three PID links has its own functions. In general, the approach to reasonably setting the parameters has always been a problem that plagued the engineering community. Typically, conventional parameter tuning relies on the technicians' experience, which is time-consuming and laborious. In addition, the lag and nonlinear phenomena in actual engineering increase the difficulty in PID parameter tuning. Accordingly, this book proposes a PSO-PID active control strategy to set the PID parameters, as shown in Fig. 4.3.

The fitting function for the PSO-PID active control shown in Fig. 4.3 is selected as $\|F_{out} - F_{desired}\|_\infty$, where $F_{desired} = 0$. The goal herein is to reduce the peak force transmitted by the power equipment to the foundation.

Example 4.1 The vibration isolation system's parameters for the power equipment are: $m_1 = 1200$ kg, $m_2 = 600$ kg, $k_1 = 1 \times 10^6$N/m, $k_2 = 1.5 \times 10^4$N/m, $c_1 = 1.6 \times 10^4\text{N} \cdot \text{m/s}$, and $c_2 = 1 \times 10^3\text{N} \cdot \text{m/s}$. The excitation load amplitude is $F_0 = 1000$ N, and the vibration frequency is $f = 2.6$Hz. The PSO algorithm parameters are: the particle population is 100, the number of iterations is 200, $c_1 = 2, c_2 = 1$,

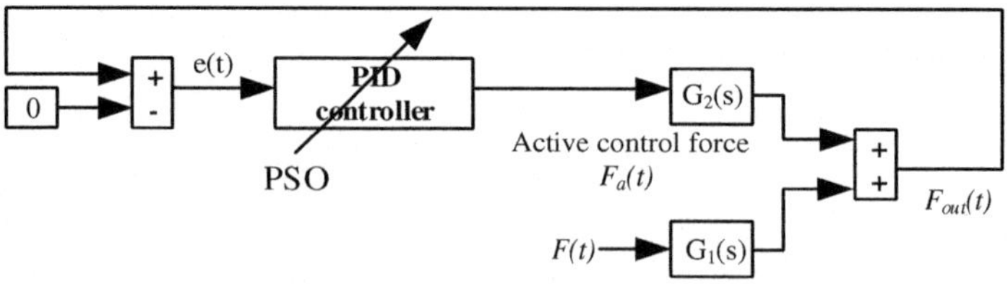

Fig. 4.3 PSO-PID active control for power equipment

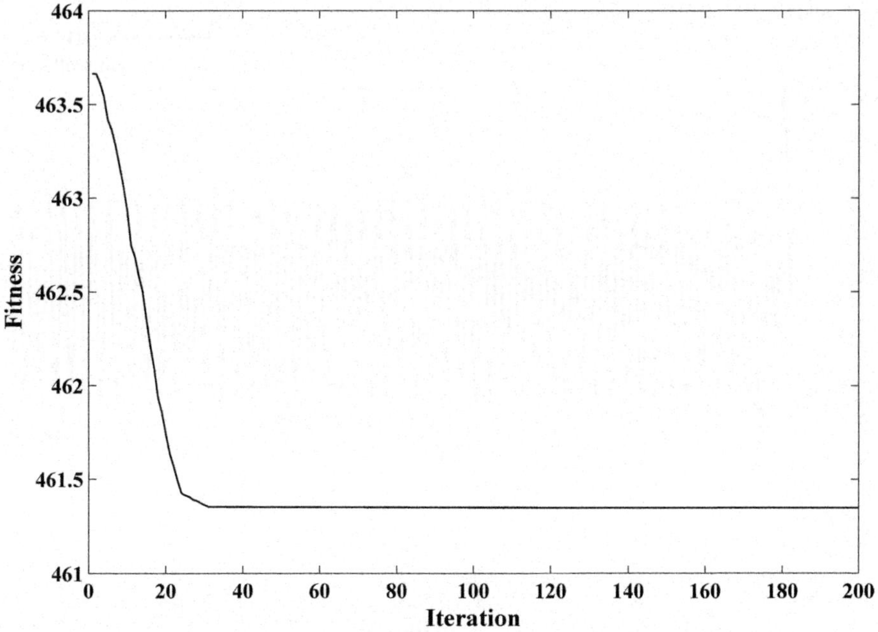

Fig. 4.4 Fitting convergence of PSO-PID active control for power equipment

and $\omega = 0.99^t$, where t is the number of iterations. The parameter search range of the proportional factor in the PID controller k_p, integration factor k_i, and differential factor k_d is arbitrarily set to $[-1000, -1000, -1000] \sim [1000, 1000, 1000]$. The fitting curve obtained herein is shown in Fig. 4.4, where the gbest solution is $k_p = -972.43$, $k_i = -983.76$, and $k_d = -850.35$.

Based on these optimal PID control parameters, the force transmitted to the foundation of the power equipment with an active control system is shown in Fig. 4.5. Compared to the uncontrolled vibration isolation system, the proposed PSO-PID active control strategy has a better control effect. In other words, the active control system improves the effect of uncontrolled vibration isolation. The control force generated by the actuator is shown in Fig. 4.6.

4.1.3 Optimized PID Active Control for Sensitive Equipment

Consider the active control system of the vibration-sensitive equipment shown in Fig. 1.5b. The external load is the interference force $F(t)$ acting on the supporting structure. This load has a simple harmonic form with a ω_{ref} frequency. The control force ($F_a(t)$) driven by the lower actuator for the active controller and the equation of motion can be expressed as follows:

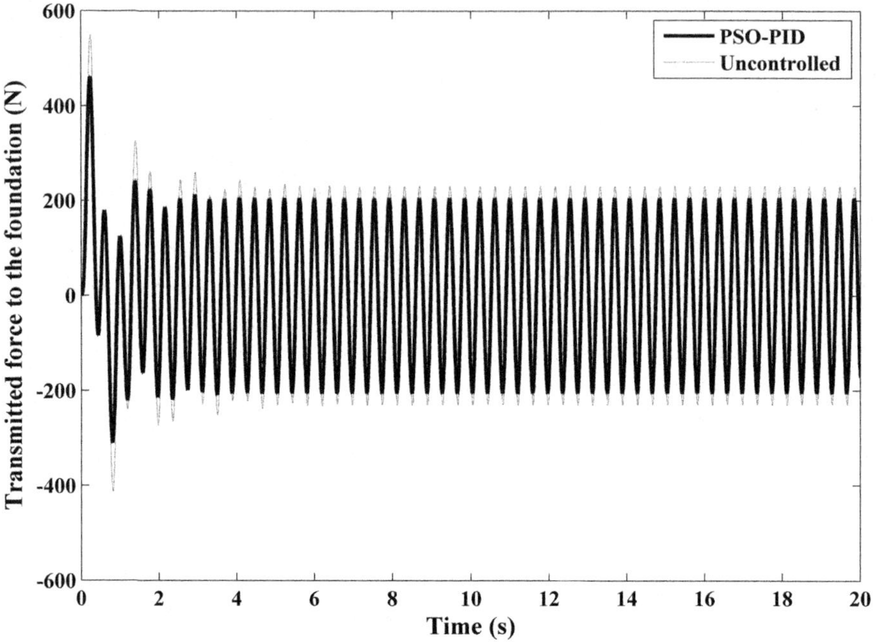

Fig. 4.5 Force transmitted from power equipment to the foundation (PSO-PID)

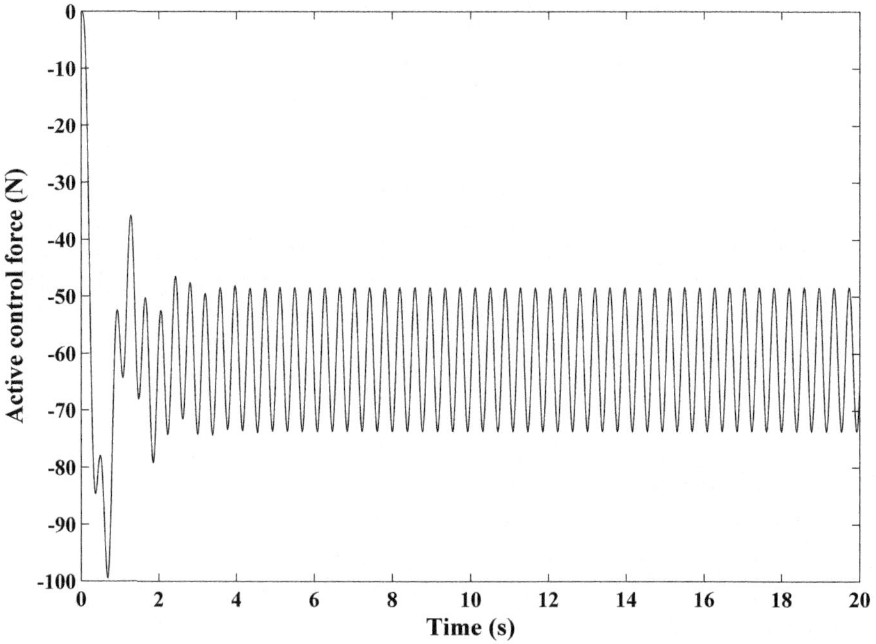

Fig. 4.6 Active control force of the PSO-PID strategy in the power equipment

$$\begin{cases} \mathrm{m}_1\ddot{x}_1 + \mathrm{k}_1 x_1 + \mathrm{c}_1\dot{x}_1 - \mathrm{c}_2(\dot{x}_2 - \dot{x}_1) - \mathrm{k}_2(x_2 - x_1) = F(t) - F_a(t) \\ \mathrm{m}_2\ddot{x}_2 + \mathrm{c}_2(\dot{x}_2 - \dot{x}_1) + \mathrm{k}_2(x_2 - x_1) = F_a(t) \end{cases} \tag{4.9}$$

By assuming the initial state to be zero and performing Laplace transform on Eq. (4.9), with a complex frequency of $s = j\omega_{ref}$, the following equation can be obtained,

$$\begin{cases} (\mathrm{m}_1 s^2 + \mathrm{k}_1 + \mathrm{k}_2 + \mathrm{c}_1 s + \mathrm{c}_2 s)X_1(s) - (\mathrm{c}_2 s + \mathrm{k}_2)X_2(s) = F_a(s) \\ (\mathrm{m}_2 s^2 + \mathrm{c}_2 s + \mathrm{k}_2)X_2(s) - (\mathrm{c}_2 s + \mathrm{k}_2)X_1(s) = F(s) - F_a(s) \end{cases} \tag{4.10}$$

Let the vibration velocity for sensitive equipment (the active control system output) be $\{\dot{x}_2\}$, and the Laplace transformation be $\{sX_2(s)\}$. Then the intermediate variables can be defined as follows:

$$\begin{cases} \Delta_1 = \mathrm{m}_1 s^2 + \mathrm{c}_1 s + \mathrm{c}_2 s + \mathrm{k}_1 + \mathrm{k}_2 \\ \Delta_2 = \mathrm{c}_2 s + \mathrm{k}_2 \\ \Delta_3 = \mathrm{m}_2 s^2 + \mathrm{c}_2 s + \mathrm{k}_2 \end{cases} \tag{4.11}$$

(1) The transfer function can be obtained given $F_a(s) = 0$ as follows:

$$G_1(s) = \frac{sX_2(s)}{F(s)} = \frac{s\Delta_1}{\Delta_1\Delta_3 - \Delta_2^2} \tag{4.12}$$

(2) The following transfer function can be obtained given $F(s) = 0$ as follows:

$$G_2(s) = \frac{sX_2(s)}{F_a(s)} = \frac{s(\Delta_2 - \Delta_1)}{\Delta_1\Delta_3 - \Delta_2^2} \tag{4.13}$$

Equations (4.12) and (4.13) can be used to develop a PID active control system with a basic principle similar to that in Fig. 4.2. The constituted PSO-PID active control has the same principle as that shown in Fig. 4.3. Moreover, the fitting function for the PSO-PID control is taken as $\left\| \dot{x}_2 - x_{2,desired} \right\|_\infty$, where $\dot{x}_{2,desired} = 0$.

Example 4.2 The vibration isolation system parameters of the sensitive equipment are: $\mathrm{m}_1 = 1200$ kg, $\mathrm{m}_2 = 100$ kg, $\mathrm{k}_1 = 1 \times 10^6$N/m, $\mathrm{k}_2 = 1.5 \times 10^4$N/m, $\mathrm{c}_1 = 1.6 \times 10^4$N · m/s, and $\mathrm{c}_2 = 1 \times 10^3$N · m/s. The excitation amplitude is $\mathrm{F}_0 = 1$N, and the frequency is $f = 0.6$Hz. The PSO algorithm parameter is the same as Example 4.1. The parameter search range for the proportional factor in the PID controller k_p, the integration factor k_i, and the differential factor k_d is arbitrarily set to [−1000, −1000, −1000] ~[1000, 1000, [1000, 1000,1000]. The fitting curve is shown in Fig. 4.7, and the obtained gbest solution is $k_p = -990.27$, $k_i = -942.35$, and $k_d = -899.16$.

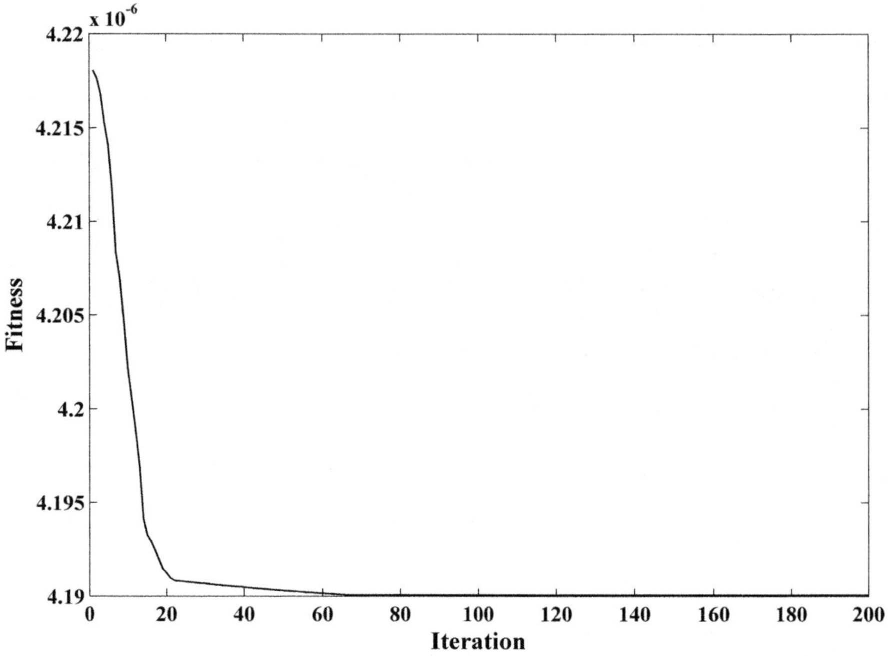

Fig. 4.7 Fitting convergence of a PSO-PID active control in sensitive equipment

Based on these optimal PID control parameters, the velocity of the sensitive equipment upon control is shown in Fig. 4.8. It can be seen that, compared to the uncontrolled vibration isolation system, the PSO-PID active control strategy effectively improves the isolation effect. The active control force generated by the actuator is shown in Fig. 4.9.

4.2 Optimized LQR Active Control

4.2.1 LQR Control Algorithm

In a linear system, if the integral of the quadratic function in a state variable $z(t)$ and the control variable $U(t)$ is taken as the functional index, then the optimal control problem of the system can be called the optimal control problem of the optimal quadratic performance index. This issue is typically referred to as a linear quadratic problem. The advantage of this problem is that it results in a unified analytical expression of the optimal control solution $U^*(t)$ as well as an optimal state feedback law that is simple to implement and easy to use in actual engineering [2].

Assume the state space equation of the controlled system is:

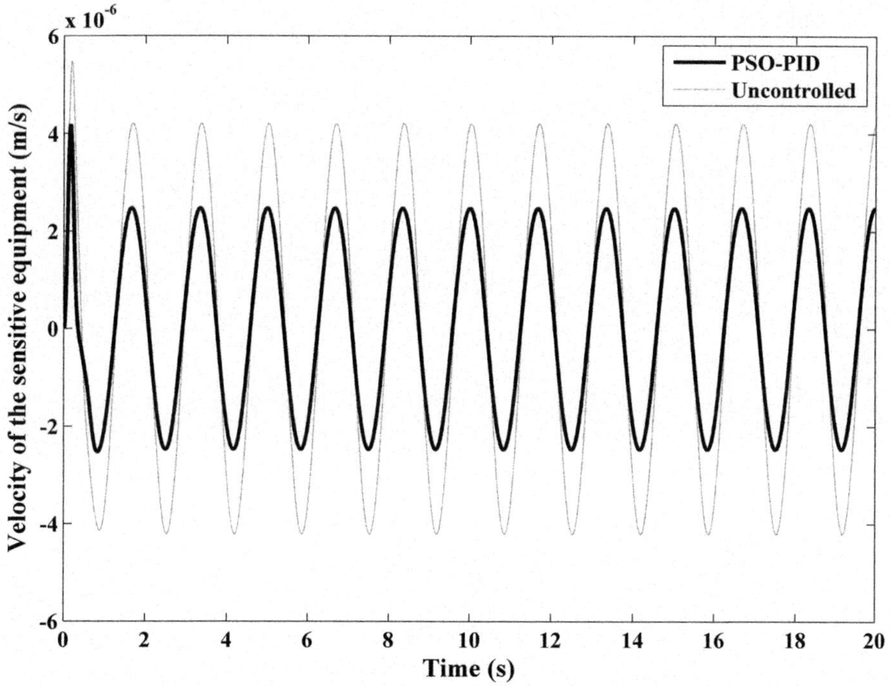

Fig. 4.8 Vibration velocity of sensitive equipment (PSO-PID)

$$\dot{z}(t) = \mathrm{A}z(t) + \mathrm{B}U(t) \tag{4.14a}$$

$$y(t) = Cz(t) \tag{4.14b}$$

The functional quadratic performance of the control system can be defined as follows:

$$J = \frac{1}{2}\int_{0}^{\infty}\left[z^{\mathrm{T}}(t)Qz(t) + U^{\mathrm{T}}(t)RU(t)\right]\mathrm{d}t \tag{4.15}$$

where Q is the semi-definite positive matrix, and R is the definite positive matrix.

The goal of system control is to induce a load in the system when the state of the system deviates from the equilibrium state (or zero state). Additionally, it aims to achieve system equilibrium without consuming too much energy. The problem herein is to search for the optimal control $U(t)$ in the infinite time interval $[t_0, \infty)$ so that the system transitions from the initial state z_0 to the zero state, while minimizing Eq. (4.15). The mathematical description of the optimal control problem is the solution of $U(t)$ $(t_0 \leq t < \infty)$, its minimized objective function is defined in Eq. (4.15), and its constraint conditions are given in Eq. (4.14a). When the

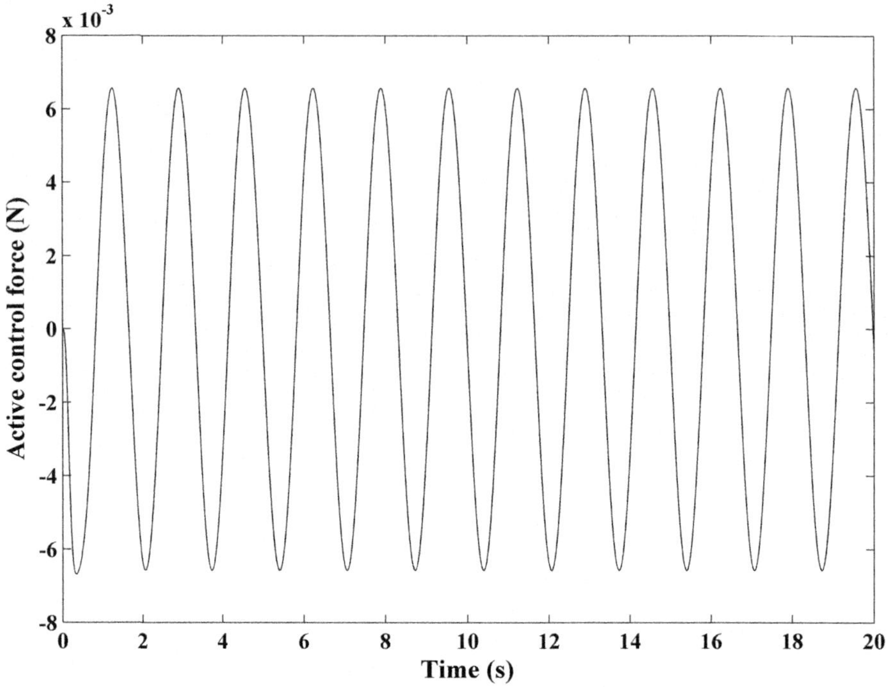

Fig. 4.9 Active control force of the PSO-PID strategy for sensitive equipment

value of $U(t)$ is not restricted, the constraint that must be imposed in the system state is given in Eq. (4.14b). The above-mentioned equation-constrained functional extremum problem can be equivalently transformed into an unconstrained functional extremum problem by using the Lagrangian multiplier and defining the sub-vector $\lambda(t) \in R^n$ to take the following Lagrange function [3]:

$$L_{fun} = \int_{t_0}^{\infty} \left[\frac{1}{2}\left(z^{\mathrm{T}} Q z + U^{\mathrm{T}} R U\right) + \lambda^{\mathrm{T}}(Az + BU - \dot{z}) \right] \mathrm{d}t \tag{4.16}$$

The above optimal control problem can be transformed into an unconditional functional extreme value problem (solving $U(t)(t_0 \leq t < \infty)$) by minimizing the objective function L_{fun}.

Consider the variational method, and perform the following transformation first:

$$\int_{t_0}^{\infty} \lambda^{\mathrm{T}} \dot{z} \mathrm{d}t = \lambda^{\mathrm{T}} z \Big|_{t_0}^{\infty} - \int_{t_0}^{\infty} \dot{\lambda}^{\mathrm{T}} z \mathrm{d}t \tag{4.17}$$

Substitute Eq. (4.17) into Eq. (4.16) to obtain the following expression:

$$L_{fun} = \int_{t_0}^{\infty} \left[H(z, U, \lambda, t) + \dot{\lambda}^{\mathrm{T}} z\right]\mathrm{d}t - \lambda^T z\big|_{t_0}^{\infty} \tag{4.18}$$

where

$$H(z, U, \lambda, t) = \frac{1}{2}(z^T Qz + U^T RU) + \lambda^T (Az + BU) \tag{4.19}$$

Given the variational form of $z(t)$, $U(t)$, and $\lambda(t)$: δz,δU and $\delta\lambda$, and the functional increment they cause: $\delta L_{fun,z}$,$\delta L_{fun,U}$ and $\delta L_{fun,\lambda}$. The variational increment relationship can be obtained by considering the first-order derivative:

$$\delta L_{fun,z} = \delta z^{\mathrm{T}} \frac{\partial L_{fun}}{\partial z} = \int_{t_0}^{\infty} \delta z^{\mathrm{T}} \left(\frac{\partial H}{\partial z} + \dot{\lambda} \right) dt - \delta z^{\mathrm{T}} \lambda\big|_{t_0}^{\infty} \tag{4.20a}$$

$$\delta L_{fun,U} = \delta U^{\mathrm{T}} \frac{\partial H}{\partial U} \mathrm{d}t \tag{4.20b}$$

$$\delta L_{fun,\lambda} = \delta \lambda^{\mathrm{T}} \frac{\partial L_{fun}}{\partial \lambda} = \int_{t_0}^{\infty} \delta \lambda^{\mathrm{T}} \left(\frac{\partial H}{\partial \lambda} - \dot{z} \right) \mathrm{d}t \tag{4.20c}$$

$$\delta L_{fun} = \delta L_{fun,z} + \delta L_{fun,U} + \delta L_{fun,\lambda} \tag{4.20d}$$

Since δz,δU, and $\delta\lambda$ are arbitrarily set, and $\delta L_{fun} = 0$, the required conditions to minimize L_{fun} are as follows:

$$\frac{\partial H}{\partial z} + \dot{\lambda} = 0 \tag{4.21a}$$

$$\frac{\partial H}{\partial U} = 0 \tag{4.21b}$$

$$\frac{\partial H}{\partial \lambda} - \dot{z} = 0 \tag{4.21c}$$

$$\delta z^T \lambda\big|_{t_0}^{\infty} = 0 \tag{4.21d}$$

Due to fixed initial conditions, if $\delta z\big|_{t_0} = 0$, $t \to \infty$, and $\delta z \neq 0$.

$$\lambda(t)\big|_{t \to \infty} = 0 \tag{4.22}$$

Substituting Eq. (4.19) into Eq. (4.21b) gives:

$$\frac{\partial H}{\partial U} = RU + B^{\mathrm{T}}\lambda = 0 \tag{4.23}$$

Since R is positive, the following expression is true:

$$U(t) = -R^{-1}B^{T}\lambda(t) \tag{4.24}$$

From Eqs. (4.19) and (4.21a), Eq. (4.25) can be obtained:

$$\dot{\lambda} = -\frac{\partial H}{\partial z} = -Qz - A^{\mathrm{T}}\lambda \tag{4.25}$$

The $U(t)$ in Eq. (4.24) is the linear function of $\lambda(t)$. In order to enable $U(t)$ as the state feedback, a linear transformation relationship between $\lambda(t)$ and $z(t)$ should be established as follows:

$$\lambda(t) = P(t)z(t) \tag{4.26}$$

Substituting Eq. (4.26) into Eq. (4.24) results in the following:

$$U(t) = -R^{-1}B^{\mathrm{T}}P(t)z(t) \tag{4.27}$$

Equation (4.27) can be rewritten into:

$$U(t) = -Gz(t), \quad G = R^{-1}B^{T}P(t) \tag{4.28}$$

where G is the optimal state feedback gain matrix.

From Eq. (4.26), the left part of Eq. (4.25) can be rewritten as:

$$\dot{\lambda} = \dot{P}(t)\, z(t) + P(t)\, \dot{z}(t) \tag{4.29}$$

Moreover, the right part of Eq. (2.25) can be rearranged as follows:

$$-Qz(t) - A^{\mathrm{T}}\lambda(t) = -\left[Q + A^{\mathrm{T}}P(t)\right]z(t) \tag{4.30}$$

Equation (4.31) can be obtained by substituting Eqs. (4.29) and (4.30) into Eq. (4.25) and combining Eqs. (4.12a) and (4.27).

$$\dot{P}(t) = -P(t)A - A^{\mathrm{T}}P(t) + P(t)BR^{-1}B^{\mathrm{T}}P(t) - Q \tag{4.31}$$

Equation (4.31) is called the Riccati equation. If P is the solution of the Riccati equation, then (4.27) and (4.28) can be changed to:

$$U(t) = -R^{-1}B^{\mathrm{T}}Pz(t) \tag{4.32}$$

Finally, substituting Eq. (4.32) into Eq. (4.14a) gives the following:

$$\dot{z}(t) = (A - BR^{-1}B^{\mathrm{T}}P)z(t) \tag{4.33}$$

4.2.2 *Optimized LQR Active Control for Power Equipment*

Consider the active control system for power equipment as shown in Fig. 1.5 (a). The system's dynamic response is expressed in Eq. (4.4). The state variables are: $z_1 = x_1$, $z_2 = x_2, z_3 = \dot{x}_1$, and $z_4 = \dot{x}_2$. and a state vector $z(t) = [z_1, z_2, z_3, z_4]^{\mathrm{T}}$ is formed here. Additionally, Eq. (4.4) can be transformed into the following state-space equation form:

$$\begin{cases} \dot{z}(t) = \mathrm{A}z(t) + \mathrm{b}_1F(t) + \mathrm{b}_2F_a(t) \\ y(t) = \mathrm{C}z(t) + \mathrm{d}_1F(t) + \mathrm{d}_2F_a(t) \end{cases} \tag{4.34}$$

where $\mathrm{A} = \begin{bmatrix} 0 & 0 & 1 & 0 \\ 0 & 0 & 0 & 1 \\ -\frac{k_1+k_2}{m_1} & \frac{k_2}{m_1} & -\frac{c_1+c_2}{m_1} & \frac{c_2}{m_1} \\ \frac{k_2}{m_2} & -\frac{k_2}{m_2} & \frac{c_2}{m_2} & -\frac{c_2}{m_2} \end{bmatrix}$, $\mathrm{b}_1 = \begin{bmatrix} 0 \\ 0 \\ 0 \\ \frac{1}{m_2} \end{bmatrix}$, $\mathrm{b}_2 = \begin{bmatrix} 0 \\ 0 \\ \frac{1}{m_1} \\ -\frac{1}{m_2} \end{bmatrix}$, $\mathrm{C} = \begin{bmatrix} 1\,0\,0\,0 \\ 0\,1\,0\,0 \\ 0\,0\,1\,0 \\ 0\,0\,0\,1 \end{bmatrix}$, $\mathrm{d}_1 = \begin{bmatrix} 0 \\ 0 \\ 0 \\ 0 \end{bmatrix}$, $\mathrm{d}_2 = \begin{bmatrix} 0 \\ 0 \\ 0 \\ 0 \end{bmatrix}$, and $y(t)$ is the observed output, and the C is the full-state observation matrix.

The LQR active control has the following indicators:

$$J = \int_0^\infty [z^T Qz + F_a^T RF_a]\mathrm{d}t \tag{4.35}$$

where Q and R are weight matrices.

The LQR algorithm solves the Riccati equation to obtain P, calculates the optimal feedback gain G, and obtains the optimal active control force $F_a(t) = -Gz(t)$. Equation (4.36) can be obtained by substituting $F_a(t)$ into Eq. (4.34).

$$\begin{cases} \dot{z}(t) = (\mathrm{A} - \mathrm{b}_2G)z(t) + \mathrm{b}_1F(t) \\ y(t) = (\mathrm{C} - \mathrm{d}_2G)z(t) + \mathrm{d}_1F(t) \end{cases} \tag{4.36}$$

Based on the optimal feedback gain G obtained from the LQR algorithm, an LQR active control system for power equipment can be established, as shown in Fig. 4.10,

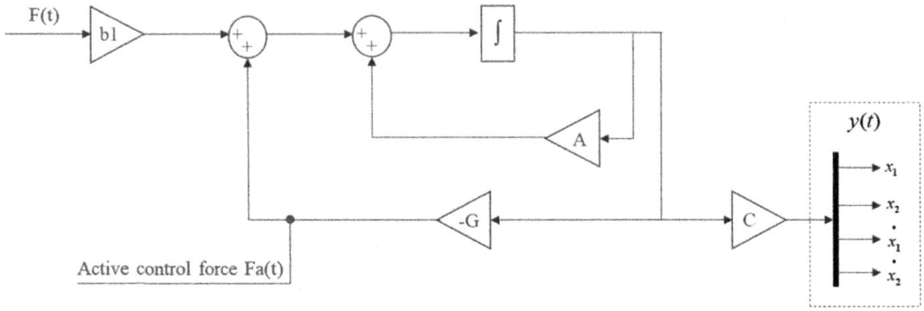

Fig. 4.10 LQR active control for power equipment

where the weight matrix in the LQR calculation is defined as $Q_{4\times 4} = \begin{bmatrix} q_1 & 0 & 0 & 0 \\ 0 & q_2 & 0 & 0 \\ 0 & 0 & q_3 & 0 \\ 0 & 0 & 0 & q_4 \end{bmatrix}$, and $R_{1\times 1}$ is positive real number.

In the LQR optimal control design, the selection of the weight matrices Q and R significantly impacts the control effect and force. Accordingly, the way to set it properly is directly related to the ideality of the control effect. For this reason, this paper establishes a PSO-LQR active control strategy based on the PSO algorithm, as shown in Fig. 4.11.

For the PSO-LQR active control shown in Fig. 4.11, the fitting function is selected as $\|F_{out} - F_{desired}\|_{\infty}$, where $F_{desired} = 0$, and $F_{out} = \mathrm{k}_1 x_1 + \mathrm{c}_1 \dot{x}_1 = \mathrm{k}_1 z_1 + \mathrm{c}_1 z_3$.

Example 4.3 The power equipment's vibration isolation system parameter settings and the PSO algorithm are the same as those in Example 4.1. The search range of the weight matrix in the LQR controller

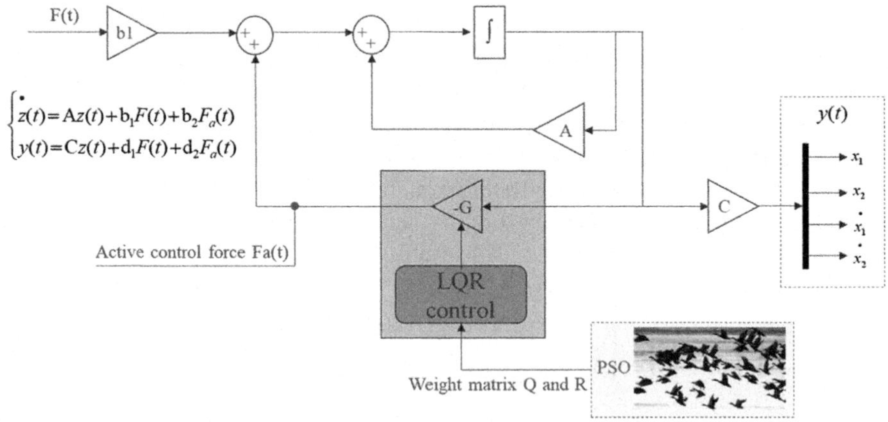

Fig. 4.11 Schematic diagram of the PSO-LQR active control for power equipment

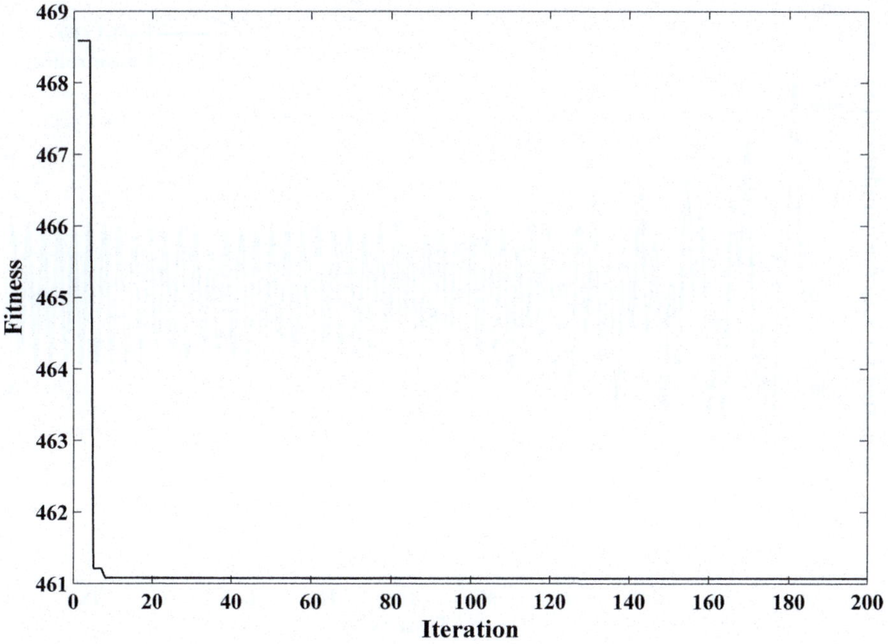

Fig. 4.12 Fitting convergence of the PSO-LQR active control for power equipment

is arbitrarily set to $\left[1 \times 10^{-7}, 1 \times 10^{-7}, 1 \times 10^{-7}, 1 \times 10^{-7}, 1 \times 10^{-7}\right] \sim \left[1 \times 10^{7}, 1 \times 10^{7}, 1 \times 10^{7}, 1 \times 10^{7}, 1\right]$. The convergence curve of the fitting value is shown in Fig. 4.12, and the gbest solution is

$$Q = 1 \times 10^6 \times \begin{bmatrix} 6.121 & 0 & 0 & 0 \\ 0 & 7.281 & 0 & 0 \\ 0 & 0 & 4.476 & 0 \\ 0 & 0 & 0 & 1.308 \end{bmatrix}, R = 0.985.$$

This optimal weight matrix is used to achieve the LQR active control, the force transmitted to the foundation after the power equipment is controlled, as shown in Fig. 4.13, and the active control force generated by the actuator is defined as shown in Fig. 4.14.

4.2.3 Optimized LQR Active Control for Sensitive Equipment

Consider the active control system of the vibration-sensitive equipment shown in Fig. 1.5b. The dynamic system is expressed in Eq. (4.9), where the state variables are

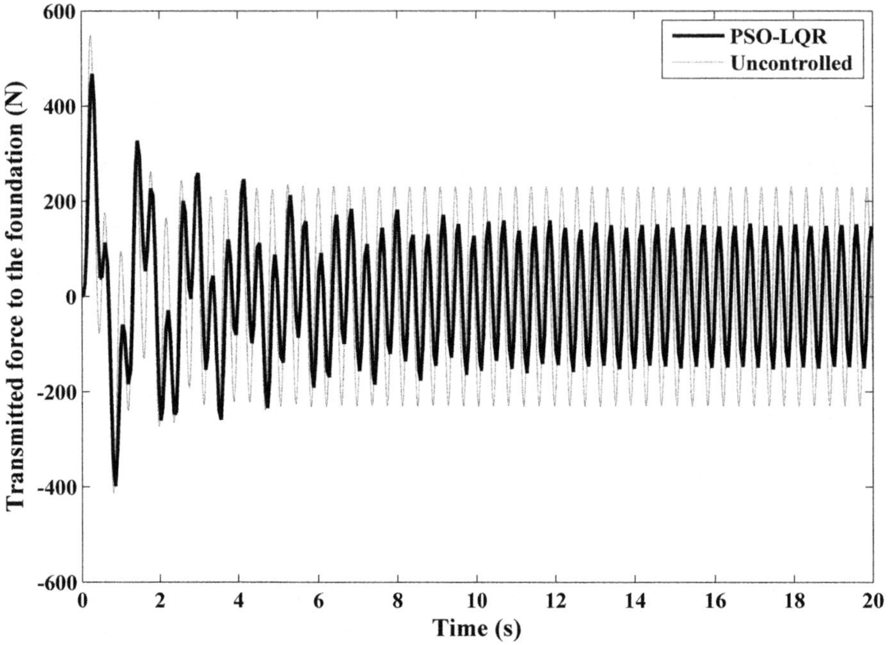

Fig. 4.13 Force transmitted from the power equipment to the foundation (PSO-LQR)

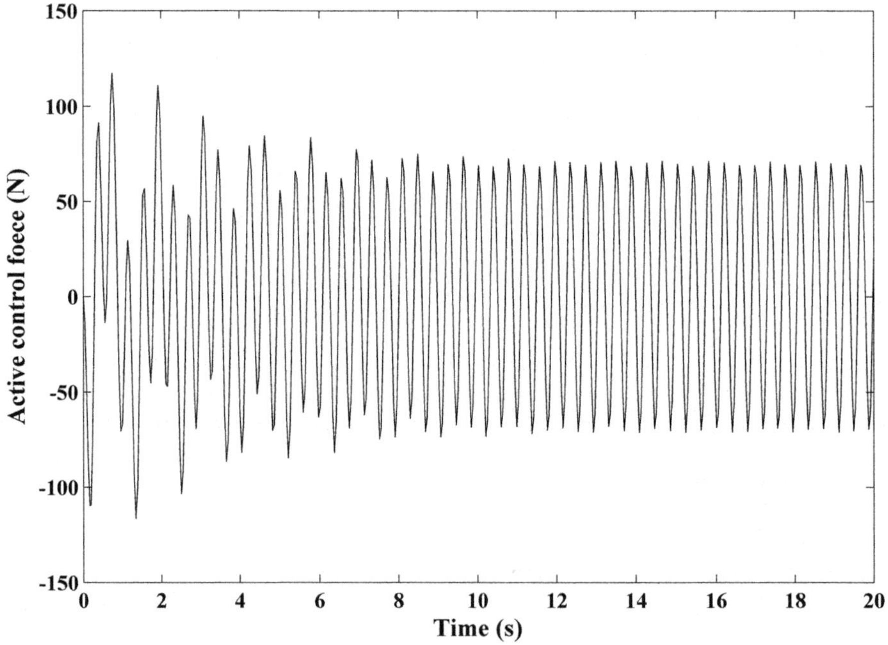

Fig. 4.14 Active control force of the PSO-LQR strategy for power equipment

$z_1 = x_1, z_2 = x_2, z_3 = \dot{x}_1$, and $z_4 = \dot{x}_2$, and the state vector is $z(t) = [z_1, z_2, z_3, z_4]^T$. Equation (4.9) can be transformed into the following state space equation form:

$$\begin{cases} \dot{z}(t) = \mathrm{A}z(t) + \mathrm{b}_1 F(t) + \mathrm{b}_2 F_a(t) \\ y(t) = \mathrm{C}z(t) + \mathrm{d}_1 F(t) + \mathrm{d}_2 F_a(t) \end{cases} \tag{4.37}$$

In which, $\mathrm{A} = \begin{bmatrix} 0 & 0 & 1 & 0 \\ 0 & 0 & 0 & 1 \\ -\frac{k_1+k_2}{m_1} & \frac{k_2}{m_1} & -\frac{c_1+c_2}{m_1} & \frac{c_2}{m_1} \\ \frac{k_2}{m_2} & -\frac{k_2}{m_2} & \frac{c_2}{m_2} & -\frac{c_2}{m_2} \end{bmatrix}$, $\mathrm{b}_1 = \begin{bmatrix} 0 \\ 0 \\ \frac{1}{m_1} \\ 0 \end{bmatrix}$, $\mathrm{b}_2 = \begin{bmatrix} 0 \\ 0 \\ -\frac{1}{m_1} \\ \frac{1}{m_2} \end{bmatrix}$,

$\mathrm{C} = \begin{bmatrix} 1\,0\,0\,0 \\ 0\,1\,0\,0 \\ 0\,0\,1\,0 \\ 0\,0\,0\,1 \end{bmatrix}$, $\mathrm{d}_1 = \begin{bmatrix} 0 \\ 0 \\ 0 \\ 0 \end{bmatrix}$, $\mathrm{d}_2 = \begin{bmatrix} 0 \\ 0 \\ 0 \\ 0 \end{bmatrix}$.

In this section, a PSO-LQR active control strategy based on the PSO algorithm for the vibration-sensitive equipment shown in Fig. 4.11 will be developed. The fitting function is $\left\| \dot{x}_2 - \dot{x}_{2,desired} \right\|_\infty$, where $\dot{x}_{2,desired} = 0$, and the state space form of $\dot{x}_2$ is z_4.

Example 4.4 The parameter settings of the vibration isolation system and the PSO algorithm of the vibration-sensitive equipment are the same as those in Example 4.2. The search range of the weight matrix in the LQR controller is arbitrarily set to $\left[1 \times 10^{-4}, 1 \times 10^{-4}, 1 \times 10^{-4}, 1 \times 10^{-4}, 1 \times 10^{-4}\right] \sim \left[1 \times 10^4, 1 \times 10^4, 1 \times 10^4, 1 \times 10^4, 1\right]$. The convergence curve of the fitting value is shown in Fig. 4.15, and the gbest solution is

$$Q = 1 \times 10^3 \times \begin{bmatrix} 9.341 & 0 & 0 & 0 \\ 0 & 2.144 & 0 & 0 \\ 0 & 0 & 4.557 & 0 \\ 0 & 0 & 0 & 6.902 \end{bmatrix}, R = 1 \times 10^{-4}.$$

Based on the optimal weight matrix for the LQR active control calculation, it can be seen that the vibration velocity of the sensitive equipment with control is shown in Fig. 4.16, and the active control force generated by the actuator is shown in Fig. 4.17.

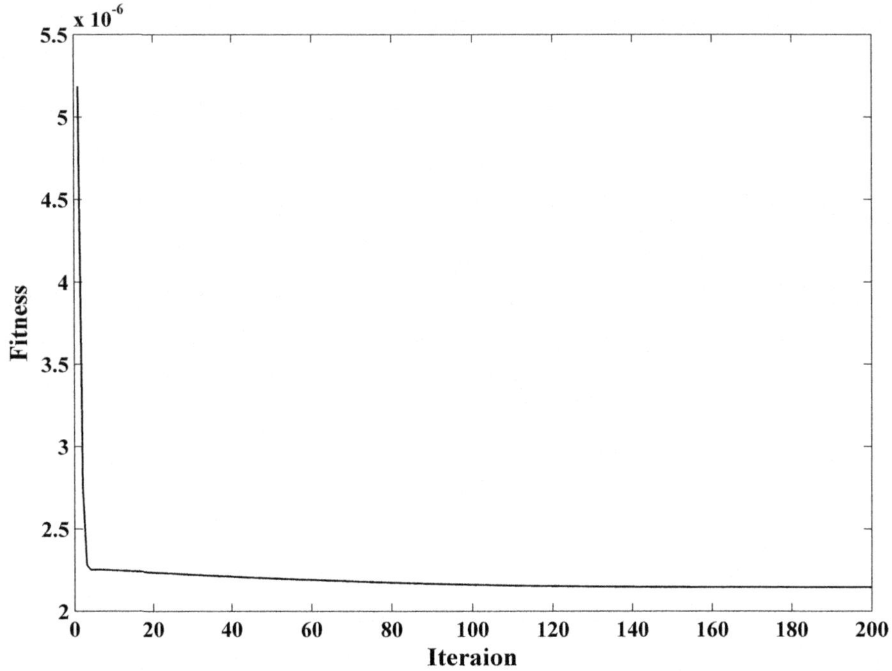

Fig. 4.15 Fitting convergence of the PSO-LQR active control for sensitive equipment

4.3 Optimized LQG Active Control

4.3.1 LQG Control Algorithm

For the state space equation of a linear stationary control system:

$$\begin{cases} \dot{z}(t) = \mathrm{A}z(t) + \mathrm{B}U(t) + \varepsilon_1(t) \\ y(t) = \mathrm{C}z(t) + \varepsilon_2(t) \end{cases} \tag{4.38}$$

where $\varepsilon_1(t)$ and $\varepsilon_2(t)$ are the input and measurement noise of the control system, respectively. These noises are all zero-mean Gauss random white noises.

Typically, the LQR control algorithm is first used to design the optimal control force for full-state feedback $U(t)$. Thereafter, the Kalman filter is used to estimate the control plant's overall state based on the structure's observed output. For this purpose, the following indicator is selected:

$$J_e = E\Big[\{z(t) - \hat{z}(t)\}^{\mathrm{T}}\{z(t) - \hat{z}(t)\}\Big] \tag{4.39}$$

where $E[*]$ represents the mean value, and $\hat{z}(t)$ is the estimated value of $z(t)$.

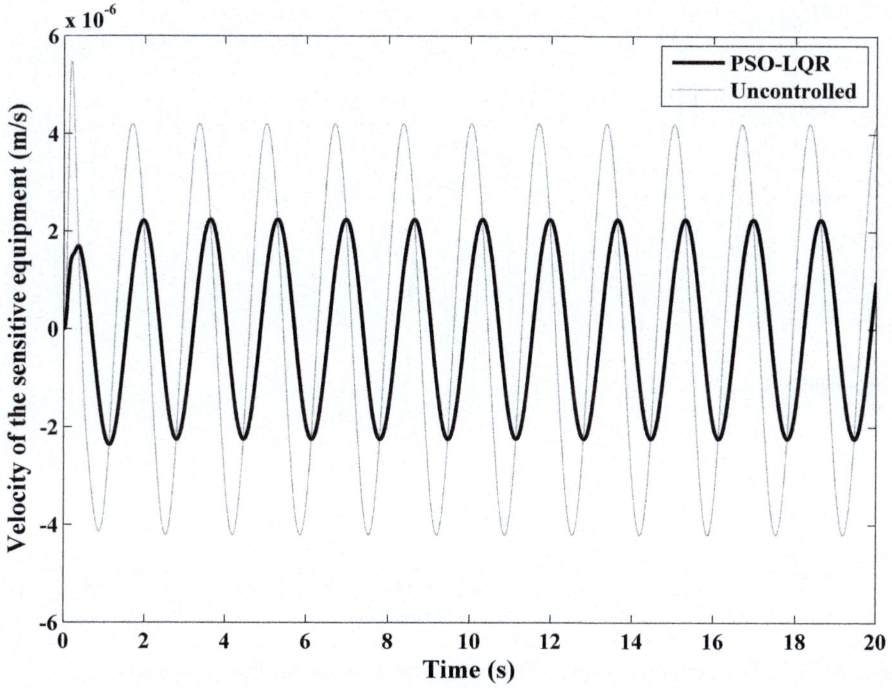

Fig. 4.16 Vibration velocity of the sensitive equipment (PSO-LQR)

The constructed Kalman filter is:

$$\dot{\hat{z}}(t) = \mathrm{A}\hat{z}(t) + \mathrm{B}U(t) + K_e[y(t) - \hat{y}(t)] \tag{4.40a}$$

$$\hat{y}(t) = \mathrm{C}\hat{z}(t) \tag{4.40b}$$

where K_e is the Kalman filter gain defined as follows:

$$K_e = P_e\mathrm{C}^{\mathrm{T}}R^{-1} \tag{4.41}$$

where P_e is the covariance matrix of steady-state filter error, which can be expressed as $P_e = \lim_{t\to\infty} E\left[\{z(t) - \hat{z}(t)\}\{z(t) - \hat{z}(t)\}^{\mathrm{T}}\right]$ and it can be solved by the following Riccati equation:

$$P_eA^{\mathrm{T}} + AP_e - P_eC^{\mathrm{T}}R_e^{-1}CP_e + Q_e = 0 \tag{4.42}$$

Assume that the estimation error is $e = z - \hat{z}$, then the Kalman filter has the following properties:

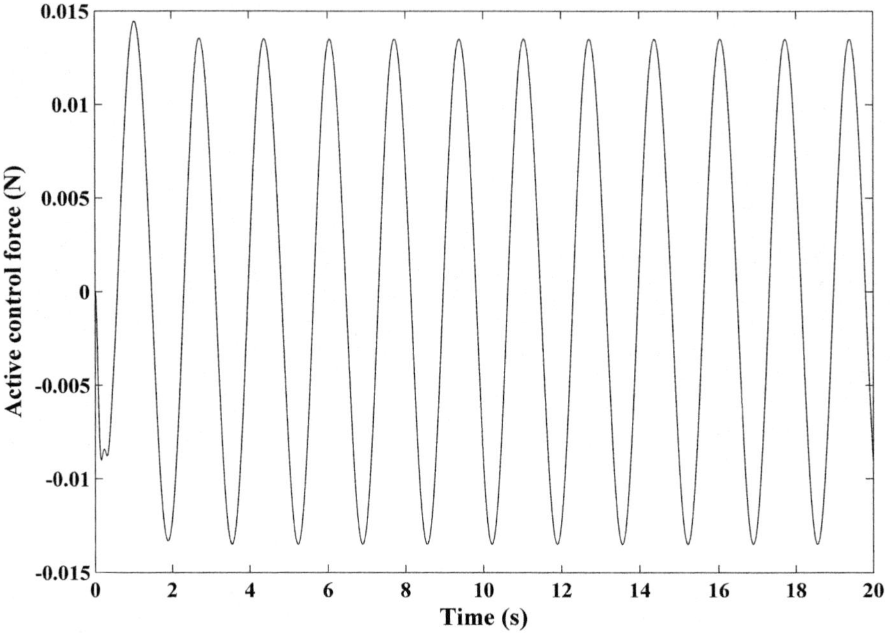

Fig. 4.17 Active control force of the PSO-LQR strategy for the sensitive equipment

$$\begin{cases} \lim_{t\to\infty} E[e(t)] = \lim_{t\to\infty} E\big[z(t) - \hat{z}(t)\big] = 0 \\ \lim_{t\to\infty} J_e = \lim_{t\to\infty} E\big[e(t)^{\mathrm{T}} e(t)\big] = trace(P_e) \end{cases} \tag{4.43}$$

The optimal control force of the system can be expressed as the feedback of the state estimation:

$$U(t) = -G\hat{z}(t) \tag{4.44}$$

The state space equation of the controlled system can be obtained by substituting Eq. (4.44) into Eq. (4.40a) as follows:

$$\dot{\hat{z}}(t) = (A - BG - K_e C)\hat{z}(t) + K_e y(t) \tag{4.45a}$$

$$\hat{y}(t) = C\hat{z}(t) \tag{4.45b}$$

4.3.2 Optimized LQG Active Control for Power Equipment

The active control system of the power equipment is shown in Fig. 1.5a. In this case, the state variables are $z_1 = x_1$, $z_2 = x_2$, $z_3 = \dot{x}_1$, and $z_4 = \dot{x}_2$, and the state vector is $z(t) = [z_1, z_2, z_3, z_4]^T$. The dynamic and state space equations are given in Eq. (4.4) and (4.34), respectively. When input noise $\varepsilon_1(t)$ and measurement noise $\varepsilon_2(t)$ are considered, the state space equation becomes:

$$\begin{cases} \dot{z}(t) = \mathrm{A}z(t) + \mathrm{b}_1 F(t) + \mathrm{b}_2 F_a(t) + \varepsilon_1(t) \\ y(t) = \mathrm{C}z(t) + \mathrm{d}_1 F(t) + \mathrm{d}_2 F_a(t) + \varepsilon_2(t) \end{cases} \tag{4.46}$$

The LQG optimal control theory can be used to eliminate the noise influence and hence to obtain better active control. The optimal feedback control force herein can be expressed as:

$$F_a(t) = -G\hat{y}(t) \tag{4.47}$$

where G is the optimal feedback control gain from the LQR control algorithm, $\hat{y}(t)$ is the estimated value of $y(t)$ which requires constructing the Kalman filter as follows:

$$\begin{cases} \dot{\hat{z}}(t) = \mathrm{A}\hat{z}(t) + \mathrm{b}_1 F(t) + \mathrm{b}_2 F_a(t) + K_e\left[y(t) - \hat{y}(t)\right] \\ \hat{y}(t) = \mathrm{C}\hat{z}(t) \end{cases} \tag{4.48}$$

The constructed LQG control with the Kalman filter are illustrated in Figs. 4.18 and Fig. 4.19.

A PSO-LQG active control strategy based on the PSO algorithm is used for the power equipment discussed in this section. The fitting function is the same as the PSO-LQR strategy.

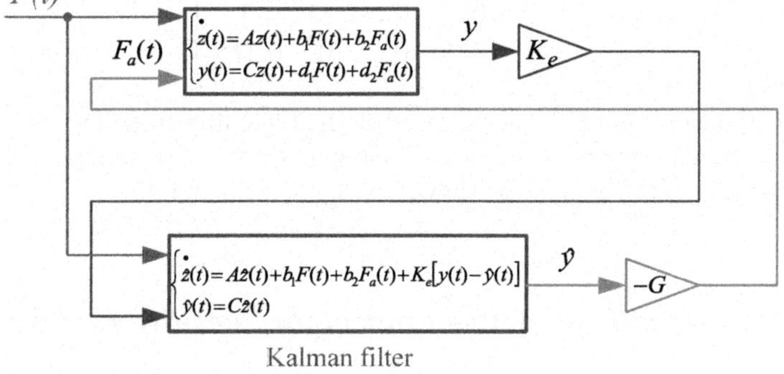

Fig. 4.18 LQG active control with a Kalman filter for power equipment

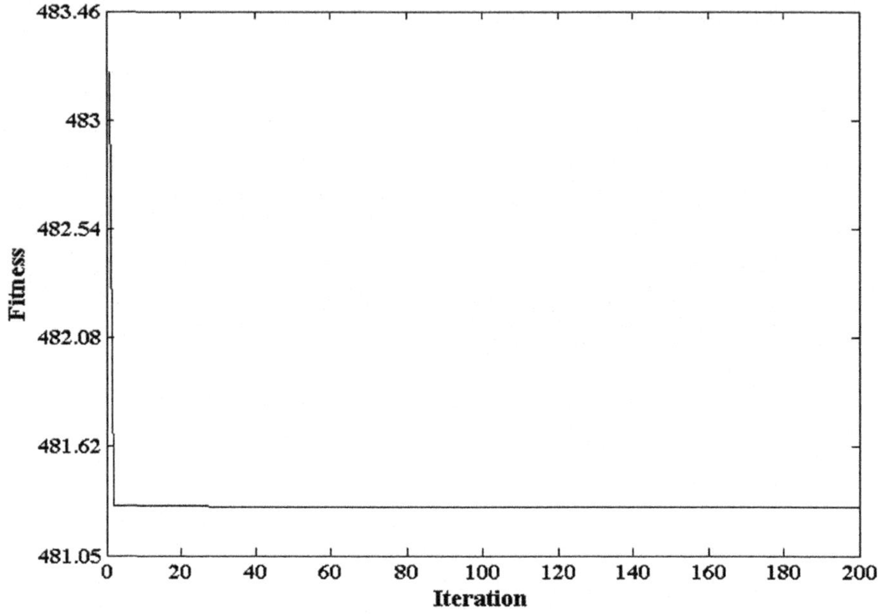

Fig. 4.19 Fitting convergence of the PSO-LQG active control for power equipment

Example 4.5 The power equipment vibration isolation system parameters and the PSO algorithm are the same as those in Example 4.3. The search range of the weight matrix in the LQR controller is arbitrarily set to $\left[1\times10^{-7},1\times10^{-7},1\times10^{-7},1\times10^{-7},1\times10^{-7}\right]\sim\left[1\times10^{7},1\times10^{7},1\times10^{7},1\times10^{7},1\right]$, the fitting curve can be obtained as shown in Fig. 2.19, and the optimal weight matrix is

$$Q=1\times10^{6}\times\begin{bmatrix}6.384&0&0&0\\0&5.752&0&0\\0&0&2.351&0\\0&0&0&0.315\end{bmatrix},R=0.993.$$

Based on the optimal LQG active control, the force transmitted to the foundation after the power equipment is controlled is shown in Fig. 4.20, and the active control force generated by the actuator is shown in Figs. 4.21 and 4.22.

4.3.3 Optimized LQG Active Control for Sensitive Equipment

Consider the active control system of the sensitive equipment shown in Fig. 1.5b. The dynamic and state space equations are expressed in Eqs. (4.9) and (4.37), respectively.

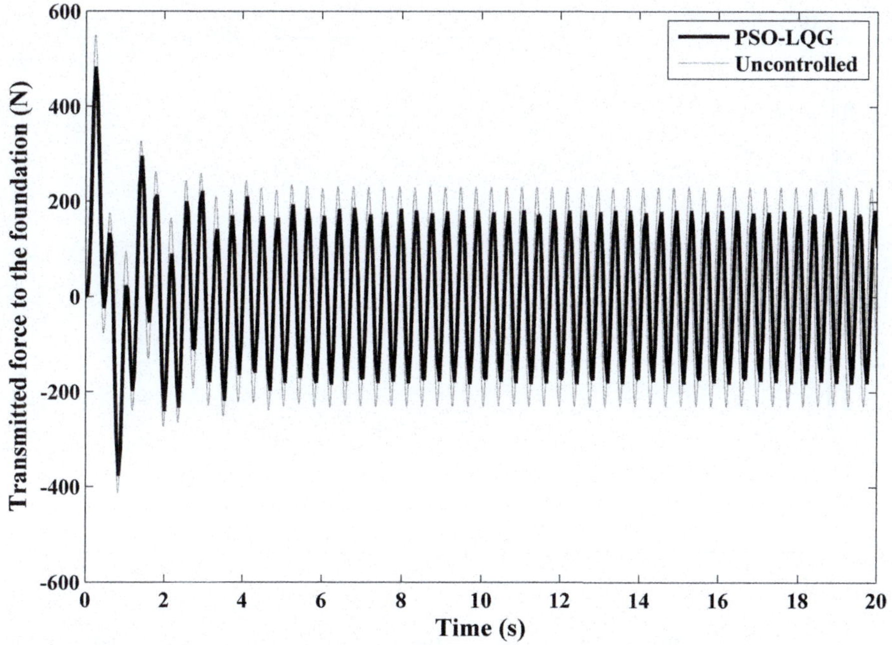

Fig. 4.20 Force transmitted from the power equipment to the foundation (PSO-LQG)

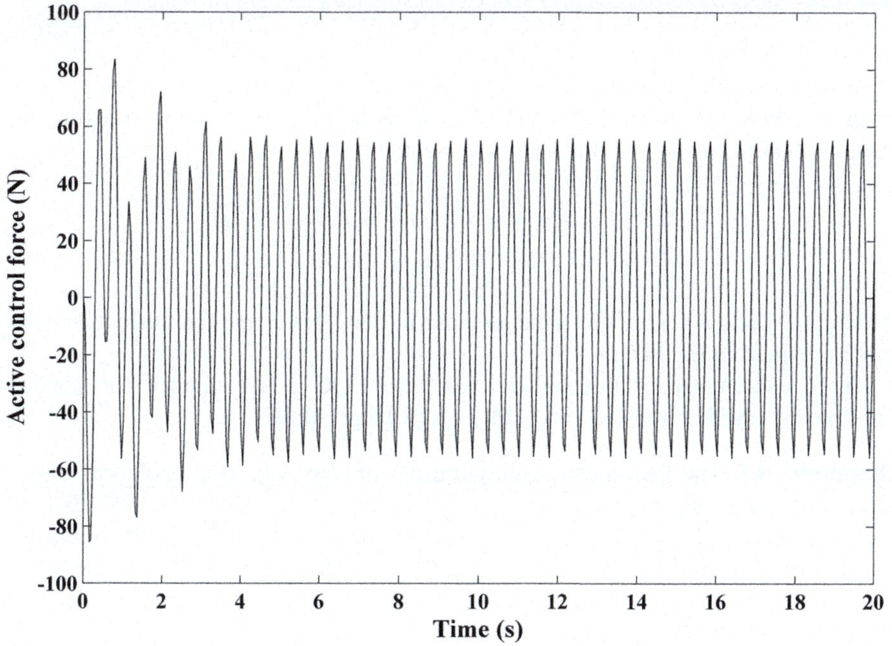

Fig. 4.21 Active control force of the PSO-LQG strategy for power equipment

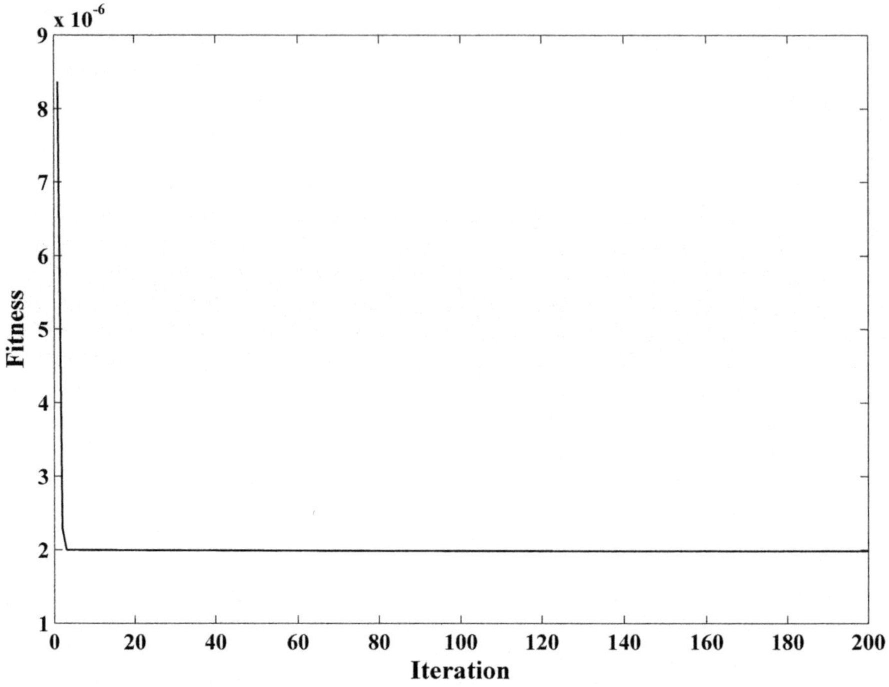

Fig. 4.22 Fitting convergence of the PSO-LQG active control for sensitive equipment

The state variables are set to $z_1 = x_1$, $z_2 = x_2$, $z_3 = \dot{x}_1$, and $z_4 = \dot{x}_2$, and the state vector is set to $z(t) = [z_1, z_2, z_3, z_4]^{\mathrm{T}}$. When the input noise $\varepsilon_1(t)$ and measurement noise $\varepsilon_2(t)$ are considered, the state space equation becomes:

$$\begin{cases} \dot{z}(t) = \mathrm{A}z(t) + \mathrm{b}_1 F(t) + \mathrm{b}_2 F_a(t) + \varepsilon_1(t) \\ y(t) = \mathrm{C}z(t) + \mathrm{d}_1 F(t) + \mathrm{d}_2 F_a(t) + \varepsilon_2(t) \end{cases} \tag{4.49}$$

where $\varepsilon_1(t)$ and $\varepsilon_2(t)$ are the same as those in Sect. 2.3.2, and the other parameters are as defined in Eq. (4.37).

This section carries out research on PSO-LQG active control for sensitive equipment based on the PSO and LQG control algorithms.

Example 4.6 The parameter configuration of the vibration isolation system and PSO algorithm of sensitive equipment is the same as in Example 4.4. The search range of the weight matrix in the LQR controller is arbitrarily set to $\left[1 \times 10^{-4}, 1 \times 10^{-4}, 1 \times 10^{-4}, 1 \times 10^{-4}, 1 \times 10^{-4}\right] \sim \left[1 \times 10^{4}, 1 \times 10^{4}, 1 \times 10^{4}, 1 \times 10^{4}, 1\right]$. The fitting convergence curve is shown in Fig. 2.22, and the obtained gbest solution is

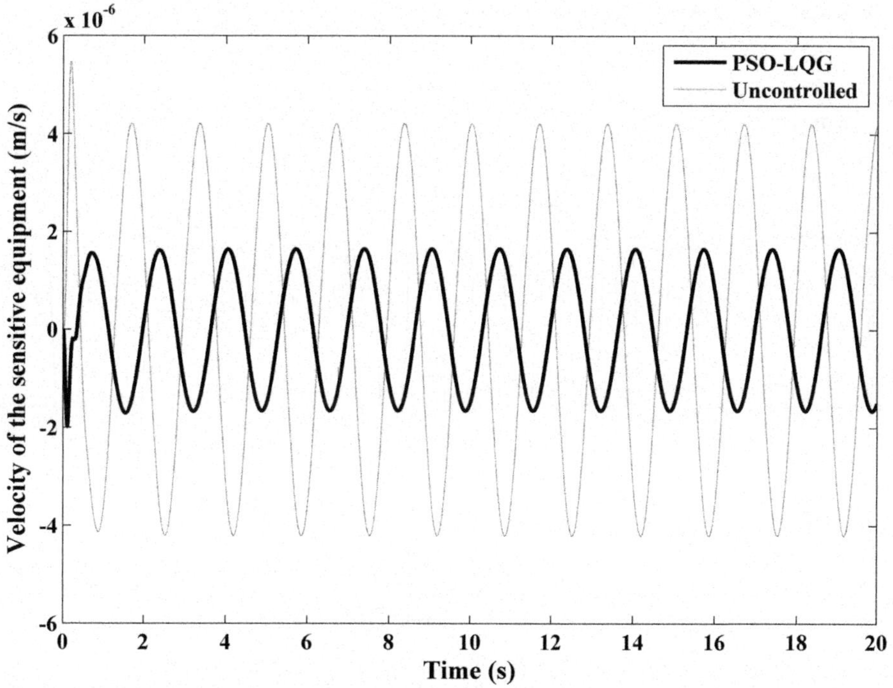

Fig. 4.23 Vibration velocity of sensitive equipment (PSO-LQG)

$$Q = 1 \times 10^4 \times \begin{bmatrix} 7.6344 & 0 & 0 & 0 \\ 0 & 3.9642 & 0 & 0 \\ 0 & 0 & 1.4577 & 0 \\ 0 & 0 & 0 & 9.4763 \end{bmatrix}, R = 1 \times 10^{-4}.$$

Based on the optimal LQG active control, the vibration velocity of sensitive equipment with active control is shown in Fig. 4.23, and the active control force generated by the actuator is shown in Fig. 4.24.

4.4 Optimized H$_\infty$ Active Control

4.4.1 H_2/H_∞ Control Algorithm

The state space equation of the dynamic closed-loop control system shown in Fig. 4.25 is as follows:

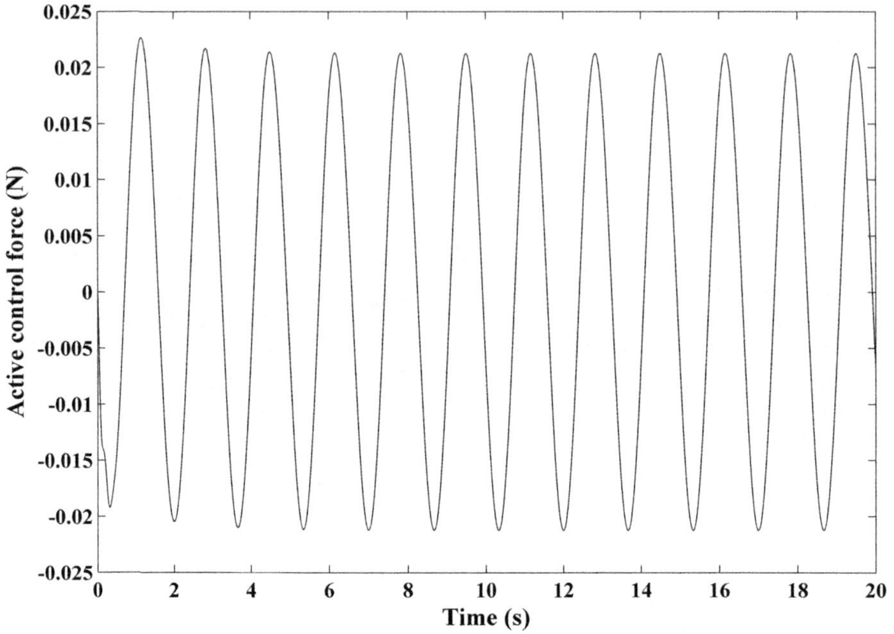

Fig. 4.24 Active control force of PSO-LQG strategy for sensitive equipment

$$\begin{cases} \dot{z}(t) = \mathrm{A}z(t) + \mathrm{b}_1 F(t) + \mathrm{b}_2 U(t) \\ Y(t) = \mathrm{C}_1 z(t) + \mathrm{d}_1 U(t) \\ y(t) = \mathrm{C}_2 z(t) + \mathrm{d}_2 F(t) \end{cases} \tag{4.50}$$

where $z(t)$ is the state variables, $y(t)$ is the observed output, $Y(t)$ is the control output, $U(t)$ is the control input, and $F(t)$ is the input variable.

In general, the H_2/H_∞ optimal control to design the controller offers a stable closed-loop control system that minimizes the H_2 norm or H_∞ norm of the proposed transfer function T_{YF}.

Let x be the controller's state variable. Its state space equation can be expressed as follows:

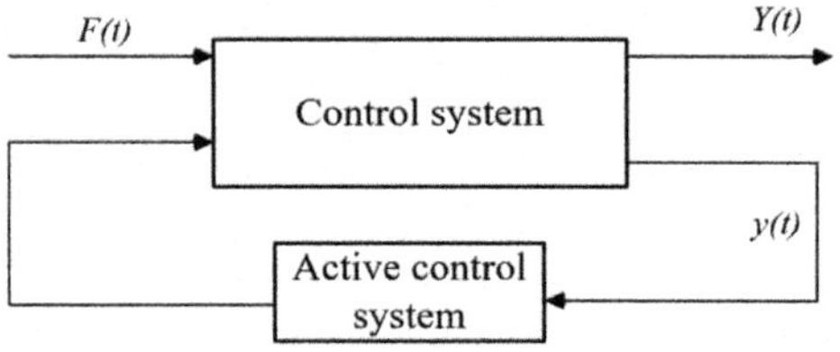

Fig. 4.25 Dynamic closed-loop control system

$$\begin{cases} \dot{x}(t) = \mathrm{A_{con}}x(t) + \mathrm{B_{con}}y(t) \\ U(t) = \mathrm{C_{con}}x(t) + \mathrm{D_{con}}y(t) \end{cases} \tag{4.51}$$

Equation (4.52) can be derived by substituting the observation output $y(t)$ in Eq. (4.50) into Eq. (4.51) as follows:

$$\dot{x}(t) = \mathrm{A_{con}}x(t) + \mathrm{B_{con}}[\mathrm{C_2}z(t) + \mathrm{d_2}F(t)] \tag{4.52}$$

On the other hand, Eq. (4.53) can be derived by substituting the control force in (4.51) into (4.50).

$$\begin{cases} \dot{z}(t) = \mathrm{A}z(t) + \mathrm{b_2}\left[\mathrm{C_{con}}x(t) + \mathrm{D_{con}}y(t)\right] + \mathrm{b_1}F(t) \\ y(t) = \mathrm{C_2}z(t) + \mathrm{d_2}F(t) \\ Y(t) = \mathrm{C_1}z(t) + \mathrm{d_1}\left[\mathrm{C_{con}}x(t) + \mathrm{D_{con}}y(t)\right] \end{cases} \tag{4.53}$$

Substituting the observation output $y(t)$ into $z(t)$ and $Y(t)$ results in the following expression:

$$\begin{cases} \dot{z}(t) = (\mathrm{A} + b_2\mathrm{D_{con}C_2})z(t) + \mathrm{b_2C_{con}}x(t) + (\mathrm{b_1} + \mathrm{b_2D_{con}d_2})F(t) \\ Y(t) = (\mathrm{C_1} + \mathrm{d_1D_{con}C_2})z(t) + \mathrm{d_1C_{con}}x(t) + \mathrm{d_1D_{con}d_2}F(t) \end{cases} \tag{4.54}$$

The following state space equation form can be obtained by rewriting Eq. (4.54).

$$\begin{aligned} \begin{bmatrix} \dot{z}(t) \\ \dot{x}(t) \end{bmatrix} &= \begin{bmatrix} \mathrm{A} + \mathrm{b_2D_{con}C_2} & \mathrm{b_2C_{con}} \\ \mathrm{B_{con}C_2} & \mathrm{A_{con}} \end{bmatrix} \begin{bmatrix} z(t) \\ x(t) \end{bmatrix} + \begin{bmatrix} \mathrm{b_1} + \mathrm{b_2D_{con}d_2} \\ \mathrm{B_{con}d_2} \end{bmatrix} F(t) \\ Y(t) &= \begin{bmatrix} \mathrm{C_2} + \mathrm{d_1D_{con}C_2} & \mathrm{d_1C_{con}} \end{bmatrix} \begin{bmatrix} z(t) \\ x(t) \end{bmatrix} + \mathrm{d_1D_{con}d_2}F(t) \end{aligned} \tag{4.55}$$

Let $Z(t) = \begin{bmatrix} z(t) \\ x(t) \end{bmatrix}$, then Eq. (4.55) can be rewritten as:

$$\begin{cases} \dot{Z}(t) = \mathrm{A_{cl}}Z(t) + \mathrm{B_{cl}}F(t) \\ Y(t) = \mathrm{C_{cl}}Z(t) + \mathrm{D_{cl}}F(t) \end{cases} \tag{4.56}$$

where

$\mathrm{A_{cl}} = \begin{bmatrix} \mathrm{A} + \mathrm{b_2D_{con}C_2} & \mathrm{b_2C_{con}} \\ \mathrm{B_{con}C_2} & \mathrm{A_{con}} \end{bmatrix}$, $\mathrm{B_{cl}} = \begin{bmatrix} \mathrm{b_1} + \mathrm{b_2D_{con}d_2} \\ \mathrm{B_{con}d_2} \end{bmatrix}$, $\mathrm{C_{cl}} = \begin{bmatrix} \mathrm{C_2} + \mathrm{d_1D_{con}C_2} & \mathrm{d_1C_{con}} \end{bmatrix}$, $\mathrm{D_{cl}} = \mathrm{d_1D_{con}d_2}$.

The transfer function T_{YF} can be expressed as:

$$T_{YF}(s) = \mathrm{C_{cl}}(s\mathrm{I} - \mathrm{A_{cl}})^{-1}\mathrm{B_{cl}} + \mathrm{D_{cl}} \tag{4.57}$$

where I is the unit diagonal matrix, and s is complex frequency.

The stability of the control system shown in Fig. 4.25 is ensured when the eigenvalues of the system matrix $\mathrm{A_{cl}}$ are all in the left half plane of $(-\infty, 0)$. In this case, the $\mathrm{H_2}$ norm of the transfer function T_{YF} is defined as:

$$\|T_{YF}\|_2 = \frac{1}{2\pi} trace\left[\int_{-\infty}^{+\infty} T_{YF}(j\omega_{ref})T_{YF}^{*}(j\omega_{ref})d\omega_{ref}\right] \tag{4.58}$$

where $T_{YF}^{*}(j\omega_{ref})$ is the complex conjugate matrix of $T_{YF}(j\omega_{ref})$.

The $\mathrm{H_\infty}$ norm of the transfer function T_{YF} is defined as

$$\|T_{YF}\|_\infty = \sup_{0 \le \omega_{ref} < \infty} \sigma_{\max}\left[T_{YF}(j\omega_{ref})\right] \tag{4.59}$$

where $\sigma_{\max}\left[T_{YF}(j\omega_{ref})\right]$ is the maximum singular value of the transfer function.

Minimizing the $\mathrm{H_2}$ norm or $\mathrm{H_\infty}$ norm of the transfer function T_{YF} causes the norm value not to exceed the given upper bound γ_1 and γ_2, respectively:

$$\|T_{YF}\|_2 < \gamma_1, \ \|T_{YF}\|_\infty < \gamma_2 \tag{4.60}$$

If the controller is considered as state feedback control, the control input $U(t)$ can be defined as follows:

$$U(t) = Ky(t) \tag{4.61}$$

where K is the state feedback controller gain, and $y(t)$ is the observed output of the control system.

For the following form of state space control equation

$$\begin{cases} \dot{z}(t) = \mathrm{A}z(t) + \mathrm{b_1}F(t) + \mathrm{b_2}U(t) \\ Y(t) = \mathrm{C_1}z(t) + \mathrm{d_{11}}F(t) + \mathrm{d_{12}}U(t) \\ y(t) = \mathrm{C_2}z(t) + \mathrm{d_{21}}F(t) + \mathrm{d_{22}}U(t) \end{cases} \tag{4.62}$$

Equation (4.62) can be transformed into the following form:

$$\begin{bmatrix} \dot{z}(t) \\ Y(t) \\ y(t) \end{bmatrix} = \begin{bmatrix} \mathrm{A} & \mathrm{b_1} & \mathrm{b_2} \\ \mathrm{C_1} & \mathrm{d_{11}} & \mathrm{d_{\cdot 12}} \\ \mathrm{C_2} & \mathrm{d_{21}} & \mathrm{d_{22}} \end{bmatrix} \begin{bmatrix} z(t) \\ F(t) \\ U(t) \end{bmatrix} \tag{4.63}$$

The control input can be expressed as follows:

$$U(t) = K(I_1 - d_{22}K)^{-1}[C_2 z(t) + d_{21}F(t)] \tag{4.64}$$

where I_1 is the diagonal unit matrix.

Equation (4.65) can be obtained by substituting Eq. (4.64) into Eq. (4.62) as follows:

$$\begin{bmatrix} \dot{z}(t) \\ Y(t) \end{bmatrix} = \begin{bmatrix} A_{cl} & B_{cl} \\ C_{cl} & D_{cl} \end{bmatrix} \begin{bmatrix} z(t) \\ U(t) \end{bmatrix} \tag{4.65}$$

where

$A_{cl} = A + b_2 K(I_1 - d_{22}K)^{-1}C_2$, $B_{cl} = b_1 + b_2 K(I_1 - d_{22}K)^{-1}d_{21}$,
$C_{cl} = C_1 + d_{12}K(I_1 - d_{22}K)^{-1}C_2$, $D_{cl} = d_{11} + d_{12}K(I_1 - d_{22}K)^{-1}d_{21}$.

It can be seen that the transfer function of the closed-loop system obtained from Eq. (4.65) is the same as that of Eq. (4.57).

4.4.2 Optimized H_∞ Active Control for Power Equipment

Consider the active control system for the power equipment shown in Fig. 1.5a. The dynamic equation is Eq. (4.4), its state variables are $z_1 = x_1$, $z_2 = x_2$, $z_3 = \dot{x}_1$, $z_4 = \dot{x}_2$, and its state vector is $z(t) = [z_1, z_2, z_3, z_4]^T$. The observation output of the control system is set to $y(t) = \left[x_1, x_2, \dot{x}_1, \dot{x}_2\right]^T$. Besides, the control output is set to the force transmitted to the foundation by the power equipment, which is $Y(t) = \left\{k_1 x_1 + c_1 \dot{x}_1\right\}$. The state space equation is as follows:

$$\begin{cases} \dot{z}(t) = Az(t) + b_1 F(t) + b_2 F_a(t) \\ Y(t) = C_1 z(t) + d_{11} F(t) + d_{12} F_a(t) \\ y(t) = C_2 z(t) + d_{21} F(t) + d_{22} F_a(t) \end{cases} \tag{4.66}$$

where

$$A = \begin{bmatrix} 0 & 0 & 1 & 0 \\ 0 & 0 & 0 & 1 \\ -\frac{k_1+k_2}{m_1} & \frac{k_2}{m_1} & -\frac{c_1+c_2}{m_1} & \frac{c_2}{m_1} \\ \frac{k_2}{m_2} & -\frac{k_2}{m_2} & \frac{c_2}{m_2} & -\frac{c_2}{m_2} \end{bmatrix},\ b_1 = \begin{bmatrix} 0 \\ 0 \\ 0 \\ \frac{1}{m_2} \end{bmatrix},\ b_2 = \begin{bmatrix} 0 \\ 0 \\ \frac{1}{m_1} \\ -\frac{1}{m_2} \end{bmatrix},\ C_1 =$$

$$[k_1\ 0\ c_1\ 0],\ d_{11} = [0],\ d_{12} = [0],\ C_2 = \begin{bmatrix} 1 & 0 & 0 & 0 \\ 0 & 1 & 0 & 0 \\ 0 & 0 & 1 & 0 \\ 0 & 0 & 0 & 1 \end{bmatrix},\ d_{21} = \begin{bmatrix} 0 \\ 0 \\ 0 \\ 0 \end{bmatrix},\ d_{22} = \begin{bmatrix} 0 \\ 0 \\ 0 \\ 0 \end{bmatrix}.$$

The transfer function matrix derived from Eq. (4.66) is shown in Eq. (4.65). For the H_∞ control theory, the feedback controller is constructed as $K = \left[K_1\ K_2\ K_3\ K_4\right]$, and the following conditions should be satisfied:

(1) The closed-loop system must be stabilized.
(2) The infinite norm $\|T_{YF}(s, K)\|_\infty$ of the transfer function $T_{YF}(s, K)$ requires meeting the condition that $\|T_{YF}(s, K)\|_\infty < \gamma_2$, where γ_2 is a positive number.

In this section, the PSO algorithm is used to optimize the feedback controller. The fitting function can be defined as:

$$\|T_{YF}(s, K)\|_\infty \tag{4.67}$$

where $s = j\omega_{ref}$ is the complex frequency, and ω_{ref} is the angular frequency of the power equipment.

The proposed PSO-H∞ active control strategy for power equipment is indicated based on the research in reference [4] in Fig. 4.26.

Example 4.7 The power equipment's vibration isolation system parameter settings and the PSO algorithm are the same as those in Example 4.5. The gain search constraints of the feedback controller in the H_∞ strategy are set to $\|K\|_\infty \leq 5 \times 10^2$. The fitting convergence curve is shown in Fig. 4.27, and the obtained gbest solution is

$$K = \begin{bmatrix} 1.913 & -4.983 & 1.974 & -5.000 \end{bmatrix} \times 10^2.$$

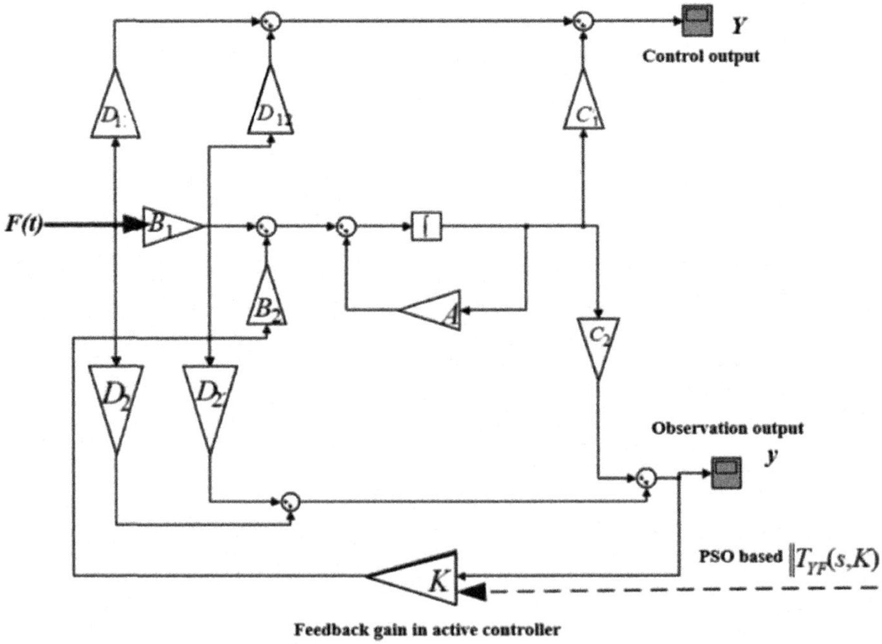

Fig. 4.26 PSO-H_∞ active control strategy for power equipment

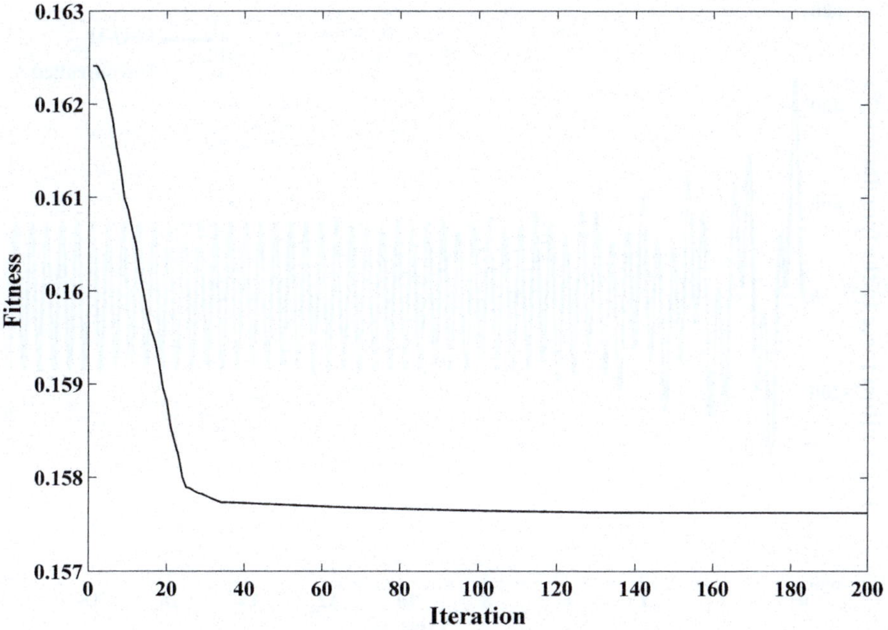

Fig. 4.27 Fitting convergence of PSO-H$_\infty$ active control for power equipment

Based on the obtained optimal state feedback controller, the force transmitted to the power equipment's foundation with active control is shown in Fig. 4.28, and the active control force generated by the actuator is shown in Fig. 4.29.

4.4.3 Optimized H$_\infty$ Active Control for Sensitive Equipment

Consider the active control system of the sensitive equipment shown in Fig. 1.5b. The dynamic equation is given Eq. (4.9), the state variables are $z_1 = x_1, z_2 = x_2, z_3 = \dot{x}_1$, and $z_4 = \dot{x}_2$, and the state vector is $z(t) = [z_1, z_2, z_3, z_4]^{\mathrm{T}}$. Let the observation output of this control system be defined as $y(t) = \left[x_1, x_2, \dot{x}_1, \dot{x}_2\right]^{\mathrm{T}}$, and the control output be the vibration velocity of the sensitive equipment expressed as $Y(t) = \left\{\dot{x}_2\right\}$. Thus, a state space equation of the form given in Eq. (4.66) can be obtained, with the coefficient matrix in the equation changing to the following:

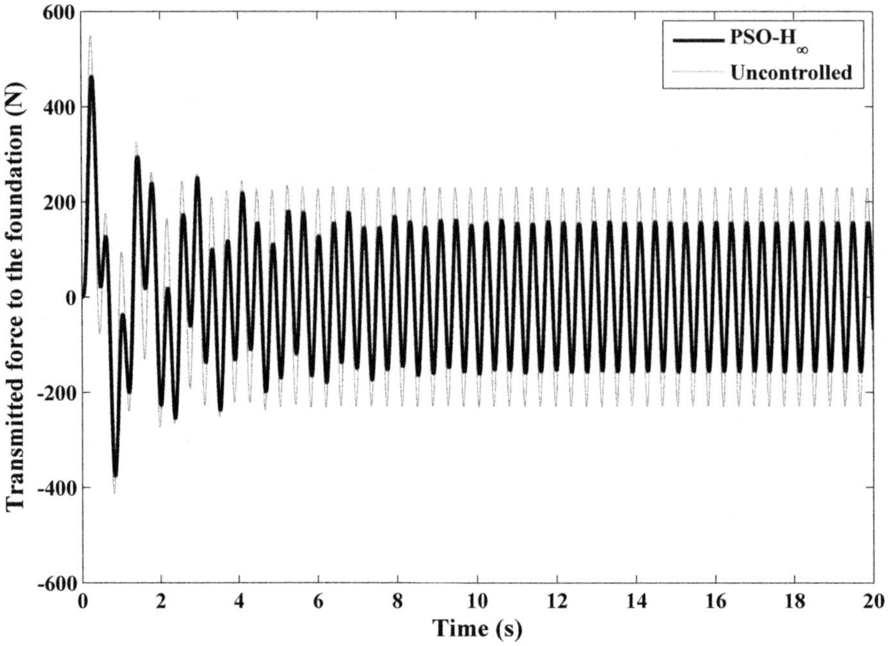

Fig. 4.28 Force transmitted from the power equipment to the foundation (PSO-H_∞)

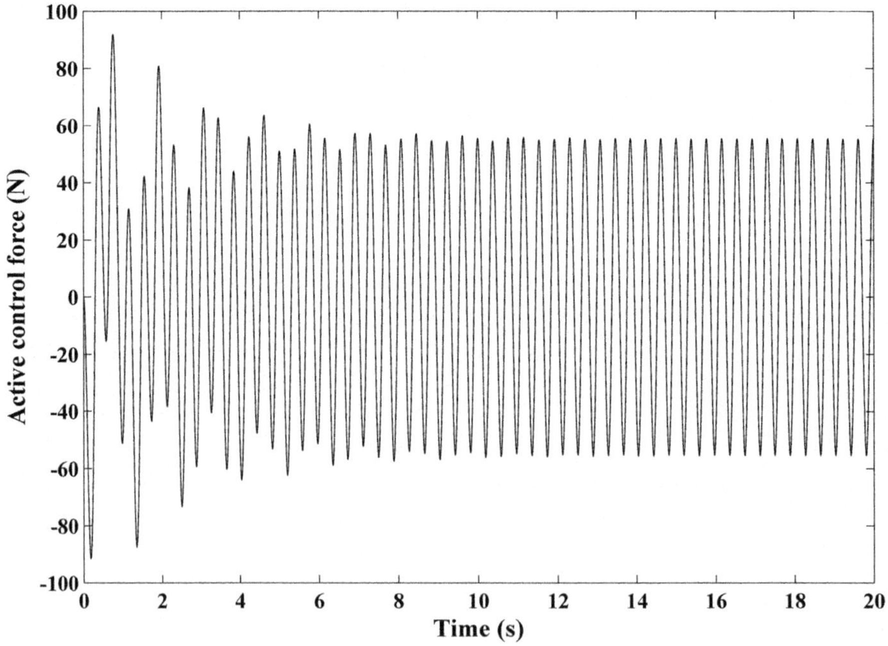

Fig. 4.29 Active control force of the PSO-H_∞ strategy for power equipment

$$A = \begin{bmatrix} 0 & 0 & 1 & 0 \\ 0 & 0 & 0 & 1 \\ -\frac{k_1+k_2}{m_1} & \frac{k_2}{m_1} & -\frac{c_1+c_2}{m_1} & \frac{c_2}{m_1} \\ \frac{k_2}{m_2} & -\frac{k_2}{m_2} & \frac{c_2}{m_2} & -\frac{c_2}{m_2} \end{bmatrix},\ b_1 = \begin{bmatrix} 0 \\ 0 \\ \frac{1}{m_1} \\ 0 \end{bmatrix},\ b_2 = \begin{bmatrix} 0 \\ 0 \\ -\frac{1}{m_1} \\ \frac{1}{m_2} \end{bmatrix},\ C_1 =$$

$$\begin{bmatrix} 0\,0\,0\,1 \end{bmatrix},\ d_{11} = [0],\ d_{12} = [0],\ C_2 = \begin{bmatrix} 1\,0\,0\,0 \\ 0\,1\,0\,0 \\ 0\,0\,1\,0 \\ 0\,0\,0\,1 \end{bmatrix},\ d_{21} = \begin{bmatrix} 0 \\ 0 \\ 0 \\ 0 \end{bmatrix},\ d_{22} = \begin{bmatrix} 0 \\ 0 \\ 0 \\ 0 \end{bmatrix}.$$

The PSO-H_∞ active control will be used for the sensitive equipment in this section.

Example 4.8 The vibration isolation system parameter setting and the PSO algorithm of the sensitive equipment is the same as that in Example 4.6. The constraints of the gain search range for the H_∞ state feedback controller are set to $\|K\|_\infty \le 4 \times 10^3$. The fitting convergence curve is shown in Fig. 4.30, and the obtained gbest solution is: K=[−3.671, −3.051, −0.997, −2.546]×10^3.

$$K = \begin{bmatrix} -3.671 & -3.051 & -0.997 & -2.546 \end{bmatrix} \times 10^3$$

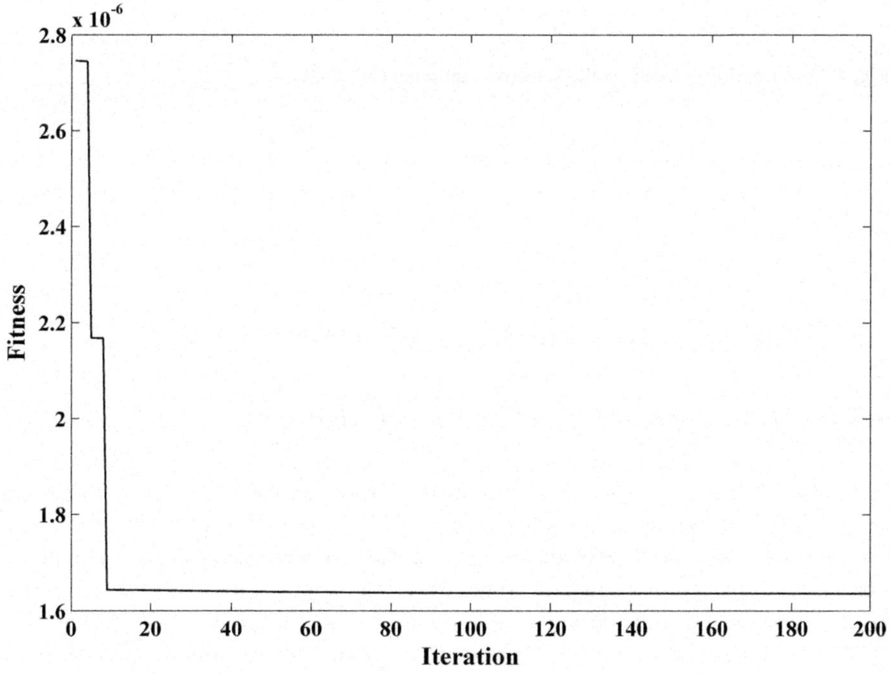

Fig. 4.30 Fitting convergence of the PSO-H_∞ active control for sensitive equipment

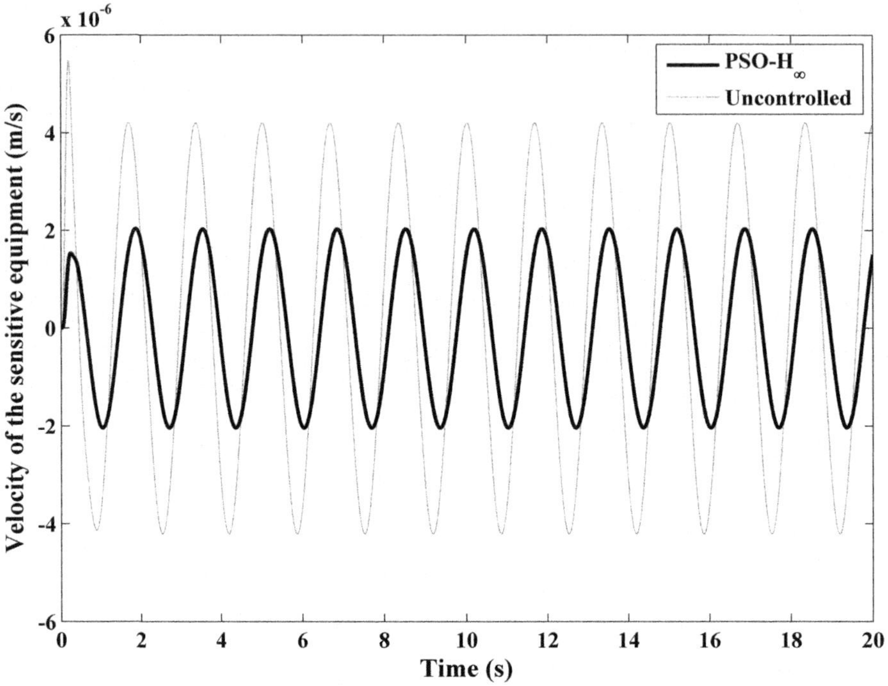

Fig. 4.31 Vibration velocity of the sensitive equipment (PSO-H_∞)

The vibration velocity of the sensitive equipment with active control is shown in Fig. 4.31, and the active control force generated by the actuator is depicted in Fig. 4.32.

4.5 Multi-objective Optimized H_2/H_∞ Active Control

4.5.1 Multi-objective H_2/H_∞ Control Algorithm

A good control system must have a certain degree of robustness to cope with internal and external uncertainty resources. It is impossible to consider the uncertainty impact on the system's control performance in a general control strategy. The H_2 and H_∞ norms are the most important indicators to reflect the robustness of the control system. H_∞ control mainly considers the system's robust stability but may sacrifice other system performances, whereas H_2 control can make up for this deficiency, but it is not robust to the uncertainty caused by model perturbations. Therefore, H_2/H_∞ hybrid control aims to achieve a comprehensive control performance and robust stability [5].

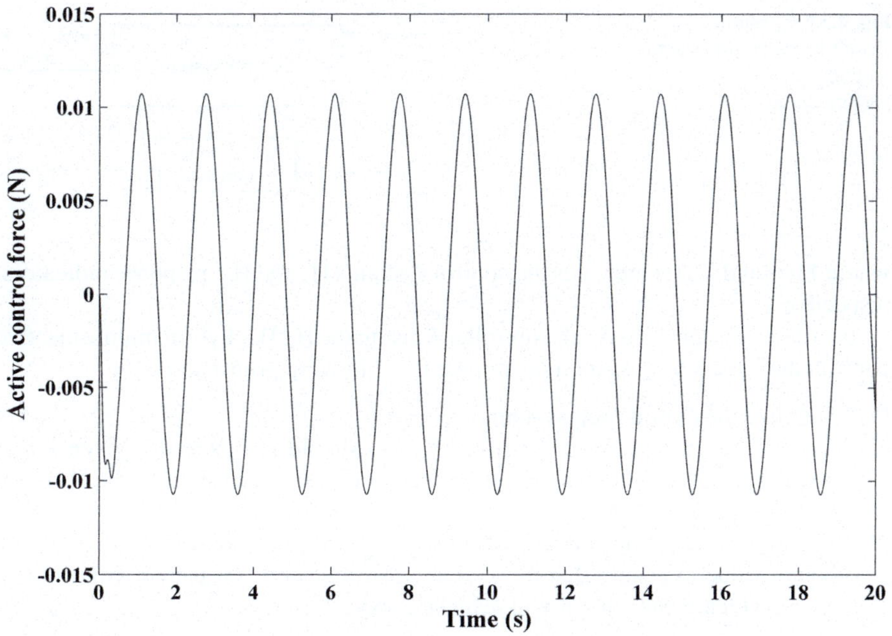

Fig. 4.32 Active control force of the PSO-H_∞ strategy for sensitive equipment

Consider the control system's H_2 and H_∞ performance indicators as a multi-objective optimization problem. Since the mixed norm control problem is usually non-convex, it is not possible to solve it using traditional mathematical methods effectively. In addition, multi-objective optimization problems are not the only solution but often the set of all non-inferior solutions, such as the Pareto optimal solution [6–9]. Traditional methods usually convert a multi-objective problem into an approximately convex single-objective problem. However, when the control system loses convexity, such methods cannot be solved, and there are significant limitations [6, 10]. In order to solve this problem, reference [6] used the MOPSO algorithm to study the H_2/H_∞ hybrid multi-objective control problem and compared it to the traditional linear matrix inequality method. Their results show that solving H_2/H_∞ hybrid control based on MOPSO is an effective new method.

This paper considers a dynamic closed-loop control system with H_2/H_∞ response index, as shown in Fig. 4.33.

The state space equation of the control system considering the H_2/H_∞ mixed response index can be expressed as:

$$\begin{cases} \dot{z}(t) = \mathrm{A}z(t) + \mathrm{b}_1 F(t) + \mathrm{b}_2 U(t) \\ Y_2(t) = \mathrm{C}_{2,1} z(t) + \mathrm{d}_{2,11} F(t) + \mathrm{d}_{2,12} U(t) \\ Y_\infty(t) = \mathrm{C}_{\infty,1} z(t) + \mathrm{d}_{\infty,11} F(t) + \mathrm{d}_{\infty,12} U(t) \\ y(t) = \mathrm{C}_2 z(t) + \mathrm{d}_{21} F(t) + \mathrm{d}_{22} U(t) \end{cases} \tag{4.68}$$

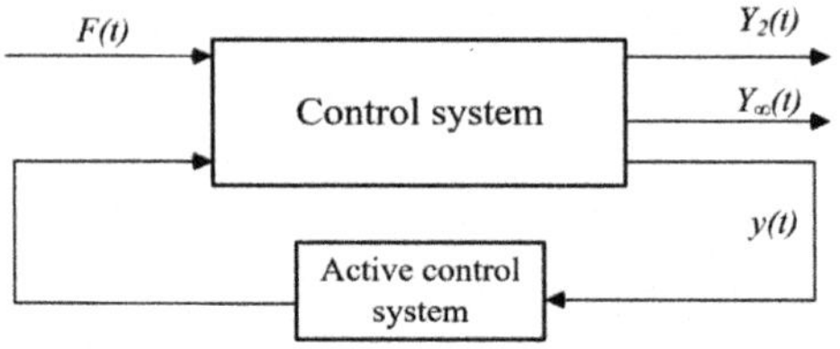

Fig. 4.33 Dynamic closed-loop control system with mixed H_2/H_∞ performances

where $Y_2(t)$ and $Y_\infty(t)$ indicates the control system's H_2 and H_∞ response indicators, respectively.

In order to design a feedback controller K using the H_2/H_∞ hybrid multi-objective control, the closed-loop system must meet the following properties:

(1) Stabilize the closed-loop system.
(2) The closed-loop transfer function $\|T_{Y_2F}\|_2$ should not exceed a given upper bound value to ensure that the system performance is measured by the H_2 norm to make it at a good level.
(3) The closed-loop transfer function $\|T_{Y_\infty F}\|_\infty$ does not exceed the given upper bound value to ensure that the system performance is measured with the H_∞ norm, which means that it will achieve good robustness.

Equation (4.69) indicates the multi-objective optimization problem and the fitting function to be satisfied.

$$\begin{cases} fitness_1 : \|T_{Y_\infty F}\|_\infty \\ fitness_2 : \|T_{Y_2F}\|_2 \end{cases} \tag{4.69}$$

4.5.2 *Multi-objective Optimized H_2/H_∞ Control for Power Equipment*

The active control system of the power equipment is shown in Fig. 1.5a, the dynamic equation is expressed in Eq. (4.4), the state variables are $z_1 = x_1$, $z_2 = x_2$, $z_3 = \dot{x}_1$, $z_4 = \dot{x}_2$, and the state vector is $z(t) = [z_1, z_2, z_3, z_4]^T$. Let the observation output of the active control system be $y(t) = \left[x_1, x_2, \dot{x}_1, \dot{x}_2\right]^T$. In this study, the H_∞ output response index is defined as the force transmitted by the power equipment to the foundation, which can bwritten as $Y_\infty(t) = \left\{k_1x_1 + c_1\dot{x}_1\right\}$, and the H_2 output response index is defined as the vibration velocity of power equipment, which can be written as $Y_2(t) = \left\{\dot{x}_2\right\}$. It is worth noting that in actual engineering, the H_2 and H_∞ response indicators must be determined according to the actual characteristics of the control system to ensure that the two indicators can accurately reflect the control performance and robustness of the specific system. Nevertheless, this study is hypothetical, hence the following state space equation can be obtained:

$$\begin{cases} \dot{z}(t) = \mathrm{A}z(t) + \mathrm{b}_1 F(t) + \mathrm{b}_2 F_a(t) \\ Y_2(t) = \mathrm{C}_{2,1}z(t) + \mathrm{d}_{2,11}F(t) + \mathrm{d}_{2,12}F_a(t) \\ Y_\infty(t) = \mathrm{C}_{\infty,1}z(t) + \mathrm{d}_{\infty,11}F(t) + \mathrm{d}_{\infty,12}F_a(t) \\ y(t) = \mathrm{C}_2 z(t) + \mathrm{d}_{21}F(t) + \mathrm{d}_{22}F_a(t) \end{cases} \tag{4.70}$$

where A, b_1, b_2, C_2, d_{21}, and d_{22} are the same as those in Eq. (4.66), $\mathrm{C}_{\infty,1}$, $\mathrm{d}_{\infty,11}$, $\mathrm{d}_{\infty,12}$ are identical to C_1, d_{11}, d_{12}, respectively,$\mathrm{C}_{2,1} = \begin{bmatrix} 0 & 0 & 0 & 1 \end{bmatrix}$, $\mathrm{d}_{2,11} = [0]$, and $\mathrm{d}_{2,12} = [0]$.

Example 4.9 The power equipment's vibration isolation system parameter setting is the same as that in Example 4.7. The parameters of the MOPSO algorithm are: the number of iterations is 200, the inertia weight coefficient is $\omega = 0.99^t$, where t is the iteration step, and the acceleration factors are $c_1 = 2$ and $c_2 = 1$. The constraints of the feedback controller gain search range in the hybrid control H_2/H_∞ is $\|K\|_\infty \leq 7 \times 10^2$.

The Pareto frontier is shown in Fig. 4.34, and the obtained gbest solution is $K = \begin{bmatrix} -4.790 & -6.998 & 7.000 & -3.698 \end{bmatrix} \times 10^2$.

Based on this optimal feedback controller, the force transmitted to the foundation of the power equipment with active control is shown in Fig. 4.35, and the active control force generated by the actuator is shown in Fig. 4.36.

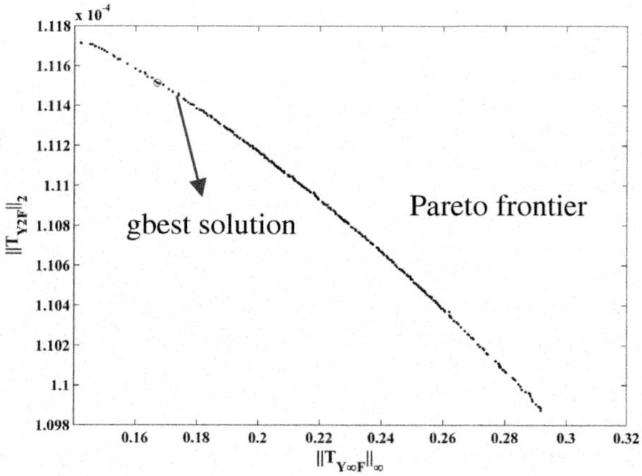

Fig. 4.34 Pareto frontier of the MOPSO-H_2/H_∞ active control for power equipment

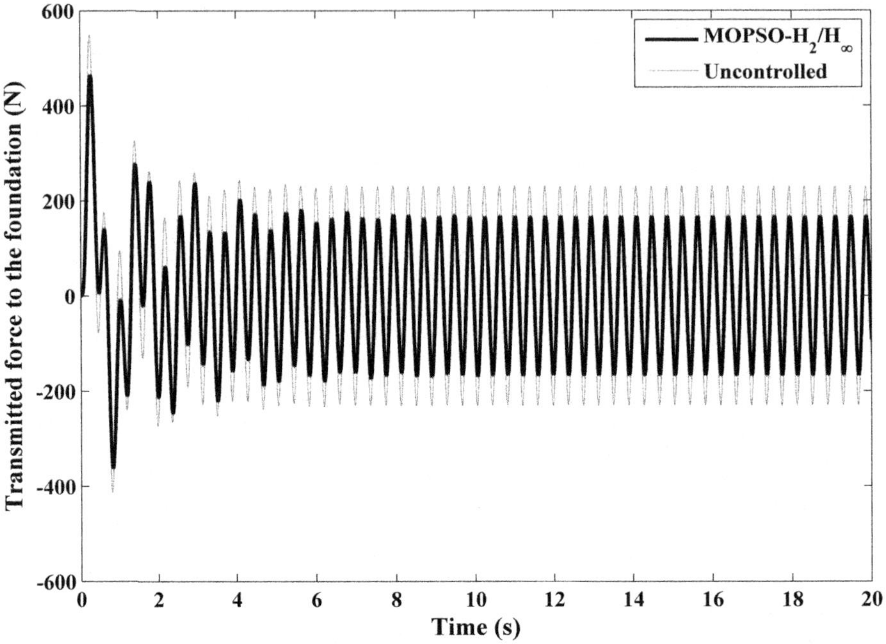

Fig. 4.35 Force Transmitted from the power equipment to the foundation (MOPSO-H_2/H_∞)

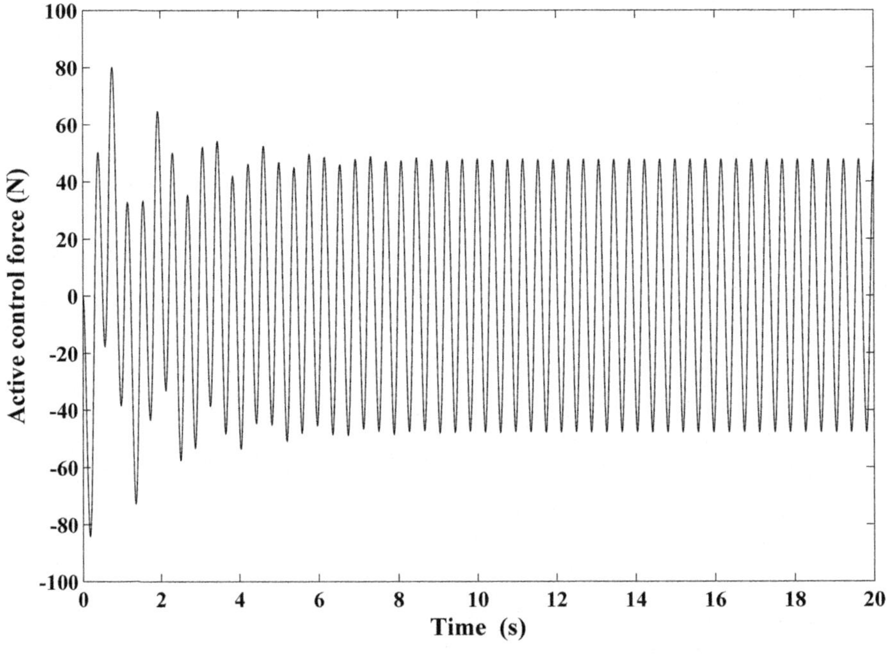

Fig. 4.36 Active control force of the MOPSO-H_2/H_∞ strategy for power equipment

4.5.3 Multi-objective Optimized H_2/H_∞ Control for Sensitive Equipment

The active control system of the sensitive equipment is shown in Fig. 1.5b, its dynamic equation is given in Eq. (4.9), the state variables are $z_1 = x_1$, $z_2 = x_2$, $z_3 = \dot{x}_1$, and $z_4 = \dot{x}_2$, and the state vector is $z(t) = [z_1, z_2, z_3, z_4]^{\mathrm{T}}$. Let the observation output of the active control system be $y(t) = \left[x_1, x_2, \dot{x}_1, \dot{x}_2\right]^{\mathrm{T}}$. In this study, the H_∞ output response index is defined as the sensitive equipment's vibration velocity $Y_\infty(t) = \left\{\dot{x}_2\right\}$, and the H_2 output response index is defined as the vibration velocity of the foundation or placement platform $Y_2(t) = \left\{\dot{x}_1\right\}$. Thus, it is possible to obtain the same state space equation as that in Eq. (2.70) can be obtained, where A, b_1, b_2, C_2, d_{21}, and d_{22} are the same as the ones in Sect. 4.4.3, and $\mathrm{C}_{\infty,1}$, $\mathrm{d}_{\infty,11}$, and $\mathrm{d}_{\infty,12}$ are the same as the C_1, d_{11}, d_{12} in Sect. 2.4.3, respectively. Besides, $\mathrm{C}_{2,1} = \left[0\ 0\ 1\ 0\right]$, $\mathrm{d}_{2,11} = [0]$, and $\mathrm{d}_{2,12} = [0]$.

Example 4.10 The sensitive equipment's vibration isolation system parameter setting is the same as that of Example 4.8, and the basic parameter setting of the MOPSO algorithm is the same as that of Example 4.9. The gain search range constraints of the feedback controller in $\mathrm{H}_2/\mathrm{H}_\infty$ hybrid control are set to $\|K\|_\infty \leq 4 \times 10^3$. The Pareto frontier results are shown in Fig. 4.37, and the gbest solution is $K = \left[-3.994\ 3.891\ -3.132\ -3.999\right] \times 10^3$.

Based on the obtained optimal feedback controller, the vibration velocity of the sensitive equipment with active control is shown in Fig. 4.38, and the active control force generated by the actuator is shown in Fig. 4.39.

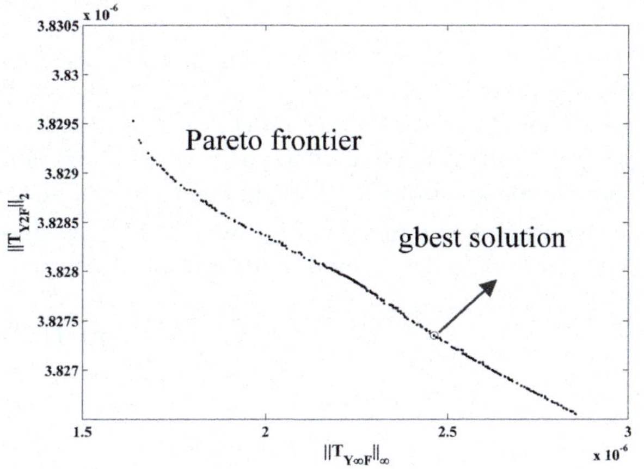

Fig. 4.37 Pareto frontier of the MOPSO-$\mathrm{H}_2/\mathrm{H}_\infty$ active control for sensitive equipment

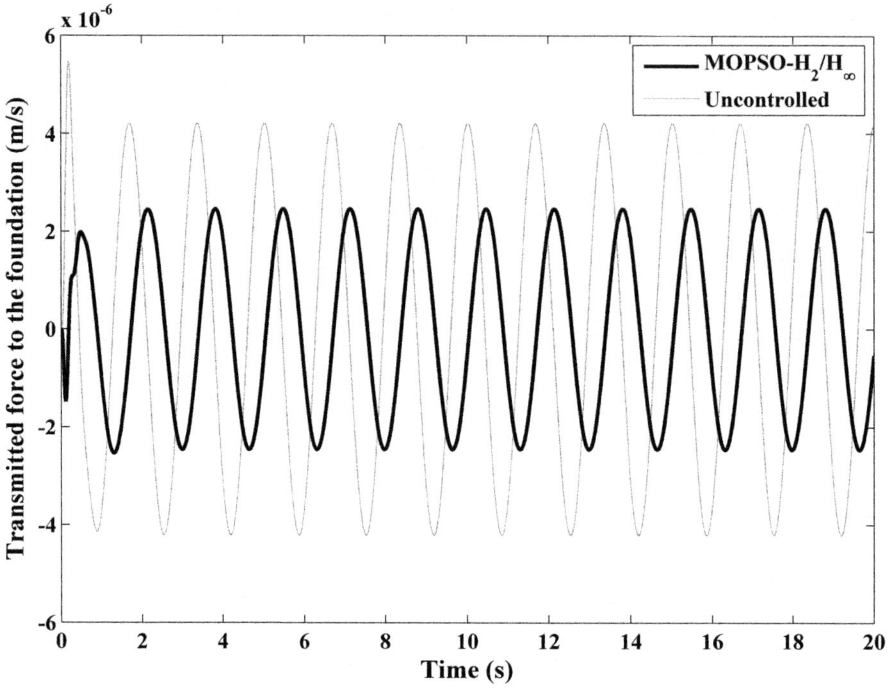

Fig. 4.38 Vibration velocity of the sensitive equipment (MOPSO-H_2/H_∞)

4.6 Optimized VUFLC Active Control

4.6.1 *Fuzzy Logic Control*

When there are several uncertainties in the controlled object or the control process, the traditional control method may not provide satisfactory control performance or even may cause the control process to fail. Zadeh [11] first proposed the concept of fuzzy sets and defined fuzzy subsets in 1965. He also assumed the universe U, and the mapping of Eq. (4.71) to determine the fuzzy subset $\tilde{A}$ of U, where $\mu_{\tilde{A}}(u)$ is the membership function of $\tilde{A}$, which indicates the degree to which the element u belongs to $\tilde{A}$, and it is referred to as *Membership*.

$$\begin{aligned} &\mu_{\tilde{A}} : U \to [0, 1] \\ &u \to \mu_{\tilde{A}}(u) \end{aligned} \tag{4.71}$$

The fuzzy subsets on the universe U are usually expressed in the Zadeh notation as follows:

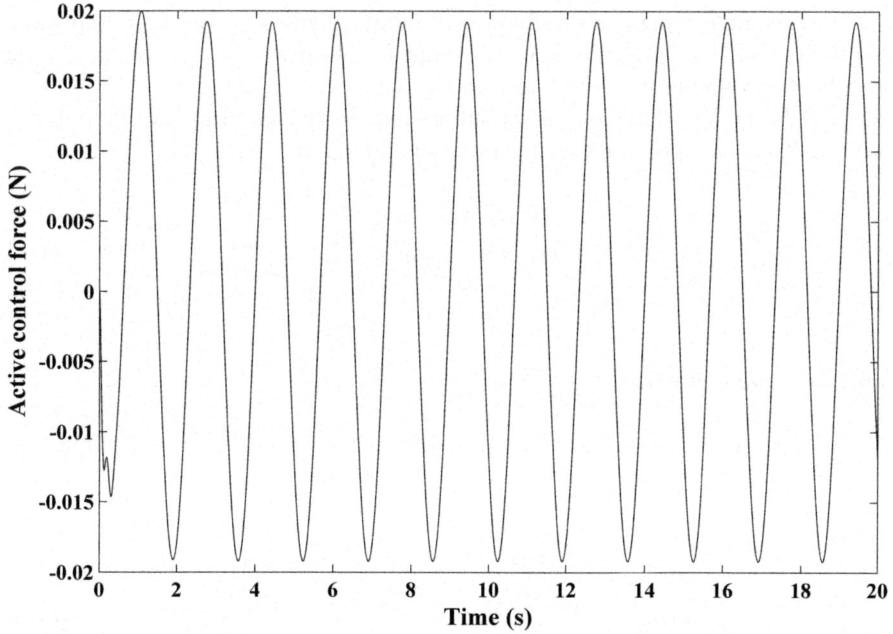

Fig. 4.39 Active control force of the MOPSO-H_2/H_∞ strategy for sensitive equipment

(1) If the universe U is a discrete finite set $\{x_1, x_2, \ldots, x_n\}$, $\tilde{A}(x_i) = a_i$ $(i = 1, 2, \ldots, n)$ can be expressed as:

$$\tilde{A} = \frac{a_1}{x_1} + \frac{a_2}{x_2} + \ldots\ldots \frac{a_n}{x_n} \tag{4.72}$$

(2) If the universe U is continuous, $\tilde{A}$ can be expressed as:

$$\tilde{A} = \int_U \tilde{A}(u)/u \tag{4.73}$$

4.6.1.1 Fuzzy Relationship

In ordinary collections, when $(x, y) \in R$, x is said to be related to y and is recorded as xRy when $(x, y) \notin R$, x is said to not be related to y. Use the characteristic function to represent R:

$$\chi_R(x, y) = \begin{cases} 0, (x, y) \in R \\ 1, (x, y) \notin R \end{cases} \tag{4.74}$$

Generalizing ordinary sets to fuzzy sets, letting X and Y be two non-empty sets, and solving fuzzy relations from X to Y with a fuzzy subset of $X \times Y$ is called a, and the whole fuzzy relations from X to Y can be represented by $Fuzzy(X \times Y)$. Assume that $\tilde{R} \in Fuzzy(X \times Y)$ $\tilde{R}$ is a fuzzy subset, and the degree of membership $\tilde{R}(x, y)$ $(x, y) \in R$ can be represented by the membership function of Eq. (2.75)

$$\begin{aligned} &\mu_R : X \times Y \to [0, 1] \\ &(x, y) \to \mu_{\tilde{R}}(x, y) \end{aligned} \tag{4.75}$$

4.6.1.2 Fuzzy Matrix

Assume that $X = \{x_1, x_2, \ldots\ldots, x_m\}$ and $Y = \{y_1, y_2, \ldots\ldots, y_n\}$ are finite sets, and then the fuzzy relationship $\tilde{R}$ of $X \times Y$ can be expressed as a $m \times n$-order fuzzy matrix

$$\tilde{R} = \left(r_{ij}\right)_{m\times n} = \begin{bmatrix} \tilde{R}(x_1, y_1) & \tilde{R}(x_1, y_2) & \ldots & \tilde{R}(x_1, y_n) \\ \tilde{R}(x_2, y_1) & \tilde{R}(x_2, y_2) & \ldots & \tilde{R}(x_2, y_n) \\ \ldots & \ldots & \ldots & \ldots \\ \tilde{R}(x_m, y_1) & \tilde{R}(x_m, y_2) & \ldots & \tilde{R}(x_m, y_n) \end{bmatrix} \tag{4.76}$$

Fuzzy matrices can be used to represent fuzzy relations for the given sets A, B, and C, assuming that $\tilde{S}$ is the fuzzy relationship of $A \times B$, $\tilde{Q}$ is the fuzzy relationship of $B \times C$, and $\tilde{R} = \tilde{S} \circ \tilde{Q}$ is the synthetic relationship of $\tilde{S}$ and $\tilde{Q}$, and the fuzzy relation from A to C.

4.6.1.3 Fuzzy Transformation

Fuzzy transformation is an important calculation process in fuzzy control. With the help of fuzzy transformation, the fuzzy output can be obtained based on the fuzzy input. The fuzzy transformation is briefly explained as follows:

Let $\tilde{A} = (a_1, a_2, \ldots, a_m)$ be an m-dimensional vector, $\tilde{R} = \begin{bmatrix} r_{11} & r_{12} & \ldots & r_{1n} \\ r_{21} & r_{22} & \ldots & r_{2n} \\ \ldots & \ldots & \ldots & \ldots \\ r_{m1} & r_{m2} & \ldots & r_{mn} \end{bmatrix}$ be a fuzzy relationship represented by an $m \times n$-dimensional matrix, and $\tilde{A} \circ \tilde{R} = \tilde{B}$ be the fuzzy transformation, where $\tilde{R}$ represents the inference rules' synthesis. Then, the unique n-dimensional fuzzy vector $\tilde{B}(b_1, b_2, \ldots, b_n)$ can be determined. If these rules exist: (1) $\tilde{A}$ is the fuzzy vector of the input universe, (2) $\tilde{B}$ is the fuzzy vector of the output control universe, and (3) $\tilde{R}$ is a relationship between the input universe and output universe, then $\tilde{A} \circ \tilde{R} = \tilde{B}$ is the fuzzy transformation process from input to output. Hence, the process of solving the fuzzy control output from a fuzzy input is carried out through some kinds of rules, and finally, the fuzzy logic control is developed.

4.6.1.4 Fuzzy Inference

There are three main forms of fuzzy statements commonly used in fuzzy control algorithms:

(1) If $\tilde{A}$ then $\tilde{B}$.
(2) If $\tilde{A}$ then $\tilde{B}$, else $\tilde{C}$.
(3) If $\tilde{A}$ and $\tilde{B}$ then $\tilde{C}$.

In these cases, the third statement is the most common. This is because the error between the given value and the actual value and the rate of error change must be considered in most control systems. 'Martini minimum algorithm' is commonly used in fuzzy inference, and the maximum-minimum operation is directly used in the inference process. In short, if $x = x_0$ and $y = y_0$, a new membership can be defined as follows:

$$\mu_{\tilde{C}}(z) = \left[w_1 \wedge \mu_{\tilde{C}_1}(z)\right] \vee \left[w_2 \wedge \mu_{\tilde{C}_2}(z)\right] \tag{4.77}$$

where $w_1 = \mu_{\tilde{A}_1}(x_0) \wedge \mu_{\tilde{B}_1}(y_0)$, and $w_2 = \mu_{\tilde{A}_2}(x_0) \wedge \mu_{\tilde{B}_2}(y_0)$.

4.6.1.5 Fuzzy Control

Based on the above descriptions, in order to perform fuzzy control, the following issues must be solved:

(1) Fuzzification of exact quantities, transforming linguistic values into fuzzy subsets of the universe.
(2) Proposing the fuzzy control rule and determining how to calculate and implement the fuzzy relationship.
(3) Judge the fuzzy output and then complete the conversion from fuzzy quantities to exact quantities.

Exact quantity fuzzification is a fuzzy set that maps real input measurements to the fuzzy universe. When designing the fuzzy level, the linguistic variables of the error, the rate of error change, the rate of control change, Positive Big (PB), Positive Small (PS), Zero (Z), Negative Small (NS), and Negative Big (PB) can be used in the description.

Linear and nonlinear are the two types of membership functions. The previous studies show that the linear and nonlinear membership functions have little effect on fuzzy control. The effect on the fuzzy control is mainly determined using the maximum membership function of the intersection between adjacent fuzzy subsets, the inner product. This indicator represents the degree of correlation between fuzzy control rules; if the inner product is too small, the rules are less relevant, and the model is not robust, and if the inner product is too large, the rules are more relevant and the model is robust but has a lower resolution. In the fuzzy control system, the input language variables are all deterministic numbers, but the input of the fuzzy controller is a fuzzy set.

Anti-fuzzification is to convert the fuzzy output generated by the fuzzy inference process into an accurate output. The common methods for anti-fuzzification are listed as follows:

(1) Maximum membership method

In this method, the exact results corresponding to the maximum degree of membership obtained by fuzzy inference are taken as the exact ones after anti-fuzzification. The average value should be taken if there are multiple maximum degrees of membership.

(2) Median method

The universe element corresponding to the membership function curve of the fuzzy set and the average point of the area surrounded by the abscissa are used in the decision to fully use the information contained in the fuzzy output set.

(3) Gravity Method

Every element in the universe x_i $(i = 1, 2, \ldots, n)$ is selected as a weighting factor of the membership $\mu_{\tilde{U}_i}(x_i)$ for adjudging a fuzzy output set $\tilde{U}_i$. In this study, the scilicet $x_i \mu_{\tilde{U}_i}(x_i)$ $(i = 1, 2, \ldots, n)$ is adopted to calculate the product sum $\sum_{i=1}^{n} x_i \mu_{\tilde{U}_i}(x_i)$ and the average value x_0 of the membership sum $\sum_{i=1}^{n} \mu_{\tilde{U}_i}(x_i)$. The obtained average value x_0 is the result of anti-fuzzification obtained by applying the gravity method.

$$x_0 = \frac{\sum_{i=1}^{n} x_i \mu_{\tilde{U}_i}(x_i)}{\sum_{i=1}^{n} \mu_{\tilde{U}_i}(x_i)} \tag{4.78}$$

4.6.2 *Variable Universe Fuzzy Logic Control*

The fuzzy logic controller is essentially an interpolator that can be divided into fuzzification, fuzzy inference, fuzzy decision, and anti-fuzzification [14]. The principle of fuzzy control shows that the linguistic variables of the fuzzy controller's input and output largely determine the quality of the fuzzy control. The fuzzy controller can approach arbitrary control functions as long as the number of linguistic variables implemented is sufficient to produce a sufficiently fine universe division, and thus the fuzzy controller can approximate any control function. However, increasing linguistic variables means that the control rules increase rapidly, while the variable universe fuzzy control solves the problem without controlling the number of rules. Variable universe fuzzy logic control (VUFLC) is essentially a strategy to achieve

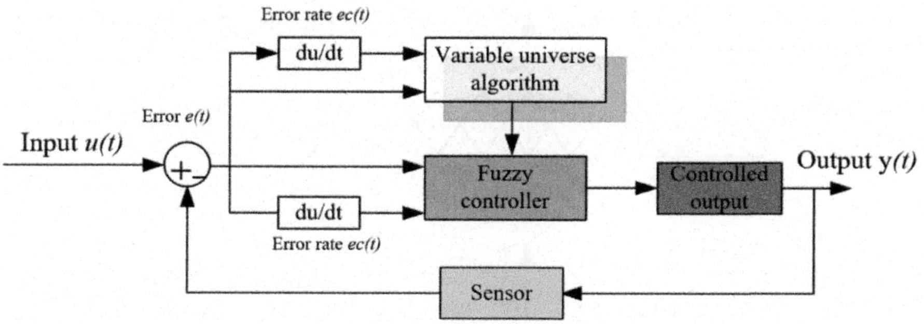

Fig. 4.40 Schematic diagram of VUFLC strategy

the change from rough control to fine control. The input and output scaling factors are generally introduced in current research without changing the fuzzy controller parameters (fuzzy rules, scale factors, quantization factors, etc.) so that the input and output universe can be dynamically scaled upon changing the fuzzy control's input [13, 14]. The basic structure of a variable universe fuzzy controller is shown in Fig. 2.40.

As shown in Fig. 4.40, the variable universe fuzzy control constantly adjusts the scaling factor according to the input error and the error's change rate so that the fuzzy universe can adaptively change. In VUFLC, the scaling factors of the input and output universes are defined as $\alpha(x)$ and $\beta(x)$, and in the two-input single-output fuzzy control, the input universe scaling factors are defined as $\alpha_1(x)$ and $\alpha_2(x)$, and the output universe scaling factor is defined as $\beta(x)$. A two-input single-output system will be used herein to illustrate the change in the universe. Let the initial universe be $[-E, E]$, and the universe after transforming be $[-\alpha(x)E, \alpha(x)E]$, as shown in Fig. 2.41. In this regard, the scaling factor has the following characteristics (Fig. 4.41):

(1) Duality: $\forall x \in X, \alpha(x) = \alpha(-x)$.
(2) Zeroing: $\alpha(0) = 0$.
(3) Monotonicity: α is strictly monotonically increasing in the universe $[-E, E]$.
(4) Coordination: $\forall x \in X, |x| \leq \alpha(x)E$.
(5) Positive regularity: $\alpha(\pm E) = 1$.

For a two-input single-output fuzzy control, the scaling factor can be either a function of the input error *e* and the error's rate of change *ec* or a function of either the input error or the error's rate of change.

In practical engineering, fuzzy control is performed in real-time. Hence, all fuzzy control depends on the time step *k*. Accordingly, the variable universe algorithm has become a function of time. As the universe is constantly changing, its membership function changes if the membership function is written as $A_i[x(k)]$, $B_j\left[y(k)\right]$, and $C_{ij}[z(k)]$, and then the original fuzzy control rules R become a set of dynamic rules $R(k)$:

$R(k)$: if $x(k)$ is $A_i(k)$ and $y(k)$ is $B_j(k)$, then $z(k)$ is $C_{ij}(k)$.

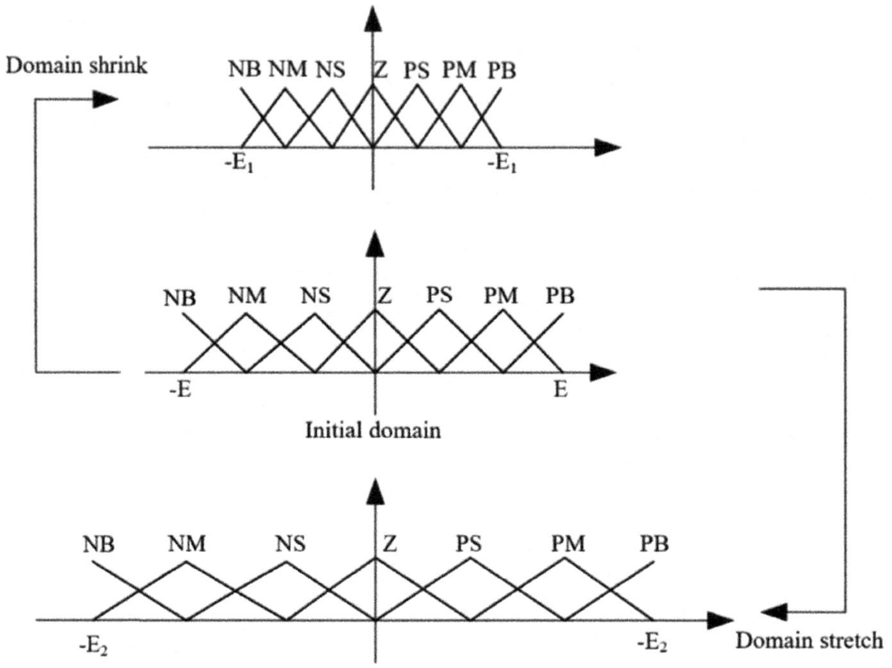

Fig. 4.41 Schematic diagram of the variable universe

The dynamic rules $R(k)$ can be expressed in the discrete form as follows:

$$\begin{array}{l} \text{if } x \text{ is } x_1(k) \text{and } y \text{ is } y_1(k), \text{ then } z \text{ is } z_{11}(k) \\ \text{if } x \text{ is } x_1(k) \text{and } y \text{ is } y_2(k), \text{ then } z \text{ is } z_{12}(k) \\ \quad \ldots \\ \text{if } x \text{ is } x_n(k) \text{and } y \text{ is } y_m(k), \text{ then } z \text{ is } z_{nm}(k) \end{array} \quad (4.79)$$

Equation (4.79) is equivalent to the interpolation function of the fuzzy control upon performing the variable universe. The equivalent control function of the fuzzy control changing over time is shown in Fig. 4.42.

It can be seen that the trend and basic characteristics of the initial control function can be inherited into the control function after performing the variable universe, allowing the controller to achieve higher accuracy. Currently available algorithms for scaling factors in the input universe are as follows [12–14]:

$$\alpha(x) = 1 - ce^{-kx^2}, c \in (0, 1), k > 0 \quad (4.80a)$$

$$\alpha(x) = \left[\frac{|x|}{E}\right]^{\tau}, \tau \in (0, 1) \quad (4.80b)$$

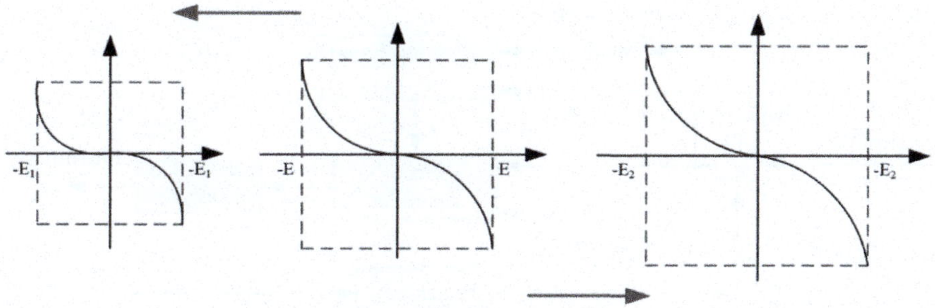

Fig. 4.42 Schematic variation of the control function of FLC

In the two-input single-output system, x in Eqs. (4.80a) and (4.80b) indicate the error e and its change rate ec, with c, k, τ being adjustable parameters.

The common scaling factors for output universes are:

$$\beta(t) = k\sum_{i=1}^{n} p_i \int_0^t e_i(\tau)\mathrm{d}_\tau + \beta(0) \tag{4.81a}$$

$$\beta(x, y) = \left[\frac{|x|}{E}\right]^{T_1}\left[\frac{|y|}{EC}\right]^{T_2} \tag{4.81b}$$

$$\beta(x, y) = \frac{1}{2}\left\{\left[\frac{|x|}{E}\right]^{T_1} + \left[\frac{|y|}{EC}\right]^{T_2}\right\} \tag{4.81c}$$

where k, p_i are adjustable parameters, n is the total number of input variables, $\beta(0)$ must be set according to the actual situation, with $\beta(0) = 1.0$ being desirable, and $0<T_1, T_2<1$.

If the $\beta(x, y)$ in Eqs. (4.81b) and (4.81c) is changed to $\beta(y)$, only the influence of the input error on the scaling factor is considered. This study adopts the scaling factor in Eq. (4.80b) and the output scaling factor in Eq. (4.81b). The variable universe fuzzy control is described in Fig. 4.43.

4.6.3 *Optimized VUFLC Active Control for Power Equipment*

Figure 1.5a shows the active control system of the power equipment. The dynamic equation is expressed in Eq. (4.4), the state variables are $z_1 = x_1$, $z_2 = x_2$, $z_3 = \dot{x}_1$, and $z_4 = \dot{x}_2$, and the state vector is $z(t) = [z_1, z_2, z_3, z_4]^{\mathrm{T}}$. The control output of the VUFLC $Y(t) = \left\{\mathrm{k}_1 x_1 + \mathrm{c}_1 \dot{x}_1\right\}$ is defined as the transmitted force from power equipment to the foundation, and the VUFLC input is the generated vibration disturbance

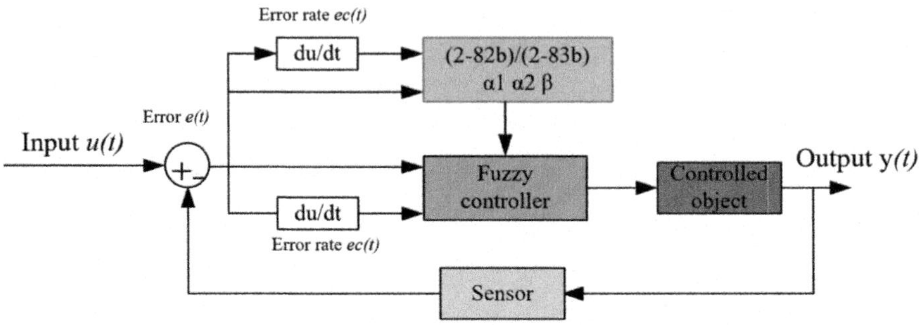

Fig. 4.43 Schematic diagram of the VUFLC with telescopic factors

$F(t)$. The state space equation of this system is defined as follows:

$$\begin{cases} \dot{z}(t) = \mathrm{A}z(t) + \mathrm{B}F_{input} \\ y(t) = \mathrm{C}z(t) + \mathrm{D}F_{input} \end{cases} \quad (4.82)$$

where $\mathrm{A} = \begin{bmatrix} 0 & 0 & 1 & 0 \\ 0 & 0 & 0 & 1 \\ -\frac{k_1+k_2}{m_1} & \frac{k_2}{m_1} & -\frac{c_1+c_2}{m_1} & \frac{c_2}{m_1} \\ \frac{k_2}{m_2} & -\frac{k_2}{m_2} & \frac{c_2}{m_2} & -\frac{c_2}{m_2} \end{bmatrix}$, $\mathrm{B} = \begin{bmatrix} 0 & 0 \\ 0 & 0 \\ 0 & \frac{1}{m_1} \\ \frac{1}{m_2} & -\frac{1}{m_2} \end{bmatrix}$, $\mathrm{C} = \begin{bmatrix} 1\,0\,0\,0 \\ 0\,1\,0\,0 \\ 0\,0\,1\,0 \\ 0\,0\,0\,1 \end{bmatrix}$, $\mathrm{D} = \begin{bmatrix} 0\,0 \\ 0\,0 \\ 0\,0 \\ 0\,0 \end{bmatrix}$, $F_{input} = \begin{bmatrix} F(t) \\ F_a(t) \end{bmatrix}$, and $F_a(t)$ indicates the active control force generated by VUFLC.

Figure 4.44 shows the VUFLC control optimized using the PSO algorithm.

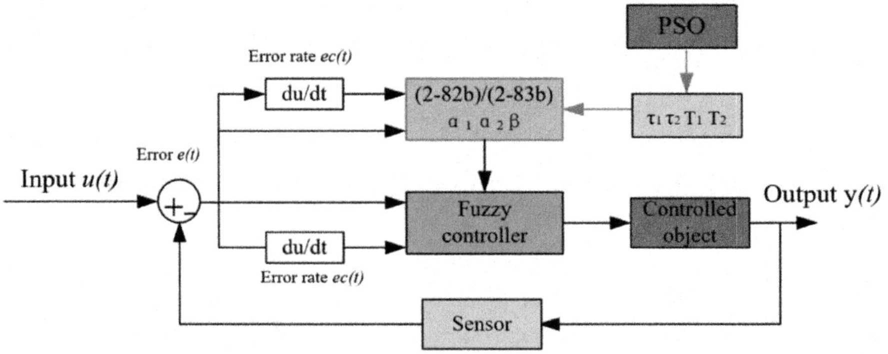

Fig. 4.44 Schematic diagram of the PSO-VUFLC active control

Table 4.1 Fuzzy inference rules (power equipment)

Inference rule		e				
		NB	NS	Z	PS	PB
ec	NB	PB	PB	PS	PS	PS
	NS	PS	PS	PS	Z	Z
	Z	Z	Z	Z	Z	NS
	PS	Z	Z	Z	Z	NS
	PB	Z	NS	NS	NS	NS

Example 4.11 The parameters of the power equipment's vibration isolation system and the PSO algorithm are the same as those in Example 4.7. The basic fuzzy controller parameters are the error *e*, the error's change rate *ec*, and the initial basic universe *u* ranging between -6 and 6. The fuzzy subsets of *e*, *ec*, and *u* are all set of {NB, NS, Z, PS, PB}. Moreover, the membership function is defined as ladder and triangle type, respectively. Fuzzy inference rules are shown in Table 4.1.

The scaling factors of the VUFLC's input and output are considered in the form of Eqs. (4.80b) and (4.81b), which are expressed as follows:

$$\begin{aligned} LY(e) &= [-\alpha_1 \times 6, \alpha_1 \times 6] \\ LY(ec) &= [-\alpha_2 \times 6, \alpha_2 \times 6] \\ LY(u) &= [-\beta \times 6, \beta \times 6] \end{aligned} \tag{4.83}$$

where $\alpha_1 = \left(\frac{|e|}{6}\right)^{\tau_1}$, $\alpha_2 = \left(\frac{|ec|}{6}\right)^{\tau_2}$, $\beta = \left(\frac{|e|}{6}\right)^{T_1}\left(\frac{|ec|}{6}\right)^{T_2}$, $\tau_1, \tau_2, T_1, T_2 \in (0,1)$ and *e* and *ec* are the dynamic error and error changing rate, respectively.

In the PSO, the arbitrary search range of τ_1, τ_2, T_1, T_2 is as follows:

$$0.01 \le \tau_1, \tau_2, T_1, T_2 \le 0.5.$$

The fitting curve is shown in Fig. 4.45. The gbest solution is:

$$\tau_1 = 0.0932, \tau_2 = 0.448, T_1 = 0.500, T_2 = 0.384.$$

The scaled fuzzy inference rule surface and membership function are shown in Fig. 4.46.

Based on the obtained optimal scaling factor and fuzzy logic controller and the research on VUFLC active control of power equipment, the equipment's active control force transmitted to the foundation is shown in Fig. 4.47, in which FLC is the primary fuzzy logic controller that does not consider variable universe strategy and its control effect ranges between uncontrolled and PSO-VUFLC. The active control force generated by the actuator in PSO-VUFLC is shown in Fig. 4.48.

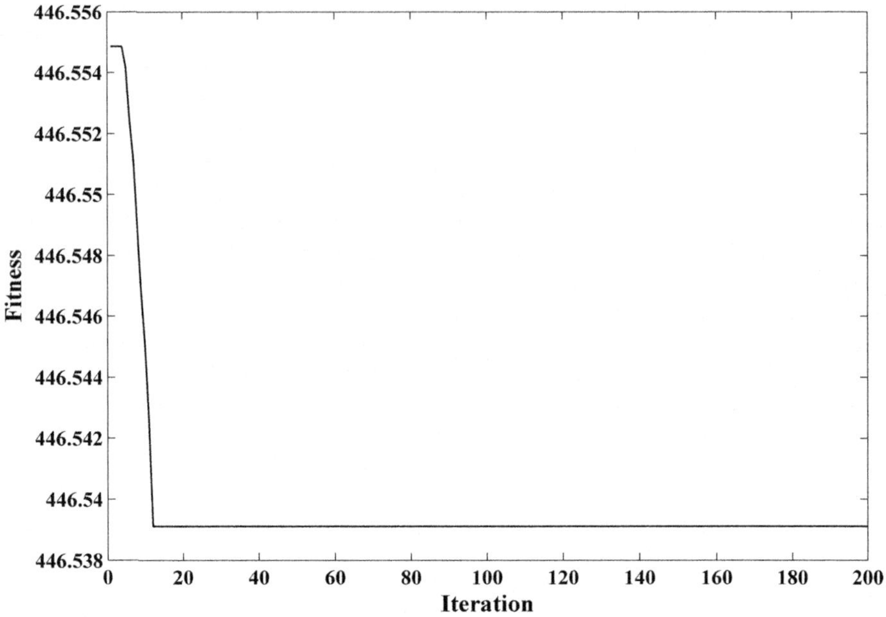

Fig. 4.45 Fitting convergence of the PSO-VUFLC active control for power equipment

4.6.4 Optimized VUFLC Active Control for Sensitive Equipment

Figure 1.5b shows the sensitive equipment's active control system. The dynamic equation is given in Eq. (4.4), the variables are $z_1 = x_1$, $z_2 = x_2$, $z_3 = \dot{x}_1$, and $z_4 = \dot{x}_2$, and the state vector is $z(t) = [z_1, z_2, z_3, z_4]^{\mathrm{T}}$. The VUFLC control output is defined as the vibration velocity of the sensitive equipment, and it is expressed as $Y(t) = \left\{\dot{x}_2\right\}$. The VUFLC input is defined as the vibration disturbance on the placement platform or foundation $F(t)$. The state space equation herein has the same form as that in Eq. (4.84), but the coefficient matrix must be changed as follows:

$$A = \begin{bmatrix} 0 & 0 & 1 & 0 \\ 0 & 0 & 0 & 1 \\ -\frac{k_1+k_2}{m_1} & \frac{k_2}{m_1} & -\frac{c_1+c_2}{m_1} & \frac{c_2}{m_1} \\ \frac{k_2}{m_2} & -\frac{k_2}{m_2} & \frac{c_2}{m_2} & -\frac{c_2}{m_2} \end{bmatrix}, B = \begin{bmatrix} 0 & 0 \\ 0 & 0 \\ \frac{1}{m_1} & -\frac{1}{m_1} \\ 0 & \frac{1}{m_2} \end{bmatrix}, C = \begin{bmatrix} 1\,0\,0\,0 \\ 0\,1\,0\,0 \\ 0\,0\,1\,0 \\ 0\,0\,0\,1 \end{bmatrix}, D = \begin{bmatrix} 0\,0 \\ 0\,0 \\ 0\,0 \\ 0\,0 \end{bmatrix}, F_{input} = \begin{bmatrix} F(t) \\ F_a(t) \end{bmatrix}.$$

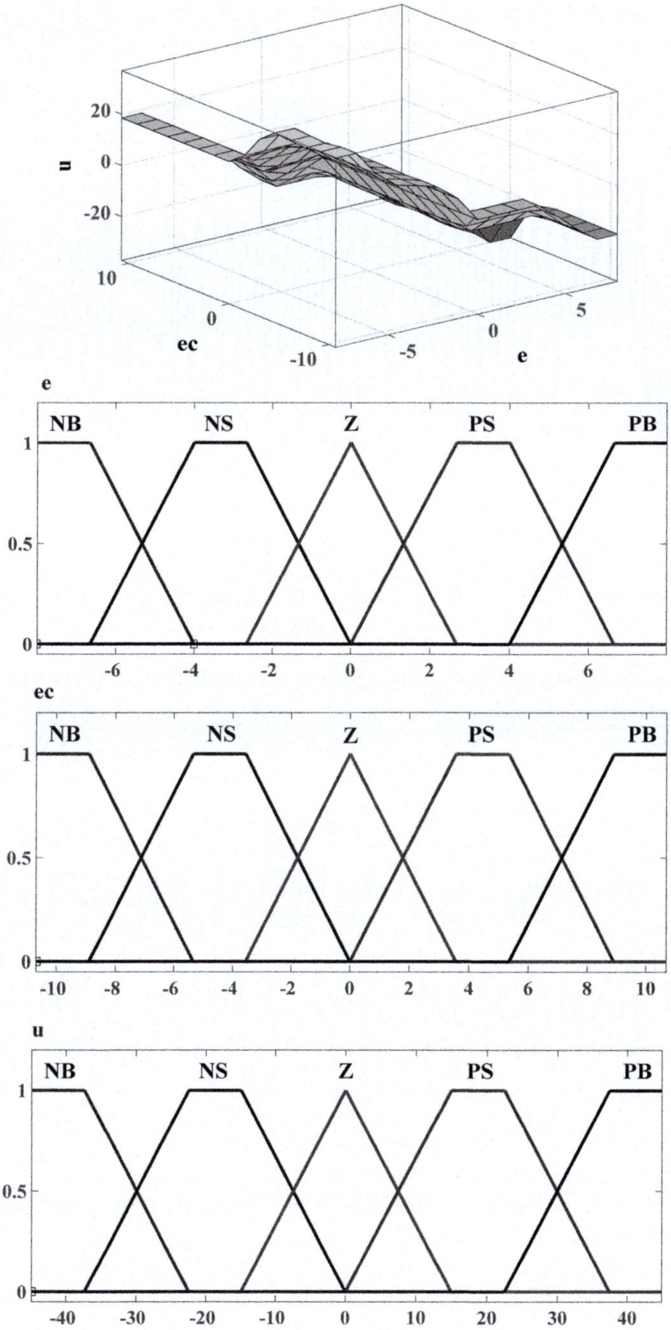

Fig. 4.46 Optimized fuzzy inference rules surface and member functions (power equipment)

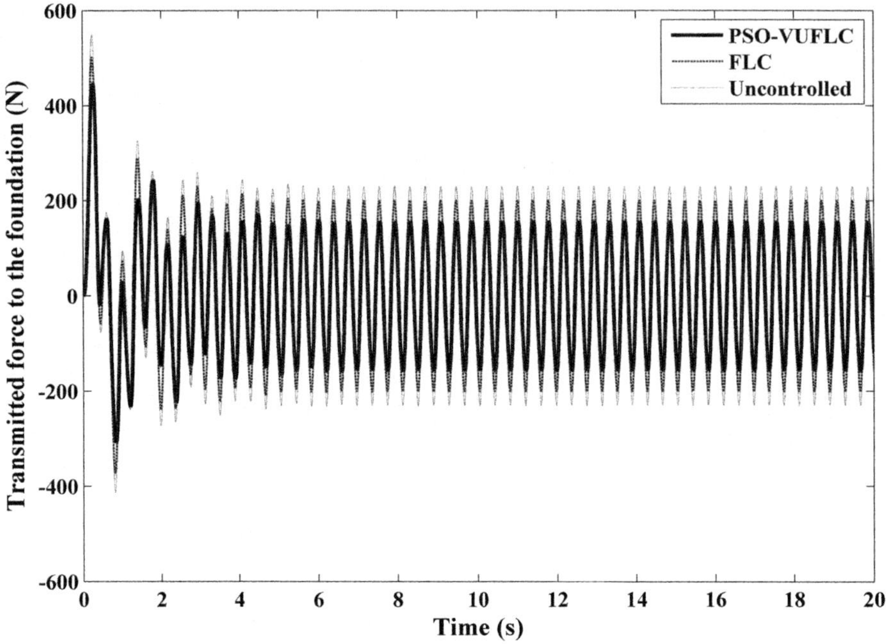

Fig. 4.47 Force transmitted from the power equipment to the foundation (PSO-VUFLC)

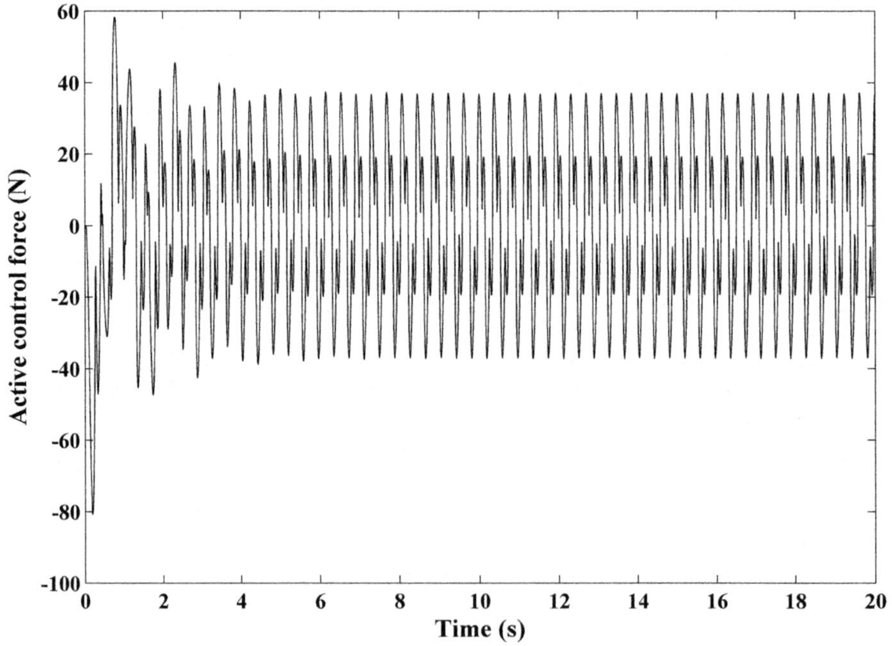

Fig. 4.48 Active control force of the PSO-VUFLC strategy for power equipment

Table 4.2 Fuzzy inference rules (sensitive equipment)

Inference rules	e					
ec	NB	NS	Z	PS	PB	
	NB	PB	PB	PS	PS	Z
	NS	PS	PS	PS	Z	Z
	Z	PS	Z	Z	Z	NS
	PS	Z	Z	NS	NS	NS
	PB	Z	NS	NS	NB	NB

The VUFLC's active control strategy for sensitive equipment based on the PSO algorithm is depicted in Fig. 4.44. However, the controlled object must be changed according to the above dynamic description.

Example 4.12 The vibration isolation system parameter configuration and PSO algorithm of sensitive equipment are the same as that in Example 4.8. Similarly, the initial basic universe of the error *e*, the error's rate of change *ec*, and the output *u* are specified in the basic fuzzy controller as [−6, 6]. The fuzzy subset of *e*, *ec*, and *u* are all {NB,NS,Z,PS,PB}. The ladder and triangle membership functions are adopted herein. Fuzzy inference rules are shown in Table 4.2. The VUFLC's telescopic factors are given in Eqs. (4.80b) and (4.81b), and the calculation form is the same as that of Eq. (4.83). In PSO, the τ_1, τ_2, T_1, T_2 search range f is defined as $0.01 \leq \tau_1, \tau_2, T_1, T_2 \leq 0.5$.

The PSO-VUFLC active control fitting convergence curve for sensitive equipment is shown in Fig. 4.49, and the obtained gbest solution is $\tau_1 = 0.495, \tau_2 = 0.112, T_1 = 0.486, T_2 = 0.0411$.

The optimized fuzzy inference rule surface and membership function are shown in Fig. 4.50. The VUFLC active control is performed on the sensitive equipment using the obtained optimal telescopic factors and fuzzy logic controller. Besides, the vibration velocity of sensitive equipment with active control is shown in Fig. 4.51. The control effect of traditional FLC is between the uncontrolled and PSO-VUFLC states. The active control force generated by the actuator in PSO-VUFLC is shown in Fig. 4.52.

4.7 Active Control Strategy Based on the Multi-objective Control Output

In the previous research on optimal active control strategies using PSO, the control output was set to a single objective, or the optimization fitting function was single. Moreover, the force transmitted to the foundation of power equipment was considered in measuring the disturbance caused by the vibration of the equipment to the surrounding environment. Besides, the sensitive equipment's vibration velocity under

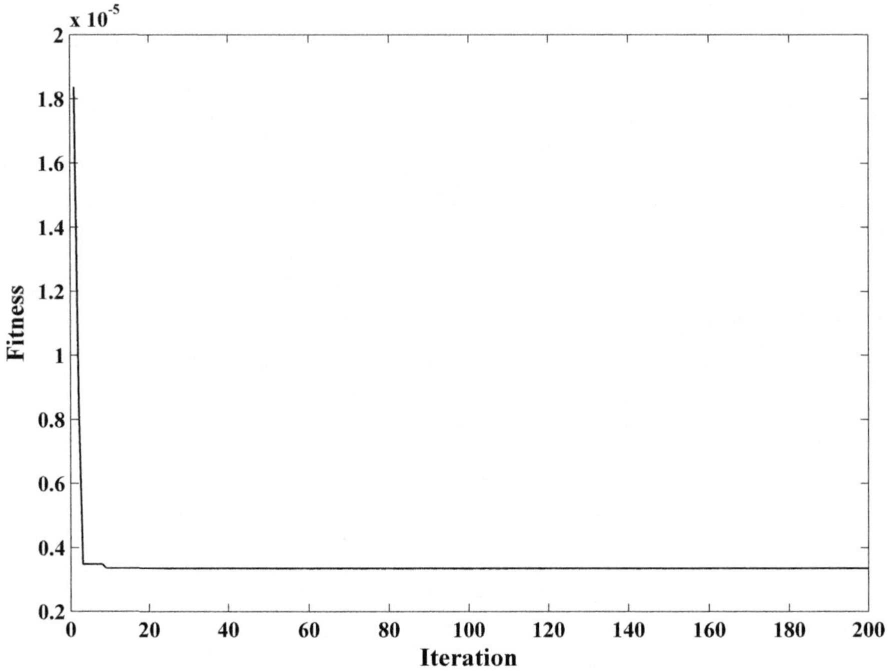

Fig. 4.49 Fitting convergence of the PSO-VUFLC active control for the sensitive equipment

external vibration hazards is considered to measure the equipment's disturbed level. In reality, a single indicator can not reflect the control effect of power or sensitive equipment. For the vibration control of power equipment, while reducing the force transmitted by the power equipment to the foundation, it is also necessary to effectively control the vibration level of the equipment itself, which is essential to overall equipment serviceability. For the vibration control of large sensitive devices, the vibration isolator's deformation directly affects the equipment's accuracy and even causes equipment failure. In such cases, an active control strategy based on multi-objective control output must be adopted.

4.7.1 Active Control Based on Multi-objective Control Output for Power Equipment

The active control system of sensitive equipment is shown in Fig. 1.5a, its dynamic equation is given in Eq. (4.4), the state variables are $z_1 = x_1$, $z_2 = x_2$, $z_3 = \dot{x}_1$, and $z_4 = \dot{x}_2$, and the state vector is $z(t) = [z_1, z_2, z_3, z_4]^{\mathrm{T}}$. The observation output of the control system is $y(t) = \left[x_1, x_2, \dot{x}_1, \dot{x}_2\right]^{\mathrm{T}}$. Besides, the control output of the control

Fig. 4.50 Optimized fuzzy inference rules surface and member functions (sensitive equipment)

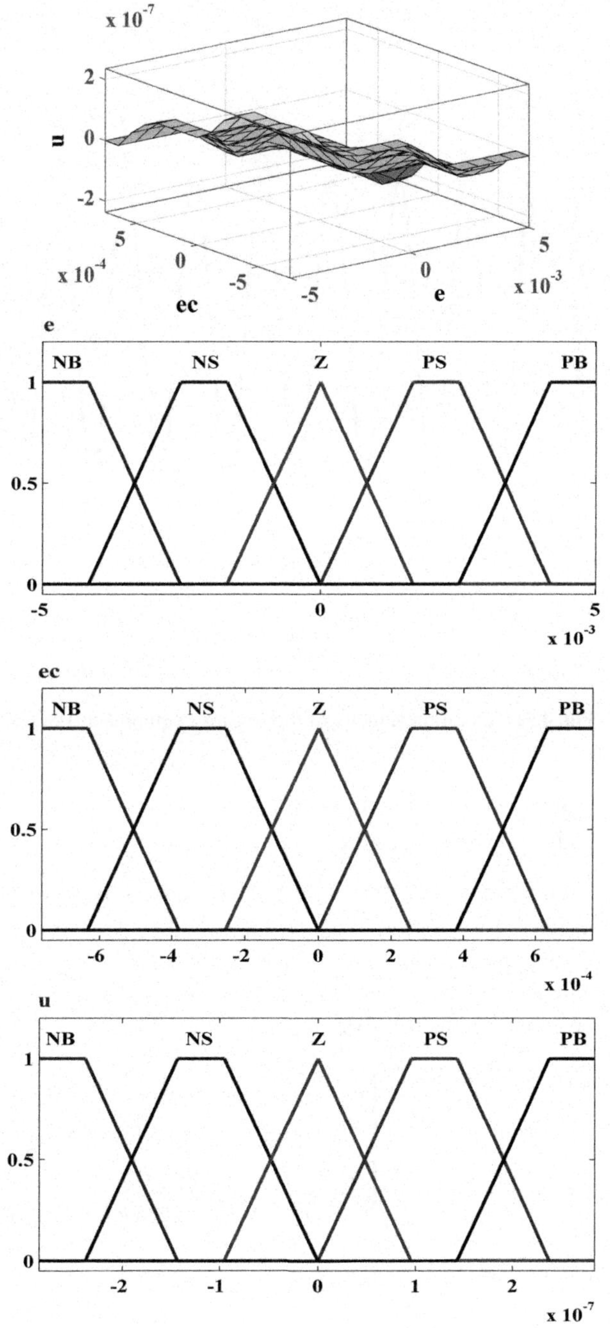

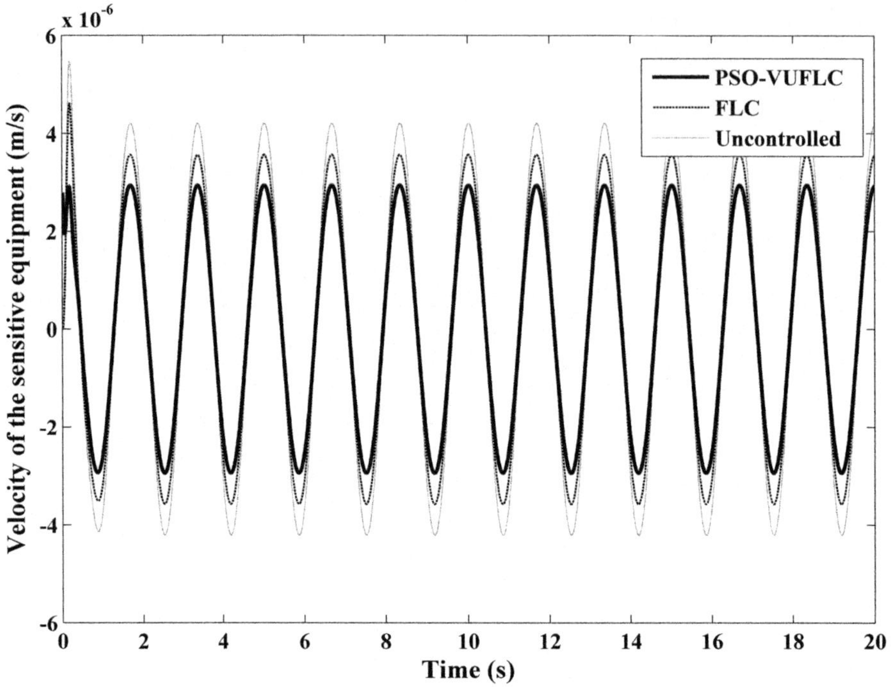

Fig. 4.51 Vibration velocity of the sensitive equipment (PSO-VUFLC)

system is considered to have two objectives: the force transmitted to the foundation by the power equipment $Y_1(t) = \left\{ \mathrm{k}_1 x_1 + \mathrm{c}_1 \dot{x}_1 \right\}$, and the vibration level of the power equipment itself $Y_2(t) = \left\{ \dot{x}_2 \right\}$. This study aims to consider both as much as possible in the active control process. Optimizing the active controller allows both control objectives to reach the best results simultaneously. For this purpose, the state space equations of the active controller design based on multi-objective control output can be expressed as follows:

$$\begin{cases} \dot{z}(t) = \mathrm{A}z(t) + \mathrm{b}_1 F(t) + \mathrm{b}_2 F_a(t) \\ Y_1(t) = \mathrm{C}_{1,1} z(t) + \mathrm{d}_{1,11} F(t) + \mathrm{d}_{1,12} F_a(t) \\ Y_2(t) = \mathrm{C}_{2,1} z(t) + \mathrm{d}_{2,11} F(t) + \mathrm{d}_{2,12} F_a(t) \\ y(t) = \mathrm{C}_2 z(t) + \mathrm{d}_{21} F(t) + \mathrm{d}_{22} F_a(t) \end{cases} \tag{4.84}$$

where

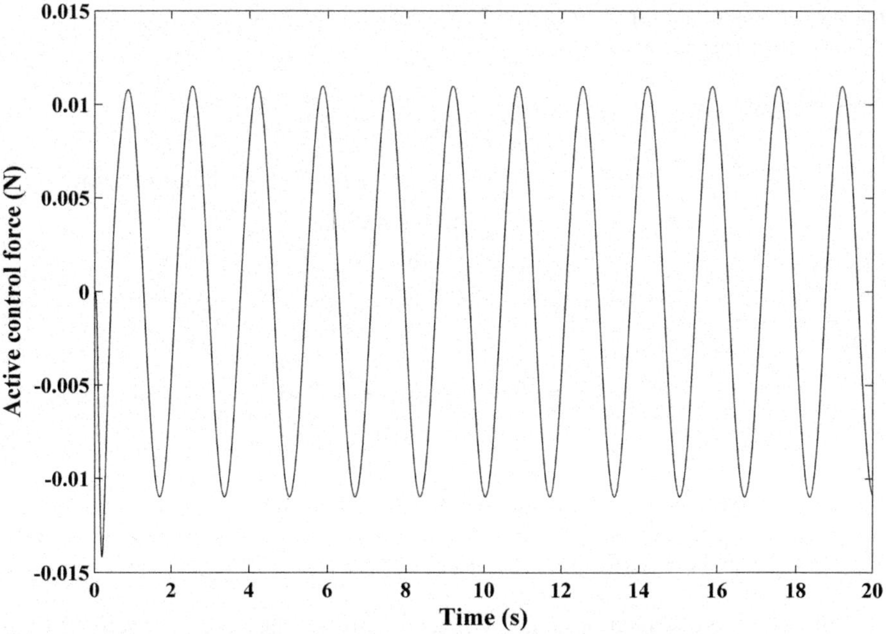

Fig. 4.52 Active control force of PSO-VUFLC strategy for sensitive equipment

$A = \begin{bmatrix} 0 & 0 & 1 & 0 \\ 0 & 0 & 0 & 1 \\ -\frac{k_1+k_2}{m_1} & \frac{k_2}{m_1} & -\frac{c_1+c_2}{m_1} & \frac{c_2}{m_1} \\ \frac{k_2}{m_2} & -\frac{k_2}{m_2} & \frac{c_2}{m_2} & -\frac{c_2}{m_2} \end{bmatrix}$, $b_1 = \begin{bmatrix} 0 \\ 0 \\ 0 \\ \frac{1}{m_2} \end{bmatrix}$, $b_2 = \begin{bmatrix} 0 \\ 0 \\ \frac{1}{m_1} \\ -\frac{1}{m_2} \end{bmatrix}$, $C_{1,1} = [k_1\ 0\ c_1\ 0]$, $d_{1,11} = [0]$, $d_{1,12} = [0]$, $C_{2,1} = [0\ 0\ 0\ 1]$, $d_{2,11} = [0]$, $d_{2,12} = [0]$,

$C_2 = \begin{bmatrix} 1\,0\,0\,0 \\ 0\,1\,0\,0 \\ 0\,0\,1\,0 \\ 0\,0\,0\,1 \end{bmatrix}$, $d_{21} = \begin{bmatrix} 0 \\ 0 \\ 0 \\ 0 \end{bmatrix}$, $d_{22} = \begin{bmatrix} 0 \\ 0 \\ 0 \\ 0 \end{bmatrix}$.

H_∞ active control theory is adopted herein, a state feedback controller that stabilizes the closed-loop system is constructed, with $K = \begin{bmatrix} K_1 & K_2 & K_3 & K_4 \end{bmatrix}$ gain, and the multi-objective fitting function is defined as follows:

$$\begin{cases} fitness_1 = \left\| T_{Y_1 F}(s, K) \right\|_\infty \\ fitness_2 = \left\| T_{Y_2 F}(s, K) \right\|_\infty \end{cases} \tag{4.85}$$

Example 4.13 The power equipment's vibration isolation system parameter configuration is the same as that in Example 4.11. Moreover, the parameter configuration of the MOPSO algorithm and the gain search range of the H_∞ state

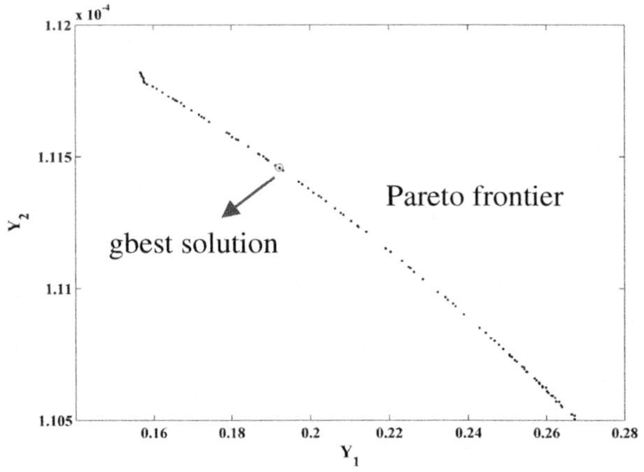

Fig. 4.53 Pareto frontier of the multi-objective control output active control for power equipment

feedback controller are the same as those in Example 2.7. The Pareto frontier results are shown in Fig. 4.53, and the obtained gbest solution is $K = \begin{bmatrix} -0.0738 & -4.998 & 5.000 & -1.301 \end{bmatrix} \times 10^2$.

Based on the obtained optimal feedback controller, the force transmitted to the foundation and the vibration velocity of the equipment with the proposed active control strategy are shown in Fig. 4.54, where PSO-H_∞ is the single-objective optimal control that only uses the force transmitted from the power equipment to the foundation as the control output in Example 4.7. It can be seen that the transmitted force to the foundation under PSO-H_∞ active control is better than the multi-objective control output and uncontrolled conditions. However, the power equipment's vibration velocity control effect is poor and even worse than the uncontrolled state. This phenomenon means that to pursue a good transmission force control effect, PSO-H_∞ active control virtually 'sacrifices' the vibration control effect of power equipment. However, it can be seen that the active control based on multi-objective control output considers both indicators, which shows the importance of implementing active control using multi-objective control output. The peak response of the two indicators is shown in Table 4.3 and Fig. 4.54.

4.7.2 *Active Control Based on the Multi-Objective Control Output for Sensitive Equipment*

The sensitive equipment's active control system is shown in Fig. 1.5b, its dynamic equation is given in Eq. (4.9), the state variables are $z_1 = x_1$, $z_2 = x_2$, $z_3 = \dot{x}_1$, and $z_4 = \dot{x}_2$, and the state vector is $z(t) = [z_1, z_2, z_3, z_4]^T$. The control system's

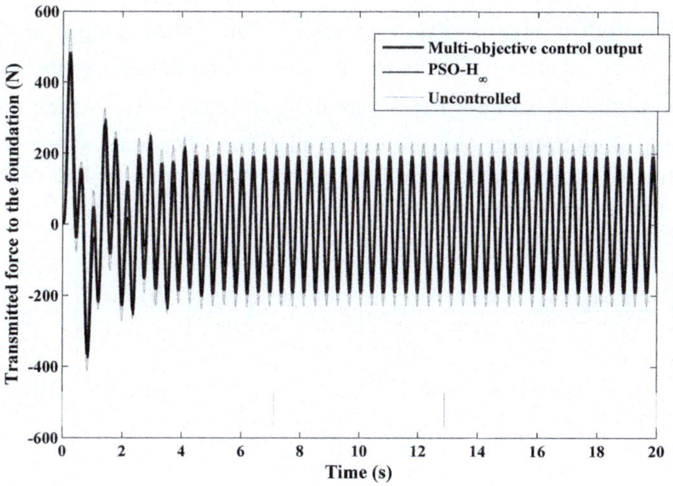

(a) transmitted force

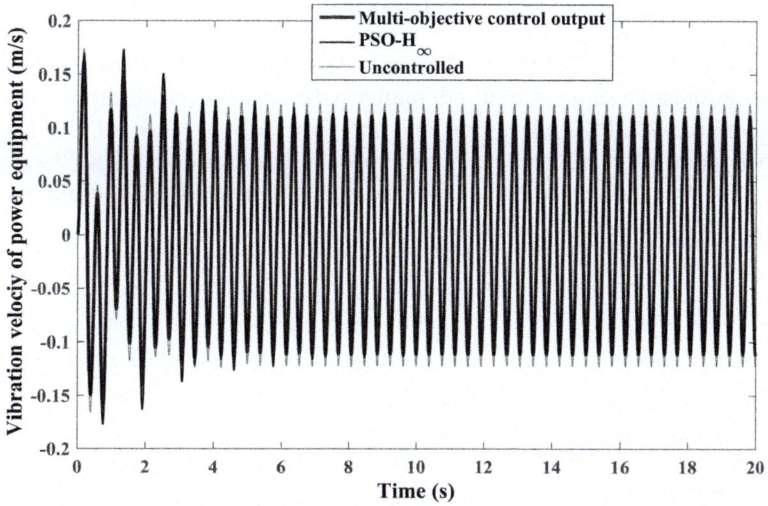

(b) vibration velocity of power equipment

Fig. 4.54 Responses of the multi-objective control output active control for power equipment

Table 4.3 Peak responses for comparison (power equipment)

Strategies	Transmitted force (N)	Vibration velocity of equipment (m/s)
Multi-objective control output	484.19	0.161
PSO-H_∞	464.27	0.176
Uncontrolled	548.62	0.173

observation output is $y(t) = \left[x_1, x_2, \dot{x}_1, \dot{x}_2\right]^T$. The control output of this control system is considered as two objectives: the sensitive equipment's vibration velocity $Y_1(t) = \left\{\dot{x}_2\right\}$, and the isolator deformation $Y_2(t) = \{x_2 - x_1\}$, which indicates the displacement of equipment and placement platform. The state space equation for active control of multi-objective control output is the same as Eq. (2.84), but the coefficient matrix has changed as follows:

$$\mathrm{A} = \begin{bmatrix} 0 & 0 & 1 & 0 \\ 0 & 0 & 0 & 1 \\ -\frac{k_1+k_2}{m_1} & \frac{k_2}{m_1} & -\frac{c_1+c_2}{m_1} & \frac{c_2}{m_1} \\ \frac{k_2}{m_2} & -\frac{k_2}{m_2} & \frac{c_2}{m_2} & -\frac{c_2}{m_2} \end{bmatrix}, \mathrm{b}_1 = \begin{bmatrix} 0 \\ 0 \\ \frac{1}{m_1} \\ 0 \end{bmatrix}, \mathrm{b}_2 = \begin{bmatrix} 0 \\ 0 \\ -\frac{1}{m_1} \\ \frac{1}{m_2} \end{bmatrix}, \mathrm{C}_{1,1} =$$

$[0\ 0\ 0\ 1]$, $\mathrm{d}_{1,11} = [0]$, $\mathrm{d}_{1,12} = [0]$, $\mathrm{C}_{2,1} = [-1\ 1\ 0\ 0]$, $\mathrm{d}_{2,11} = [0]$, $\mathrm{d}_{2,12} = [0]$,

$$\mathrm{C}_2 = \begin{bmatrix} 1\,0\,0\,0 \\ 0\,1\,0\,0 \\ 0\,0\,1\,0 \\ 0\,0\,0\,1 \end{bmatrix}, \mathrm{d}_{21} = \begin{bmatrix} 0 \\ 0 \\ 0 \\ 0 \end{bmatrix}, \mathrm{d}_{22} = \begin{bmatrix} 0 \\ 0 \\ 0 \\ 0 \end{bmatrix}.$$

The H_∞ control strategy is still considered here, and the fitting function of the multi-objective optimization is the same as that in Eq. (4.65).

Example 4.14 The sensitive equipment's vibration isolation system parameter configuration is the same as that in Example 4.12. Moreover, the parameter configuration of the MOPSO algorithm and the gain search range of the H_∞ state feedback controller are the same as those in Example 4.8. The Pareto frontier results are shown in Fig. 4.55, and the obtained gbest solution is $K = \left[\,0.149\ {-2.968}\ 2.498\ {-2.805}\,\right] \times 10^3$.

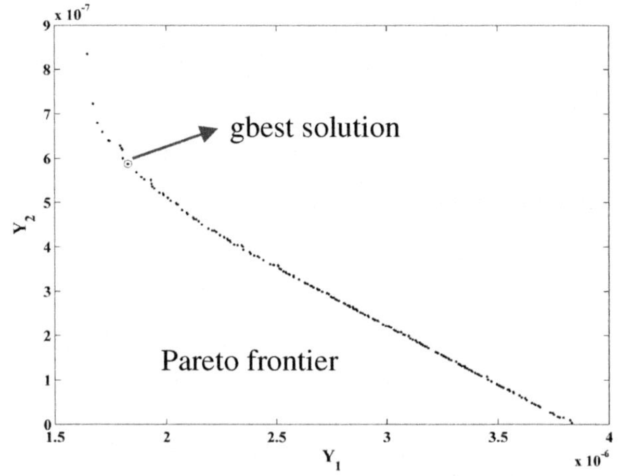

Fig. 4.55 Pareto frontier of the multi-objective control output active control for sensitive equipment

Based on the obtained optimal feedback controller, the sensitive equipment's velocity and the isolator's displacement with the proposed active control strategy are shown in Fig. 4.56, where PSO-H_∞ is the single-objective optimal control that only uses the vibration velocity of the sensitive equipment as the control output in Example 4.8. It can be seen that the control effect with the PSO-H_∞ active control is better than the multi-objective control output and uncontrolled conditions. However, the control effect of the isolator's displacement is poor and even worse than the uncontrolled state. This phenomenon means that to pursue a good isolation performance for sensitive equipment, the PSO-H_∞ active control virtually 'sacrifices' the isolator's deformation. However, it can be seen that the active control based on multi-objective control output considers both indicators, which shows the importance of implementing active control using multi-objective control output. The peak response of the both indicators is shown in Table 4.4.

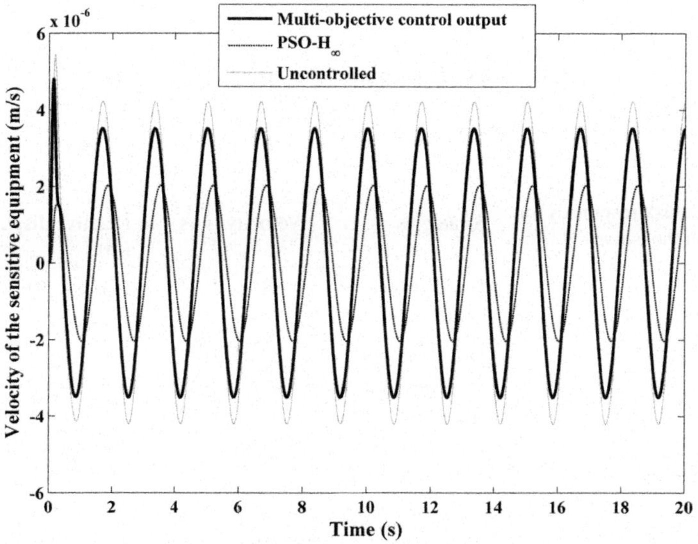

(a) velocity of the sensitive equipment

Fig. 4.56 Responses of the multi-objective control output active control for sensitive equipment

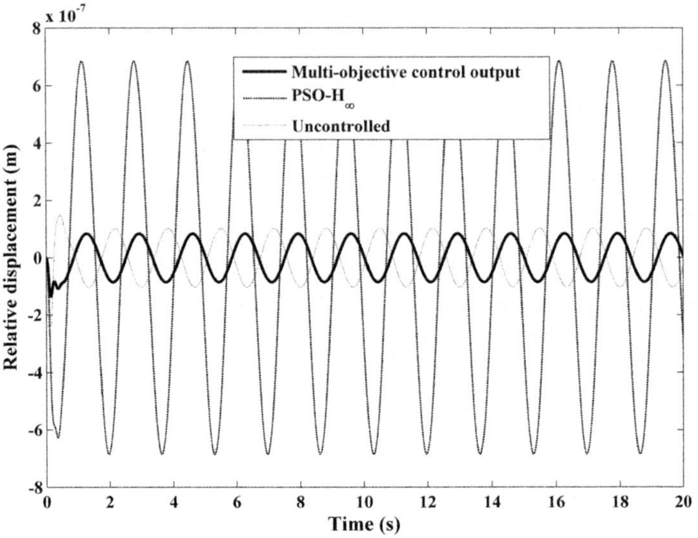

(b) Relative displacement (deformation) of isolator

Fig. 4.56 (continued)

Table 4.4 Peak responses for comparison (sensitive equipment)

Strategies	Velocity (m/s)	Relative displacement (m)
Multi-objective control output	4.831×10^{-6}	1.372×10^{-7}
PSO-H_∞	2.037×10^{-6}	6.848×10^{-7}
Uncontrolled	5.443×10^{-6}	2.373×10^{-7}

References

1. Chen YL, Fu LZ, Mei DQ et al (2003) Design of PID vibration control system for ultra-precision device [J]. J Eng Des 10(1):15–19
2. Zhao MW, Wang J, Jiang WH (2007) Modern control theory [M]. Huazhong University of Science and Technology Press, Wuhan
3. Ang KK, Wang SY, Quek ST (2002) Weighted energy linear quadratic regulator vibration control of piezoelectric composite plates [J]. Smart Mater Struct 11(1):98
4. Farshidianfar A, Saghafi A, Kalami SM et al (2012) Active vibration isolation of machinery and sensitive equipment using H∞ control criterion and particle swarm optimization method[J]. Meccanica 47(2):437–453
5. Ning XL, Liu HJ, Tan P et al (2010) Multi-objective robust H_2/H_∞ control for structural vibration based on LMI [J]. J Vibration Eng 23(2):167–172
6. He HJ, Zhang LB, Gu XS (2011) Design of H_2/H_∞ robust control based on multi-objective particle swarm optimization algorithm [J]. J East China Univ Sci Technol (Natl Sci Edition) 37(1):100–104

7. Deb K, Agrawal S, Pratap A et al (2000) A fast elitist non-dominated sorting genetic algorithm for multi-objective optimization: NSGA-II [J]. Lect Notes Comput Sci 1917:849–858
8. Abbass H, Sarker R, Newton C (2001) PDE: a Pareto-frontier differential evolution approach for multi-objective optimization problems[C]. In: Evolutionary computation. Proceedings of the 2001 Congress on IEEE, 2:971–978
9. Deb K, Saxena D (2006) Searching for Pareto-optimal solutions through dimensionality reduction for certain large-dimensional multi-objective optimization problems[C]. In: Proceedings of the World congress on computational intelligence (WCCI-2006). 2006:3352–3360
10. Baeyens E, Khargonekar P (1994) Some examples in mixed H_2/H_∞ control [J]. Proc Am Control Conf 6(21):1608–1612
11. Zadeh LA (1973) Outline of a new approach to the analysis of complex systems and decision processes [J]. Syst Man Cybernetics IEEE Trans 1:28–44
12. Li HX (1999) Variable universe adaptive fuzzy controller [J]. Sci China (Series E) 29(1):32–42
13. Peng LZ (2012) Research and application of fuzzy control strategy based on variable universe of particle swarm optimization [D]. Central South University
14. Li HX (1998) Interpolation mechanism of fuzzy control [J]. Sci China (Series E) 3(3):259–267

Chapter 5
Semi-Active Control Tracking Active Control

Abstract In this chapter, MRD semi-active control technology which can track the optimized active control is proposed. Firstly, an improved 6-segment cubic polynomial model is proposed for MRD, and an open-loop control strategy for MRD is also proposed. Based on this, nonlinear damping force tracking based on MRD is carried out including cubic nonlinear damping force, harmonic nonlinear damping force, and optimized active control force in Chap. 4, and the numerical results have confirmed the effectiveness of the proposed control strategy for power equipment and sensitive equipment.

5.1 MRD Semi-Active Control Technology

Active control is a process that induces a continuously controlled actuator force based on a specific control law to dissipate the input energy in the system. This process has been proven to be a superior control method that provides the ability to independently adjust the control force. However, designing the actuator and the required sensing system is complicated. Besides, its implementation is challenging and expensive, often requiring a large amount of control energy, causing adverse economic effects.

Compared to active control, semi-active control has the advantages of low energy consumption and relatively simple structural design, making it a feasible alternative and a trending topic in the field of vibration control. Semi-active control reduces vibration by changing the system parameter, such as its mass, stiffness, or damping. Adjusting damping is a common way to achieve semi-active control. Generally, there are two methods to modify the system's damping in a semi-active control system: to adjust the opening of the damping liquid flow port and to use functional materials to change the fluid's viscosity coefficient. In the first method, the damper consists of a piston rod with a throttle hole and a cylinder. The damping device is installed between the base and the controlled object. When the system is subjected to a vibration excitation, the piston moves linearly to the cylinder wall and generates resistance through the orifice, where the piezoelectric valve can be used to drive the orifice's regulating valve to automatically control the orifice's opening. The other damping control method uses functional materials, such as electro-rheological fluid

W. Huang and J. Xu, *Optimized Engineering Vibration Isolation, Absorption and Control*, Springer Tracts in Civil Engineering, https://doi.org/10.1007/978-981-99-2213-0_5

(ERF) and magneto-rheological fluid (MRF). In the absence of an electric or magnetic field, the viscosity of the ERF and MRF is kept at a minimal level, and the generated damping force is small. When an external electric or magnetic field is applied, the ERF or MRF viscosity increases significantly, whereas when the electric and magnetic fields' strength reaches a critical value, a phase change and solidification occur. This liquid–solid switching is fast and reversible and can be performed in milliseconds. Electro-rheological damper (ERD) and magneto-rheological damper (MRD) [1] are made by using the rheological effects of electricity and magnetic fields, which have controllable damping, high performance, small size, simple structure, large damping force, and low power consumption.

The magnetic particles in MRF have an inherent magnetic moment and can provide a shear stress that has greater magnitude and controllable damping force than ERF, and hence is more suitable for vibration control of industrial equipment. In addition, MRF controls the strength of the magnetic field by applying low voltage and large current, while ERF controls the electric field strength by applying high voltage and low current. The voltage is often up to several hundred volts. Therefore, MRD is much safer during the serviceability stage. Lord Co. in the United States started early research on MRF and MRD and has achieved great results in MRF preparation and characteristics. MRD is currently being used across various applications, including vehicle suspension and semi-active control of bridge vibration. On the other hand, it has not been widely used in controlling industrial equipment vibrations.

The MRD mechanical model is very critical for its practical engineering application. In recent years, many scholars have researched MRD's mathematical simulation due to its special mechanical hysteresis properties, which results in highly complex modeling to achieve great accuracy. Besides, applying rheological theory to analyze the mechanical characteristics of MRD is extremely complicated; hence the results of MRD mechanical performance tests are usually used to establish its mathematical model. Stanway [2] proposed a nonlinear double-viscosity model based on the assumption that the MRF is plastic before and after yielding and the damping before yielding is much greater than the damping after yielding to fit the damping force–displacement hysteresis curve and energy dissipation in each dynamic cycle. Nevertheless, he could not accurately fit the damping force-speed hysteresis characteristics before yielding. Pang et al. [3] extended works on nonlinear double-viscosity models by proposing a nonlinear hysteretic double-viscosity model. In general, their model can simulate the hysteresis characteristics of the damper before yielding, but it cannot fit the attenuation of the MRD's restoring force at low speeds. Spencer et al. [4, 5] proposed the Bouc-Wen model, which is easy to handle numerically and can simulate the damper's hysteresis behavior at low speeds. However, the model is complex, with many parameters that are difficult to identify. The authors in reference [6] developed a polynomial to fit the damping force–velocity curve, expressing the damping force as a univariate high-order polynomial with the piston speed as the independent variable.

From a practical point of view, an ideal MRD damping force model must require simple parameters and easy identification only to accurately simulate its actual force–velocity relationship characteristics and reflect its yielding effects and hysteresis behavior. Besides, it must be able to take into account the effect of the current control intensity, vibration frequency, and amplitude. Therefore, this study presents a novel polynomial mathematical model which can accurately simulate the MRD's output.

5.2 6-Segment Cubic Polynomial Mechanical Model for MRD

The Lord company in the United States tested the dynamic characteristics of a specific MRD type under a sinusoidal excitation signal of $x(t) = A\sin(2\pi ft)$, with an amplitude of 5, 10, and 15 mm. The vibration frequency was taken as 0.5, 1.0, 2.0, and 3.0 Hz, and the MRD control current was taken as 0, 0.1, 0.2, 0.3, 0.4, 0.5, 0.6, 0.7, 0.8, 0.9, and 1.0 A. The MRD mechanical characteristics at the 10 mm excitation amplitude and 1.0 Hz frequency with the control current varying between 0 and 1.0 A are shown in Fig. 5.1.

Choi et al. [6] developed a numerical model and performed a parameter identification study on the mechanical test data of an MRD by utilizing the least-square method to fit a 6-degree polynomial with linear coefficients. Zhou et al. [7] found in the actual research process that the fitting accuracy of the 9-order polynomial is higher than that of the 6-order, and the fitting accuracy of the 12-order is higher than that of the 9-order. Hence, the fitting accuracy rises with the increase in the polynomial degree. Additionally, they pointed out that high-order polynomial fitting leads to the Runge phenomenon in the mathematical model establishment, making it hard to decide whether to choose higher-order polynomials to improve the accuracy of the MRD model.

In order to clarify this issue, this chapter first performs 6-, 12-, and 20-order polynomial fitting on the MRD experimental data, where the polynomial coefficient has a linear relationship with the control current. Within this analysis, the test data at 15 mm excitation, 1.0 Hz frequency, and 0, 0.1, 0.2, 0.3, 0.4, 0.5, 0.6, 0.7, 0.8, 0.9, and 1.0 A MRD control current were selected. MRD damping force–velocity curve can be divided into a left branch (negative acceleration) and a right branch (positive acceleration), as shown in Fig. 5.2.

The polynomial fitting model is defined as follows:

$$F_p = \sum_{i=0}^{n} a_i v^i \tag{5.1a}$$

$$a_i = b_i + c_i I\,(\text{Linear relationship}) \tag{5.1b}$$

where $i = 0, 1, 2, \ldots, n$ and n is the total order of fitting for polynomial.

Polynomial coefficients under different control currents can be identified by the least squares method as follows:

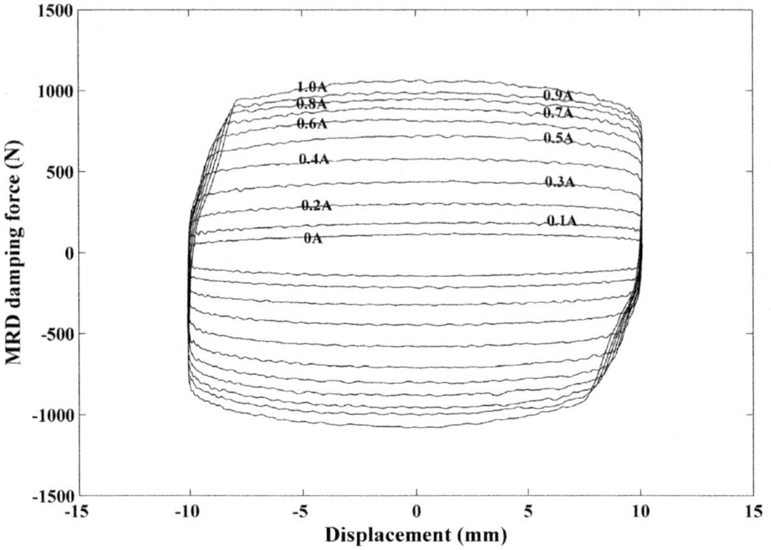

(a) MRD damping force-displacement curve

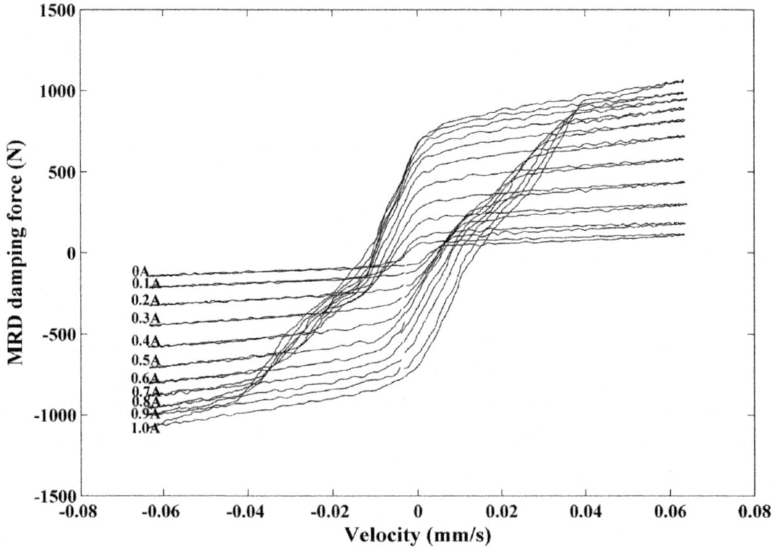

(b) MRD damping force-velocity curve

Fig. 5.1 Mechanical test data of the MRD

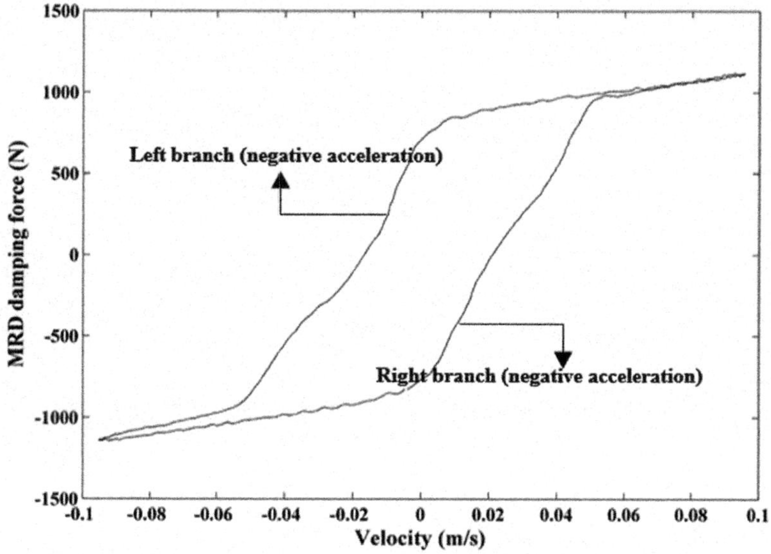

Fig. 5.2 MRD damping force–velocity curve with left and right branches

$$E(a_0, a_1, \ldots, a_n) = \min\left(\sum_{j=1}^{m} \left[F_{pj}(v_j) - F_p(v_j)\right]^2\right) \tag{5.2}$$

where a_i is the polynomial coefficient to be identified, m is the number of actual data points for participation in identification, v_j is the velocity of MRD piston to be identified, $F_{pj}(v_j)$ is the measured damping force to be identified based on the test data that has a one-to-one correspondence with velocity, $F_p(v_j)$ is the damping force calculated by the polynomial model based on the predicted data.

Firstly, a linear relationship between a_i and the current I is set, and then the intercept b_i and slope c_i are directly identified through a straight line fitting method.

5.2.1 6-Order Polynomial Fitting

The fitting polynomial coefficients a_0–a_6 were forced to have a linear relationship with the MRD control current. Figure 5.3 shows that the abscissa represents the MRD's control current I. The force characteristics of the MRD were predicted based on the obtained polynomial model under the control current I = 1.0 A. The resulting damping force–velocity characteristic curve is shown in Fig. 5.4.

Figure 5.3 indicates that it is difficult to form a linear relationship between a few coefficients a_i and the current I in a real sense (such as a_1, a_3, a_5). This is because forcing a_i and I linear fitting is not reasonable, causing the 6-order polynomial MRD damping force model's accuracy to be low with large errors.

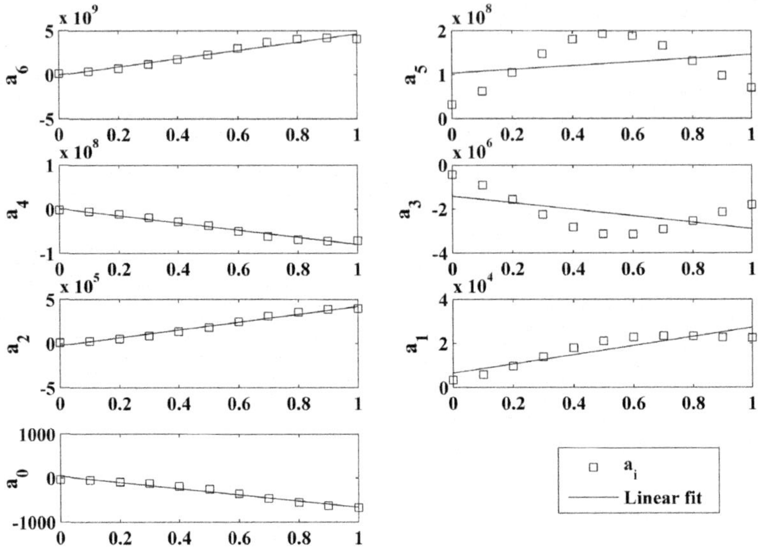

(a) right branch

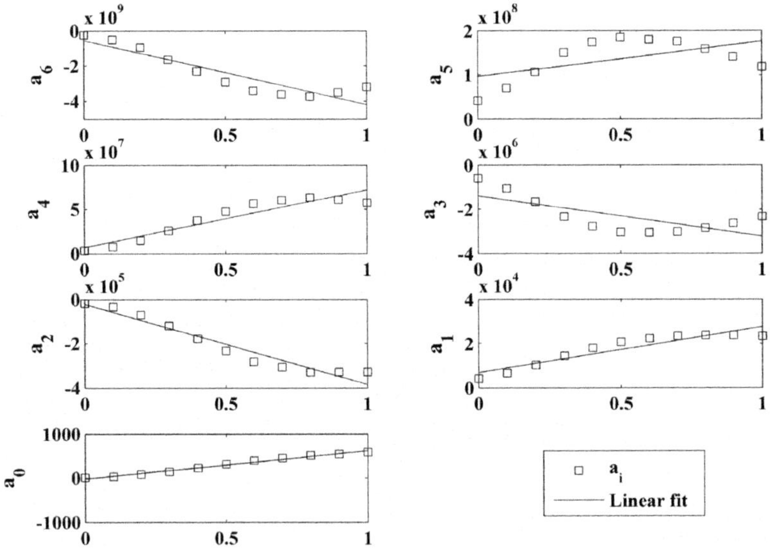

(b) left branch

Fig. 5.3 Relationship between coefficients a_i and current I using the 6-order polynomial fitting

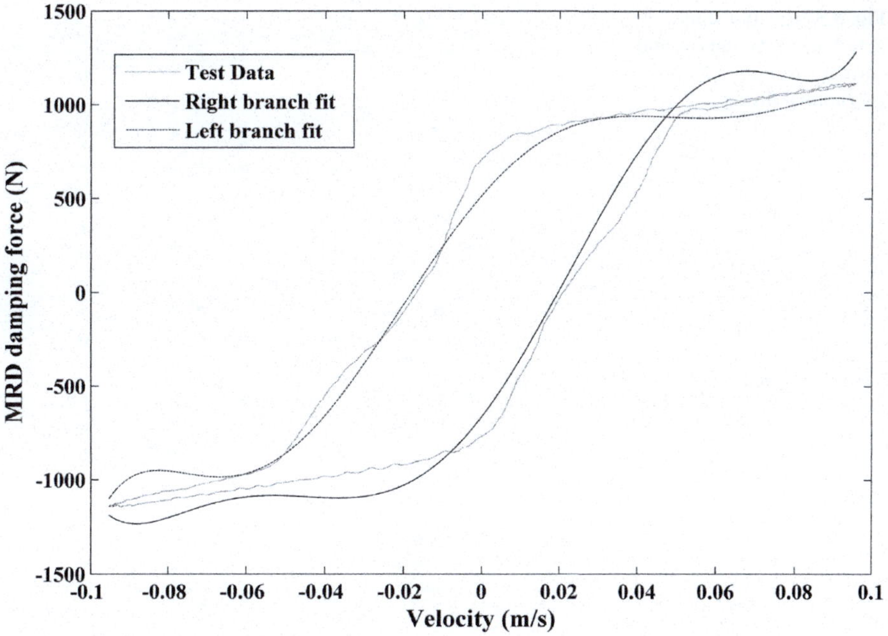

Fig. 5.4 MRD damping force–velocity predicted using the 6-order polynomial fitting

5.2.2 12-Order Polynomial Fitting

The fitting polynomial coefficients a_0–a_{12} were forced to have a linear relationship with the MRD control current. As shown in Fig. 5.5, the abscissa represents the control current I of the MRD. Based on the obtained 12-order polynomial model, the force characteristics of the MRD were predicted given the control current I = 1.0 A. The damping force–velocity characteristic curve is shown in Fig. 5.6.

Figure 5.5 shows that there is still a large difference by forcing a linear fit for the polynomial coefficients a_i and the current I, where the actual relationship was misled. The rationality of this forced linear fit will be discussed later on. Figure 5.6 shows that the adopted 12-order polynomial fitting significantly improves the prediction accuracy of the MRD damping force compared to the 6-order polynomial fitting. However, it is still not ideal, and there are certain errors. The residuals are shown in Fig. 5.7. In order to improve the fitting accuracy, the number of fittings is further increased.

Fig. 5.5 Relationship between coefficients a_i and current I using the 12-order polynomial fitting

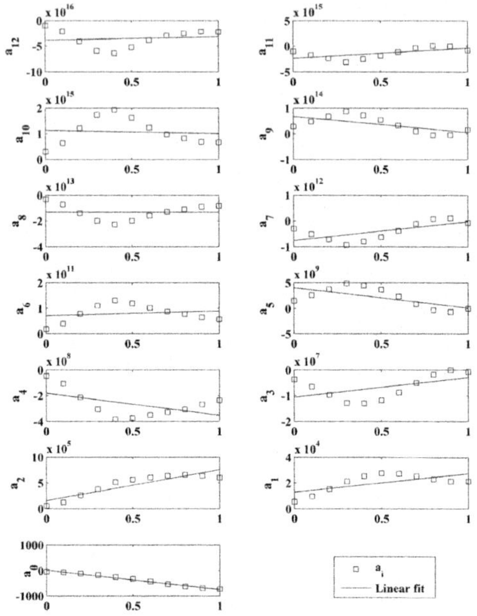

(a) right branch

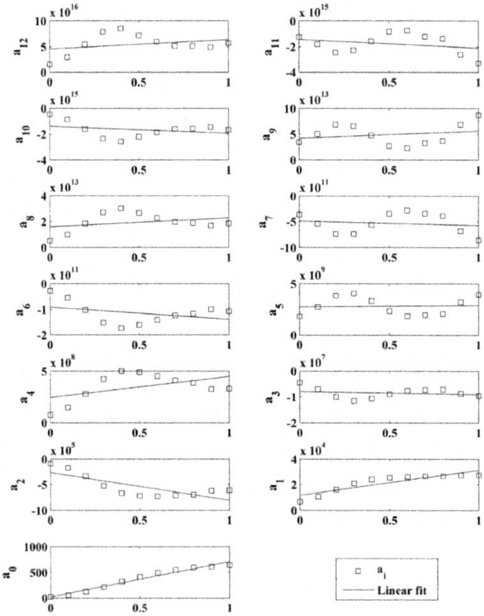

(b) left branch

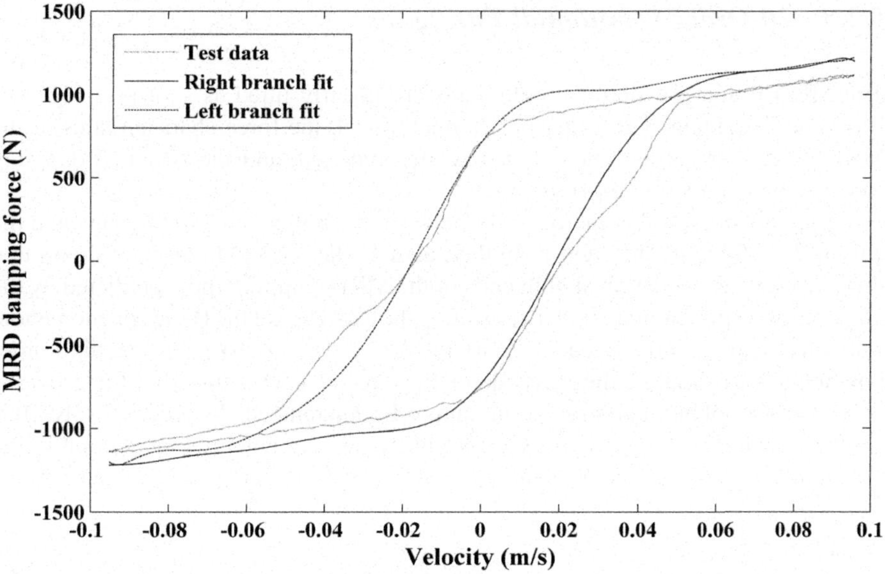

Fig. 5.6 MRD damping force–velocity predicted using the 12-order polynomial fitting

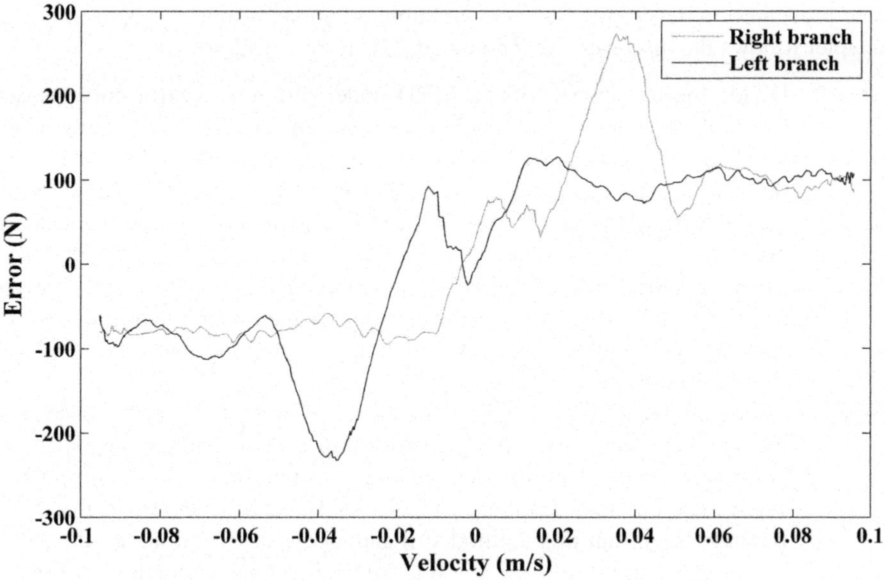

Fig. 5.7 Error of the MRD damping force using the 12-order polynomial fitting

5.2.3 20-Order Polynomial Fitting

The MRD fitting polynomial coefficients are linearly fitted as shown in Fig. 5.8. Based on the obtained 20-order polynomial model, the force characteristics of the MRD at the control current $I = 1.0$ A are predicted, and the obtained damping force–velocity curve is shown in Fig. 5.9.

Figure 5.8 depicts that the linear fit between the polynomial coefficients and the current I is still quite different from their actual relationship. Figure 5.9 shows that there is a severe oscillation at both ends of the MRD damping force predicted by the 20-order fitted polynomial model, indicating the existence of the Runge phenomenon. This phenomenon's occurrence means there is a greater risk of using high-order polynomials to simulate the MRD damping force characteristics, but the accuracy of low-order polynomial simulation cannot be guaranteed. In order to solve this contradiction, this study proposes a 6-segment cubic polynomial model that might potentially provide an accurate simulation of the MRD damping force characteristics.

5.2.4 6-Segment Cubic Polynomial Model

The procedure to develop the 6-segment cubic polynomial model proposed in this chapter follows the discussion of Zhou et al. [7] study as follows:

Step 1: Divide the test curves of the MRD under different control currents and divide the left and right branches' curves into 3 sections according to the yield velocity respectively, a total of 6 sections, as shown in Fig. 5.10. The three right sections are the AQ_1, Q_1Q_2 and Q_2B, and the three left sections are the BQ_3, Q_3Q_4, and Q_4A, where Q_1, Q_2, Q_3, and Q_4 are four yield speed points.

Step 2: The 6-segment curves divided by the left and right branches are subjected to polynomial fitting based on the least square method, with the cubic fitting selected. The polynomial coefficients under different control currents are shown in Table 5.1.

Step 3: Fit the obtained polynomial coefficients a_i separately by first considering them to be a conventional linear relationship with the control current I, then predicting the MRD's force characteristics under the control current $I = 1.0$ A. The obtained damping force–velocity curve is shown in Fig. 5.11. It can be seen that this method still cannot effectively simulate the MRD damping force. For this reason, this study abandons the method of forcing a linear fit with the control current I and adopts a higher polynomial order that is more in line with the fitting law. Moreover, this study adopts a cubic polynomial. The polynomial coefficient fitting curve is shown in Fig. 5.12, with a_{ij} representing the coefficient a_j of the cubic polynomial fitting.

Step 4: Use the obtained polynomial coefficient a_{ij} to predict the MRD damping force under different control currents. Then form a 6-segment polynomial

Fig. 5.8 Relationship between coefficients a_i and current I using the 20-order polynomial fitting

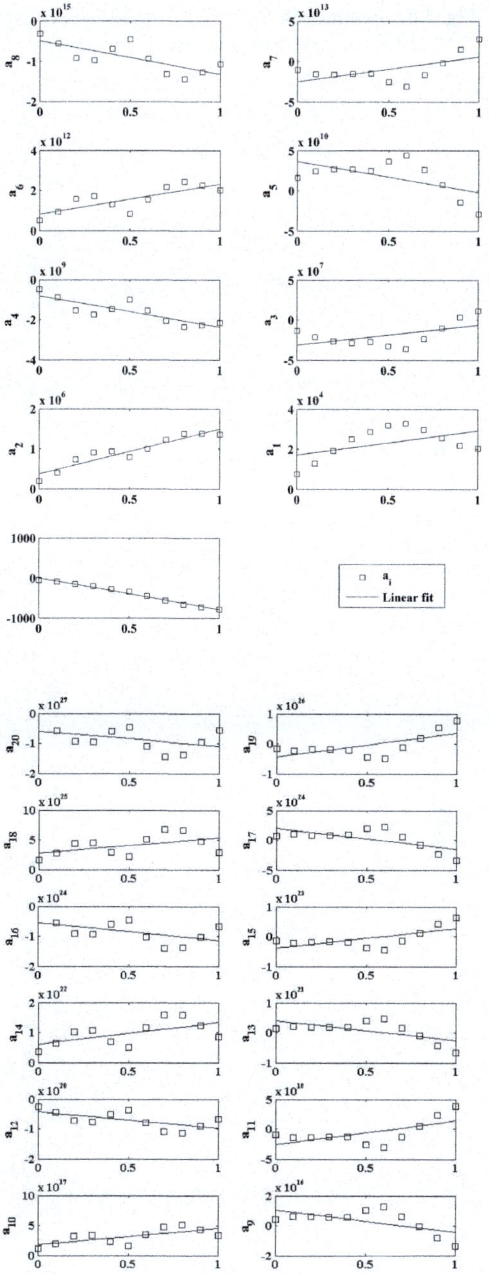

(a) right branch

Fig. 5.8 (continued)

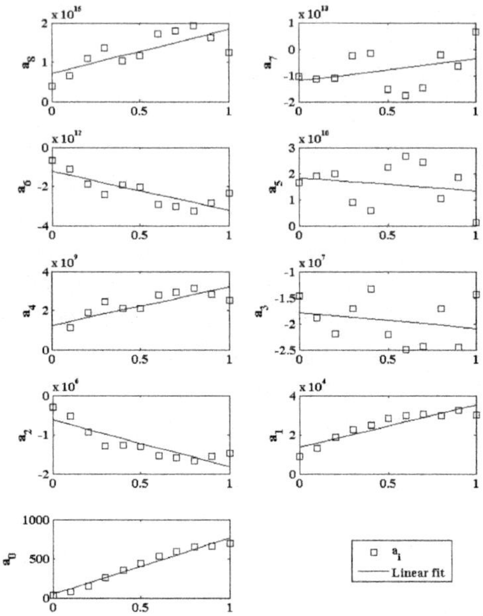

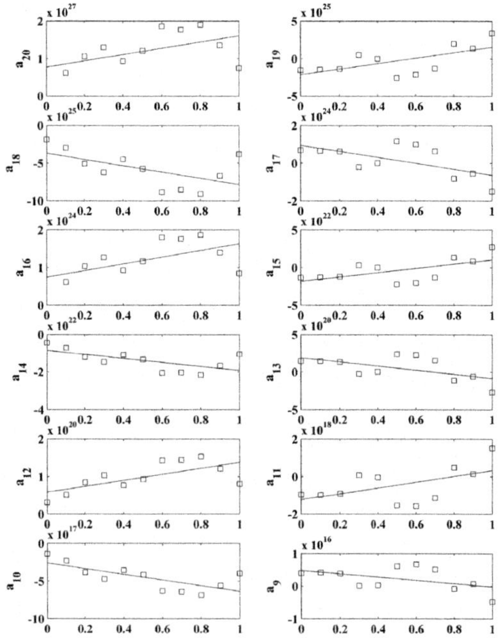

(b) left branch

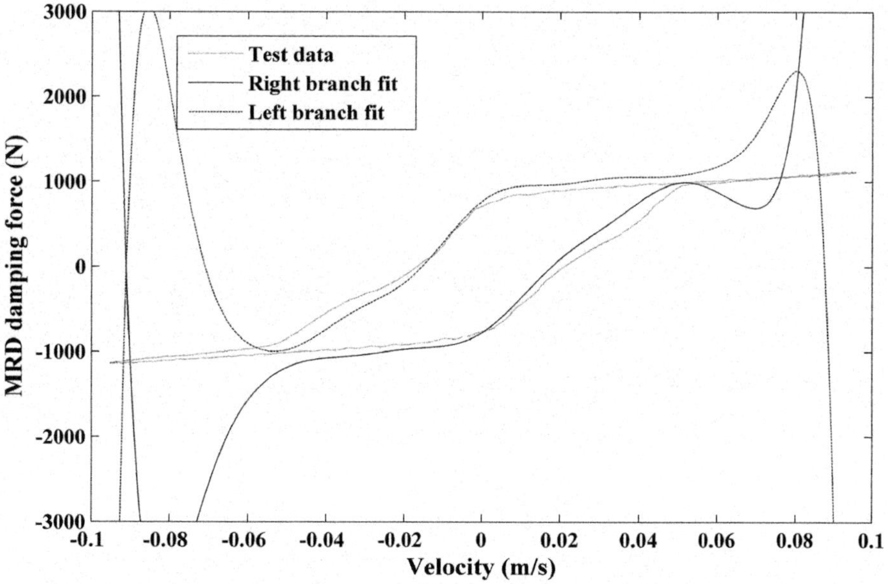

Fig. 5.9 MRD damping force–velocity predicted using the 20-order polynomial fitting

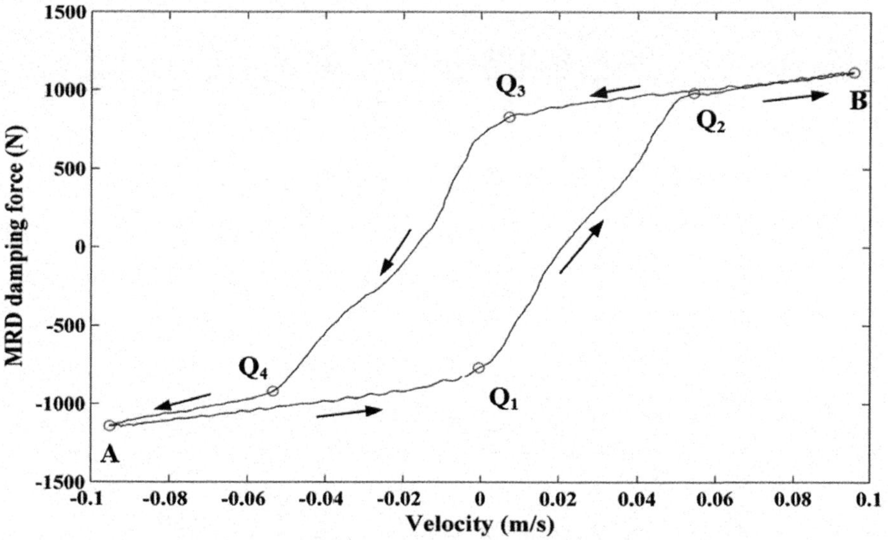

Fig. 5.10 Split the MRD test curve using the yielding velocities

Table 5.1 6-segment cubic polynomial fitting coefficients for MRD

Right branch												
Segment AQ_1					Segment Q_1Q_2				Segment Q_2B			
I(A)	$a_3\times10^6$	$a_2\times10^5$	$a_1\times10^4$	a_0	$a_3\times10^6$	$a_2\times10^5$	$a_1\times10^4$	a_0	$a_3\times10^6$	$a_2\times10^5$	$a_1\times10^4$	a_0
1.0	0.29	0.58	0.66	– 798.15	4.49	–4.53	4.64	– 846.75	–1.67	3.66	–2.26	1385.72
0.8	0.29	0.62	0.66	– 707.77	2.45	–3.58	4.47	– 699.82	–0.71	1.68	–1.00	1060.95
0.6	0.38	0.77	0.72	– 571.97	–1.09	–2.62	4.06	– 444.88	–0.31	0.55	–0.033	669.26
0.4	0.18	0.35	0.41	– 418.82	–20.6	4.93	2.85	– 263.44	0.15	–0.26	0.36	401.97
0.2	0.18	0.32	0.30	– 211.48	–69.2	14.5	1.95	– 170.00	0.37	–0.69	0.55	118.58
Left branch												
Segment BQ_3					Segment Q_3Q_4				Segment Q_4A			
电流 I (A)	a_3 $\times10^6$	a_2 $\times10^5$	a_1 $\times10^4$	a_0	a_3 $\times10^6$	a_2 $\times10^5$	a_1 $\times10^4$	a_0	a_3 $\times10^6$	a_2 $\times10^5$	a_1 $\times10^4$	a_0
1.0	0.81	– 1.44	1.04	734.68	16.1	15.5	7.04	822.14	2.59	5.88	4.84	381.34
0.8	0.48	– 0.91	0.77	697.65	11.6	11.6	6.22	718.69	1.65	3.63	2.98	–56.21
0.6	0.29	– 0.59	0.58	607.25	7.96	9.07	5.52	580.59	6.77	1.48	1.33	–366.62
0.4	0.13	– 0.28	0.35	440.40	–35.4	–11.0	2.59	352.04	0.19	0.42	0.49	–384.88
0.2	0.07	– 0.18	0.24	215.39	–46.2	–12.2	1.93	176.76	0.36	0.70	0.56	–156.72

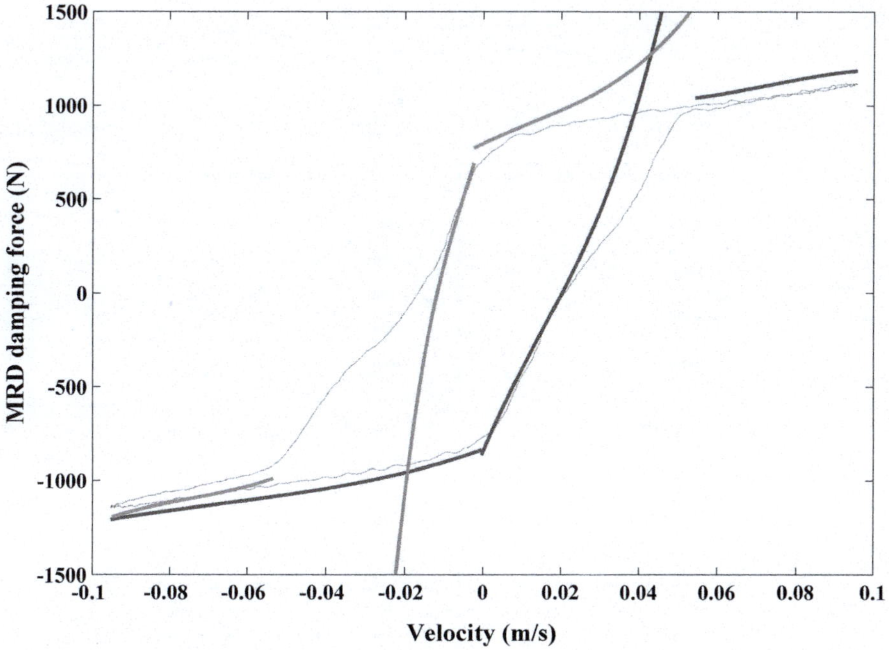

Fig. 5.11 MRD damping force–velocity predicted using the 6-segment cubic polynomial model (linear fitting for a_i and I)

curve and calculate the four intersection points J_1, J_2, J_3, and J_4, as shown in Fig. 5.13.

Step 5: Continuously process the obtained 6-segment fitting curves at the intersections to obtain the speeds V_{J1}, V_{J2}, V_{J3}, and V_{J4} at intersections J_1, J_2, J_3, and J_4, respectively, and obtain 6 velocity intervals: $V_A–V_{J1}$, $V_{J1}–V_{J2}$, $V_{J2}–V_B$, $V_B–V_{J3}$, $V_{J3}–V_{J4}$, $V_{J4}–V_A$. Thereafter, equally space each speed interval to obtain N discrete points, where the specific operation is $(V_{J1}-V_A)/N$, with N being 520 in this study.

Based on the above strategy, the obtained 6-segment cubic polynomial model was used to predict the force characteristics of MRD at the input current $I = 1.0$ A. The obtained damping force–velocity curve is shown in Fig. 5.14, where the dashed line represents the MRD test data, and the solid line represents the data predicted. The curve fitted by the model is highly consistent with the experimental curve, and it successfully overcomes the adverse effect of forcing a linear relationship with the input current I, and the fitting effect is better than the result of the 12-order polynomial fitting. The comparison in errors shown in Fig. 5.15 indicates high accuracy.

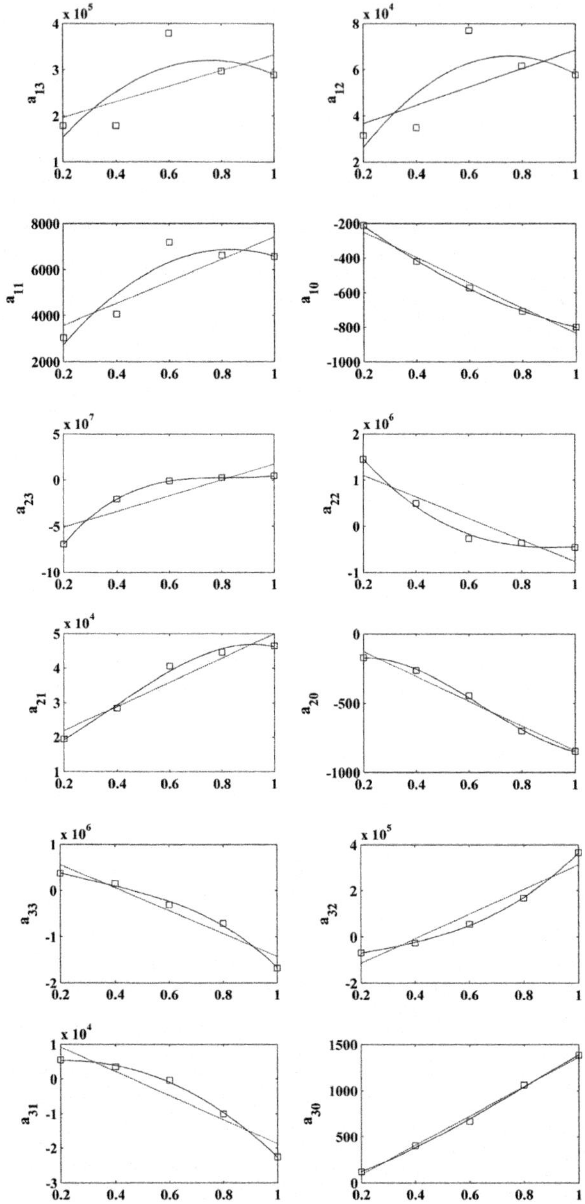

Fig. 5.12 Relationship between coefficients a_i and current I using the 6-segment cubic polynomial model

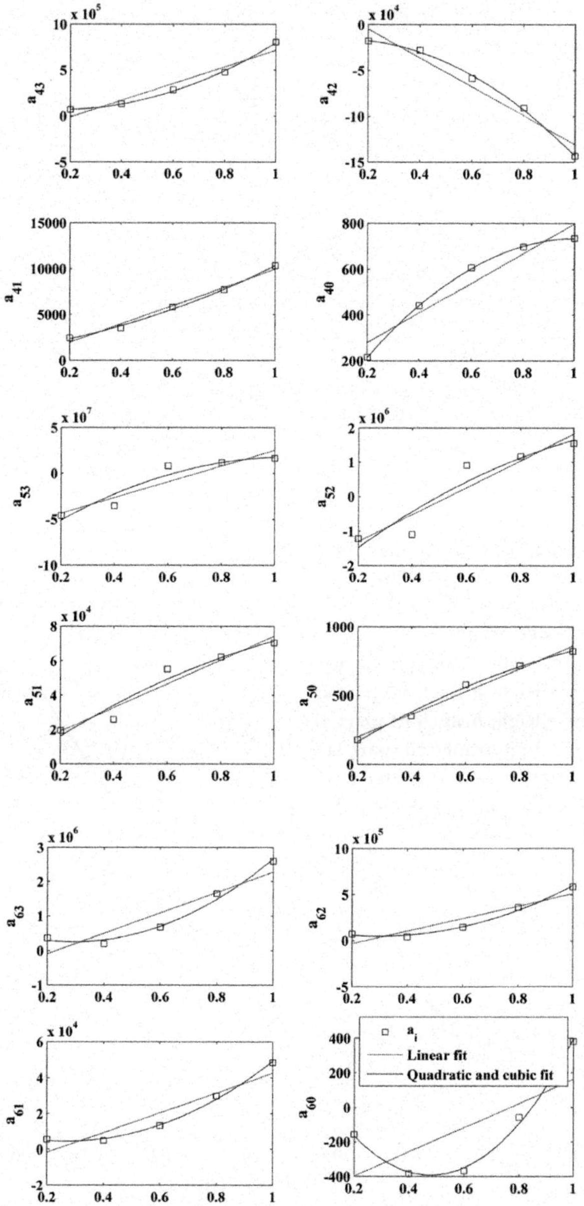

Fig. 5.12 (continued)

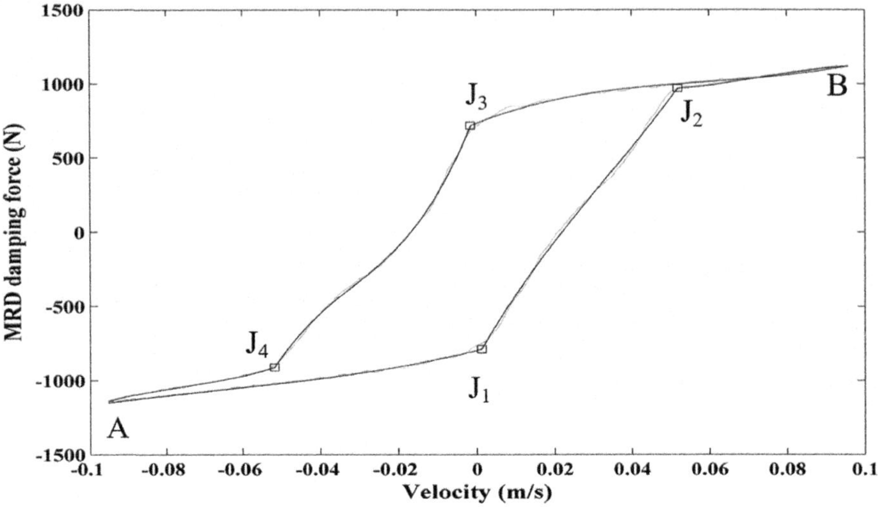

Fig. 5.13 Intersections of 6 polynomial fitted curves

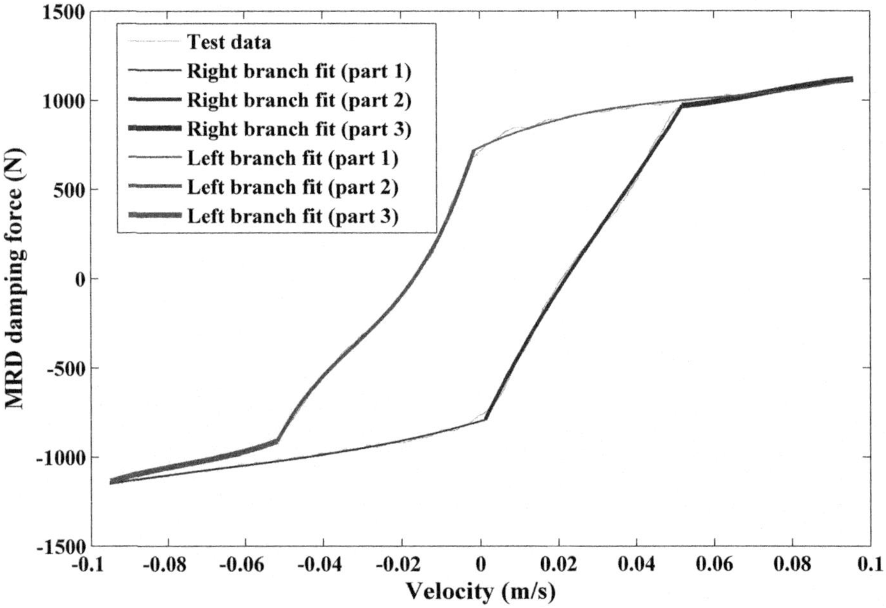

Fig. 5.14 MRD damping force–velocity curve predicted using the 6-segment cubic polynomial model (cubic fitting for a_i and I)

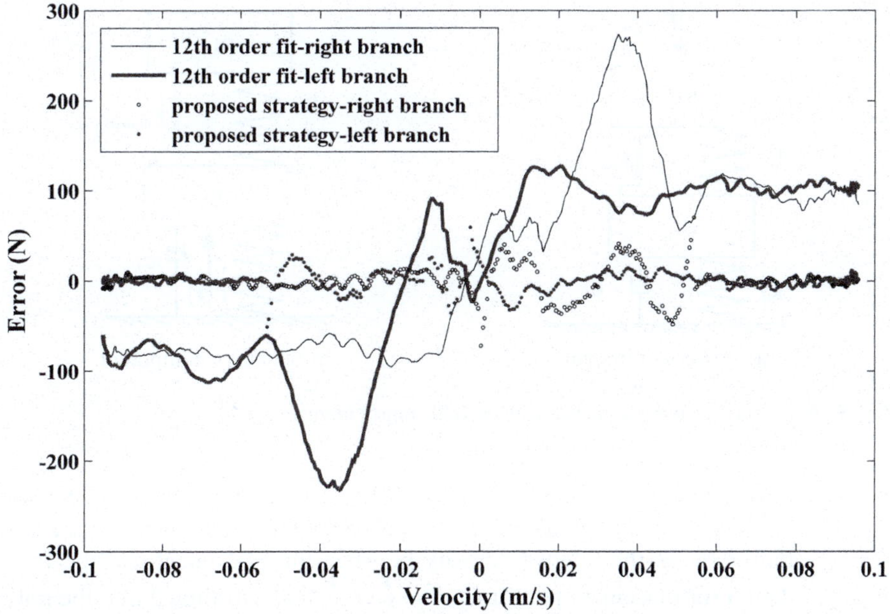

Fig. 5.15 Error in the MRD damping force predicted using the 6-segment cubic polynomial model

5.2.5 MRD Open-Loop Control Strategy

The core of an MRD semi-active control lies within the adopted control strategy. Currently, the MRD is applied to vibration control and mainly relies on semi-active control algorithms, such as the linear quadratic optimal control, independent modal optimal control, and pole configuration algorithms, including the optimal configuration, optimal predictive control, variable structure control, adaptive control, and nonlinear feedback optimal control.

This research proposes a 6-segment cubic polynomial model that can accurately predict the MRD output. Moreover, this chapter develops an open-loop control strategy to enable the proposed MRD model to be easily and effectively applied among the semi-active controls of the aforementioned two types of equipment.

Figure 5.16 shows two types of equipment vibration isolation models. $F_{desired}$ is any form of external control force used in the vibration isolation system. It can be an active control force, semi-active control force, or other nonlinear damping forces that meet different frequency or time domain control requirements. The $F_{desired}$ is the ideal force or reference force in the MRD-based semi-active control model, and it is the tracking target and critical parameters in the semi-active control process using MRD.

In general, $F_{desired}$ is substituted into the proposed 6-segment cubic polynomial model to inversely control the current I. The basic method is as follows:

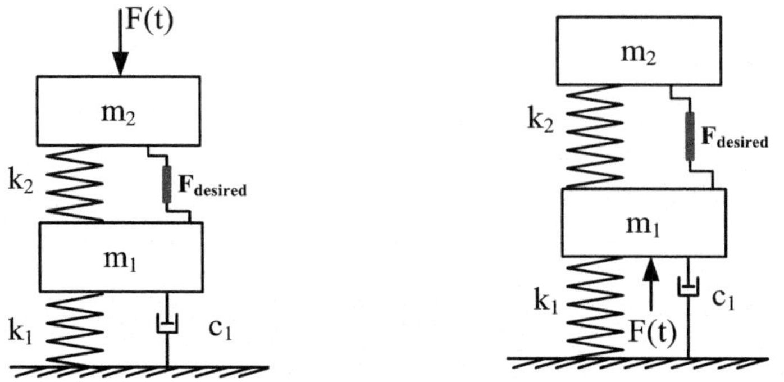

(a) Power equipment (b) Sensitive equipment

Fig. 5.16 Vibration isolation models with desired control inputs

Step 1: Calculate the input velocity to the MRD vibration isolation system (the moving velocity of the MRD piston rod). This velocity is defined, in Fig. 4.16, as the relative velocity between the equipment and the foundation (or placement platform) $v = \dot{x}_2 - \dot{x}_1$ and substituted into the polynomial fitting coefficient determined from Fig. 5.12 to get the following intermediate variables:

$$\begin{cases} \mathrm{A}_{i3} = c_{i3}v^3 + c_{i2}v^2 + c_{i1}v + c_{i0} \\ \mathrm{A}_{i2} = b_{i3}v^3 + b_{i2}v^2 + b_{i1}v + b_{i0} \\ \mathrm{A}_{i1} = d_{i3}v^3 + d_{i2}v^2 + d_{i1}v + d_{i0} \\ \mathrm{A}_{i0} = e_{i3}v^3 + e_{i2}v^2 + e_{i1}v + e_{i0} \end{cases} \tag{5.3}$$

where c_{ij}, b_{ij}, d_{ij}, and e_{ij} are respectively the fitting coefficients of a_{ij}, $i = 1$–3 or 4–6, representing three segments of the MRD's right or left branches.

Step 2: Combine the coefficients calculated using Eq. (5.3) with $\mathrm{F}_{\mathrm{desired}}$, and form the following polynomial coefficient vector:

$$p_1 = \left[\mathrm{A}_{i3}\ \mathrm{A}_{i2}\ \mathrm{A}_{i1}\ \mathrm{A}_{i0} - F_{desired} \right] \tag{5.4}$$

where p_{1k} ($k = 0$–3) represents the kth order of a polynomial, 0 is the constant term, and solve the roots of Eq. (5.4) and r_1, r_2 and r_3 are obtained, and randomly rearrange the set of roots as follows:

$$roots = randperm\left([r_1, r_2, r_3]\right) \tag{5.5}$$

Thereafter, randomly extract a real root r_{rand} in the rearranged root set *roots* and $randperm(\bullet)$ which represents the random rearrangement of elements in a vector.

It is important to point out that : The odd-numbered polynomial must have a real root. Hence, Eq. (4.4) has at least one real root. This is why the number of polynomial coefficients a_i in this study was selected to be 3. Indeed, other orders, such as 5 and 7, are also feasible, and more details will be omitted here. This conclusion is briefly explained below:

Consider this polynomial $p_{2n-1}x^{2n-1} + p_{2n-2}x^{2n-2} + \ldots + p_2x^2 + p_1x + p_0 = 0$, where the highest term coefficient satisfies $p_{2n-1} \neq 0$. Let $f(x) = p_{2n-1}x^{2n-1} + p_{2n-2}x^{2n-2} + \ldots + p_2x^2 + p_1x + p_0$, if $p_{2n-1} > 0$, and when x approaches $+\infty$, $f(x)$ approaches $+\infty$, whereas when x approaches $-\infty$, $f(x)$ approaches $-\infty$. Since $f(x)$ is continuous when $x \in R$, according to the median theorem, there must be a real root x_0 that satisfies $f(x_0) = 0$. If $p_{2n-1} < 0$, when x approaches $+\infty$, $f(x)$ approaches $-\infty$, when x approaches $-\infty$, $f(x)$ approaches $+\infty$. Since $f(x)$ is continuous when $x \in R$, according to the median theorem, there must be a real root x_0, which can satisfy $f(x_0) = 0$. Therefore, the odd-order coefficient polynomial must have a real root.

Step 3: The obtained r_{rand} is the control current I of the input MRD. Additionally, I is substituted into the coefficients c_{ij}, b_{ij}, d_{ij}, and e_{ij} to obtain the following polynomial coefficients:

$$\begin{cases} \mathrm{B}_{i3} = c_{i3}I^3 + b_{i3}I^2 + d_{i3}I + e_{i3} \\ \mathrm{B}_{i2} = c_{i2}I^3 + b_{i2}I^2 + d_{i2}I + e_{i2} \\ \mathrm{B}_{i1} = c_{i1}I^3 + b_{i1}I^2 + d_{i1}I + e_{i1} \\ \mathrm{B}_{i0} = c_{i0}I^3 + b_{i0}I^2 + d_{i0}I + e_{i0} \end{cases} \tag{5.6}$$

Step 4: Form the following polynomial coefficient vector based on the polynomial coefficients obtained from Eq. (4.6).

$$p_2 = \left[\mathrm{B}_{i3}\ \mathrm{B}_{i2}\ \mathrm{B}_{i1}\ \mathrm{B}_{i0} \right] \tag{5.7}$$

where p_{2k} $(k = 0 \sim 3)$ represents the kth order of the polynomial, with 0 being a constant term.

The actual output damping force F_{mr} of the MRD can be defined by taking the velocity input to the MRD v as an independent variable and substituting it into Eq. (5.7) to calculate the polynomial value.

The basic idea of the above strategy is summarized in Fig. 5.17. From the ideal input force $F_{desired}$ to the MRD output damping force F_{mr}, and then an open-loop control strategy is constituted here, which does not require a closed-loop control.

It should be pointed out that the proposed MRD 6-segment cubic polynomial model is based on the actual mechanical tests with a 15 mm amplitude, a 1 Hz frequency, and a 0–1.0 A control current. The velocity interval corresponding to

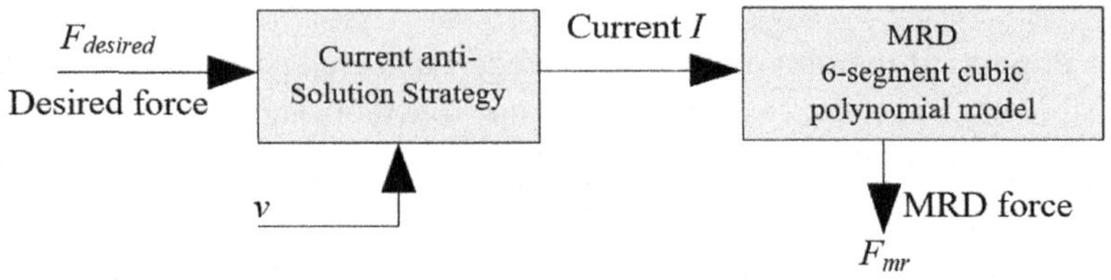

Fig. 5.17 Open-loop control strategy for MRD

the MRD damping force (the upper and lower velocity limits are determined by the amplitude of the vibration table) is determined. Therefore, the input velocity in the actual vibration system to which the MRD is applied (the piston of the MRD or rod velocity) must be in the proposed model's velocity interval, such as between v_A and v_B in Fig. 5.13. If the velocity of the input MRD is outside this interval, using the MRD model based on the test data is meaningless. Of course, if we want to obtain the MRD model required by the vibration control system under various working conditions (load excitation of different amplitude and frequency), we need to carry out specific MRD mechanical tests according to the actual dynamic characteristics of the controlled object to obtain the applicable 6-segment cubic polynomial model.

5.3 Nonlinear Damping Force Tracking Based on MRD

Indeed, the existence of nonlinear damping in the vibration isolation system is inevitable. Besides, all vibration dampers cannot be strictly linear. Accordingly, in the actual process, the existence of nonlinear damping must be considered [8]. Several studies have shown the critical role of specific nonlinear damping in vibration isolation systems, where its vibration isolation characteristics are better than linear damping, which can overcome the deficiency of linear damping. References [9–12] carried out a systematic study of cubic nonlinear damping, and the results showed that the former has better broadband vibration isolation than linear damping. Reference [13] studied novel nonlinear damping and affirmed its good vibration isolation characteristics. References [14, 15] proposed nonlinear harmonic damping [16], and references [17, 18] proposed a variety of nonlinear variable damping algorithms that can be applied to semi-active control.

It can be seen that in the actual process, the way to achieve the theoretical nonlinear damping equivalently has essential significance. Based on the proposed 6-segment cubic polynomial model, this study uses MRD to observe the nonlinear damping force in the vibration isolation system.

5.3.1 Cubic Nonlinear Damping Force

Cubic nonlinear damping was applied to the vibration isolation system of the power and sensitive equipment described in this study to replace linear damping. The considered nonlinear damping force can be expressed using the following cubic polynomial [11, 19]:

$$f_{cubic}(\bullet) = d_3(\bullet)^3 + d_2(\bullet)^2 + d_1(\bullet) \tag{5.8}$$

where d_3, d_2, and d_1 are the polynomial coefficients of the cubic nonlinear damping, d_3 and d_2 are the nonlinear characteristics, d_1 is the linear characteristics, f_{cubic} is cubic nonlinear damping force, and $(\bullet)$ is the input velocity for the nonlinear damping.

5.3.1.1 Cubic Nonlinear Damping Isolation for Power Equipment

The research on the cubic nonlinear damping vibration isolation system of power equipment is shown in Fig. 5.18, where $F(t)$, m_2, m_1, k_2, k_1, and c_1 are as before. The input velocity of the cubic nonlinear damping force f_{cubic} is the relative moving velocity between m_2 and m_1, thus $v = \dot{x}_2 - \dot{x}_1$, the dynamic equation is expressed by Eq. (4.9).

$$\begin{cases} \mathrm{m}_1 \ddot{x}_1 + \mathrm{k}_1 x_1 + \mathrm{c}_1 \dot{x}_1 - \mathrm{k}_2(x_2 - x_1) = f_{cubic}(t) \\ \mathrm{m}_2 \ddot{x}_2 + \mathrm{k}_2(x_2 - x_1) = F(t) - f_{cubic}(t) \end{cases} \tag{5.9}$$

Assume that the state variables are $z_1 = x_1$, $z_2 = x_2$, $z_3 = \dot{x}_1$, $z_4 = \dot{x}_2$. Hence, a state vector is constructed such that $z(\mathrm{t}) = [z_1, z_2, z_3, z_4]^{\mathrm{T}}$, and Eq. (5.9) is transformed into the following state space equation:

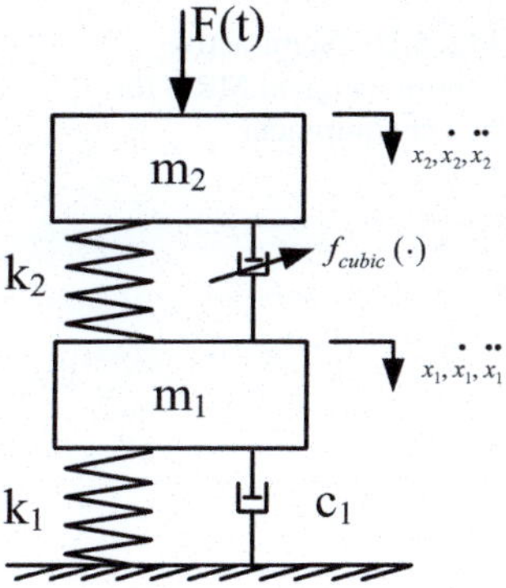

Fig. 5.18 Vibration isolation using cubic nonlinear damping for power equipment

$$\begin{cases} \dot{z}(t) = \mathrm{A}z(t) + \mathrm{B}F_{input} \\ y(t) = \mathrm{C}z(t) + \mathrm{D}F_{input} \end{cases} \tag{5.10}$$

where $F_{input} = \begin{bmatrix} F(t) \\ f_{cubic} \end{bmatrix}$,

$$\mathrm{A} = \begin{bmatrix} 0 & 0 & 1 & 0 \\ 0 & 0 & 0 & 1 \\ -\frac{\mathrm{k}_1+\mathrm{k}_2}{\mathrm{m}_1} & \frac{\mathrm{k}_2}{\mathrm{m}_1} & -\frac{\mathrm{c}_1}{\mathrm{m}_1} & 0 \\ \frac{\mathrm{k}_2}{\mathrm{m}_2} & -\frac{\mathrm{k}_2}{\mathrm{m}_2} & 0 & 0 \end{bmatrix}, \mathrm{B} = \begin{bmatrix} 0 & 0 \\ 0 & 0 \\ 0 & \frac{1}{\mathrm{m}_1} \\ \frac{1}{\mathrm{m}_2} & -\frac{1}{\mathrm{m}_2} \end{bmatrix}, \mathrm{C} = \begin{bmatrix} 1 & 0 & 0 & 0 \\ 0 & 1 & 0 & 0 \\ 0 & 0 & 1 & 0 \\ 0 & 0 & 0 & 1 \end{bmatrix}, \mathrm{D} = \begin{bmatrix} 0 & 0 \\ 0 & 0 \\ 0 & 0 \\ 0 & 0 \end{bmatrix}.$$

The nonlinear damping force $f_{cubic} = d_3v^3 + d_2v^2 + d_1v$ can be expressed in the form of state variables as follows:

$$f_{cubic} = d_3(z_4 - z_3)^3 + d_2(z_4 - z_3)^2 + d_1(z_4 - z_3) \tag{5.11}$$

Substitute Eq. (5.11) into Eq. (5.10) using the calculation strategy in Fig. 5.19.

The semi-active control system of the MRD's equivalent realization can be examined by replacing the cubic nonlinear damping with MRD, as shown in Fig. 5.20, where the current I is supplied by an external power source, and the input velocity (moving velocity of the piston rod) of the MRD is the relative moving velocity between m_2 and m_1, thus $v = \dot{x}_2 - \dot{x}_1$, the dynamic equation is shown in Eq. (5.12).

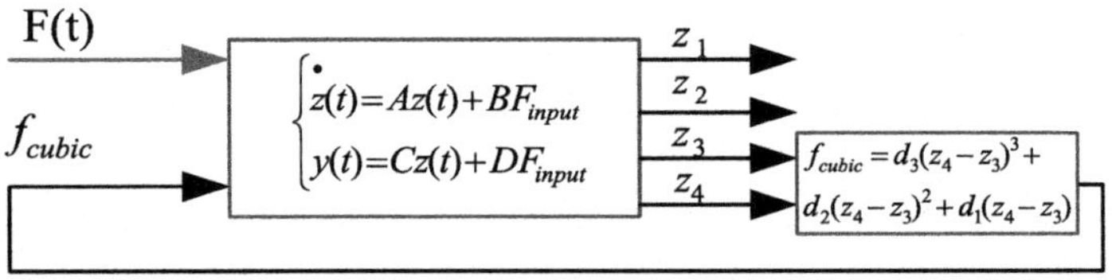

Fig. 5.19 Schematic diagram of the vibration isolation with cubic nonlinear damping

Fig. 5.20 Semi-active control using an MRD for power equipment

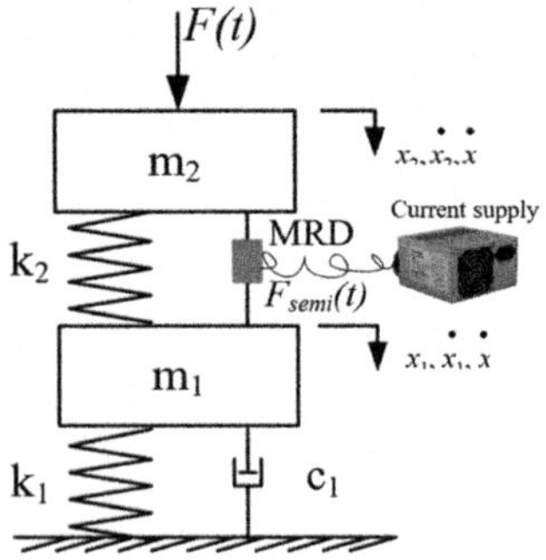

$$\begin{cases} m_1 \ddot{x}_1 + k_1 x_1 + c_1 \dot{x}_1 - k_2(x_2 - x_1) = F_{mr}(t) \\ m_2 \ddot{x}_2 + k_2(x_2 - x_1) = F(t) - F_{mr}(t) \end{cases} \tag{5.12}$$

where $F_{mr}(t)$ is the MRD output damping force.

Assume that the state variables are $z_1 = x_1, z_2 = x_2, z_3 = \dot{x}_1, z_4 = \dot{x}_2$ and the state vector is $z(t) = [z_1, z_2, z_3, z_4]^T$. Then Eq. (5.12) can be transformed into the following state space function:

$$\begin{cases} \dot{z}(t) = Az(t) + BF_{input} \\ y(t) = Cz(t) + DF_{input} \end{cases} \tag{5.13}$$

where $F_{input} = \begin{bmatrix} F(t) \\ F_{mr}(t) \end{bmatrix}$, and A and B are coefficient matrices defined in Eq. (5.10).

The output damping force $F_{mr}(t)$ of MRD is calculated based on the 6-segment cubic polynomial model proposed in this research. It should be noted that the output variable $y(t)$ must be changed to $y(t) = \begin{bmatrix} x_1 & x_2 & \dot{x}_1 & \dot{x}_2 & \ddot{x}_2 - \ddot{x}_1 \end{bmatrix}^T$, where

$$C = \begin{bmatrix} 1 & 0 & 0 & 0 \\ 0 & 1 & 0 & 0 \\ 0 & 0 & 1 & 0 \\ 0 & 0 & 0 & 1 \\ \frac{k_2}{m_2} + \frac{k_1+k_2}{m_1} & -\left(\frac{k_2}{m_2} + \frac{k_2}{m_1}\right) & \frac{c_1}{m_1} & 0 \end{bmatrix} \text{ and } D = \begin{bmatrix} 0 & 0 \\ 0 & 0 \\ 0 & 0 \\ 0 & 0 \\ \frac{1}{m_2} & -\left(\frac{1}{m_1} + \frac{1}{m_2}\right) \end{bmatrix}.$$

Additionally, the calculation of the relative acceleration $\ddot{x}_2 - \ddot{x}_1$ is used to determine the acceleration sign input to the MRD to determine whether it is located on the positive acceleration branch (right branch curve) or the negative acceleration branch (left branch curve).

The calculation strategy for semi-active control based on MRD is shown in Fig. 5.21.

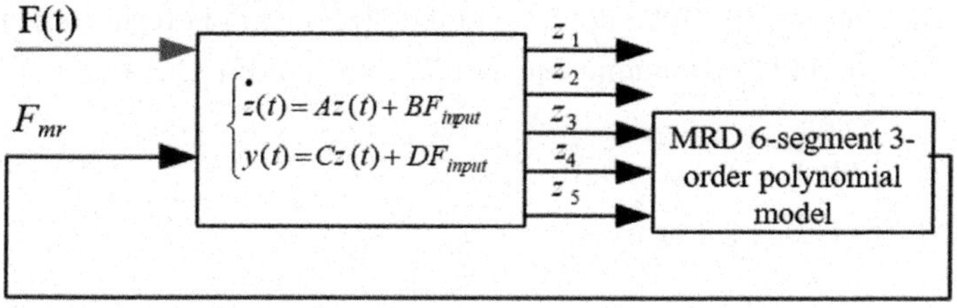

Fig. 5.21 Schematic diagram of the semi-active control with an MRD

The calculation of the output damping force of MRD is based on the 6-segment cubic polynomial model and the open-loop control strategy proposed in Sect. 5.2.5, where $F_{desired}$ is the cubic nonlinear damping force f_{cubic}.

Example 5.1 The parameters of the cubic nonlinear damping vibration isolation for power equipment and the MRD semi-active control system were set as follows: $\mathrm{m}_1 = 1200$ kg, $\mathrm{m}_2 = 600$ kg, $\mathrm{k}_1 = 1 \times 10^6$ N/m, $\mathrm{k}_2 = 1.5 \times 10^6$ N/m, $\mathrm{c}_1 = 1.6 \times 10^4$ N·m/s. Assume that the excitation load $F(t)$ generated by the power equipment is in the form of a sine wave, and the amplitude is 300 N, and the frequency is 1 Hz. The polynomial coefficients of the cubic nonlinear damping are $d_3 = 1 \times 10^6$, $d_2 = 1 \times 10^4$, and $d_1 = 1 \times 10^2$. The 6-segment cubic polynomial model used in the MRD is calculated based on the experimental data with a 15 mm amplitude, 1 Hz frequency, and 0 to 1.0 A control currents (Note: the MRD models used in this study are based on the test data of this group of operating conditions, and will not be described later).

Based on the cubic nonlinear damping force, the equivalent tracking MRD damping force is shown in Fig. 5.22. The MRD can effectively follow the cubic nonlinear damping force, where the induced MRD damping force is highly consistent with the tracked object. The effectiveness and accuracy of the proposed MRD model and open-loop control strategy are proved herein. The comparison between the force transmitted to the foundation by the power equipment under the two control conditions is shown in Fig. 5.23. It can be seen that the high-precision nonlinear damping force as a function of the MRD ensures that the vibration isolation characteristics of the cubic nonlinear damping are well copied. The MRD control current is shown in Fig. 5.24. The MRD piston rod movement velocity is shown in Fig. 5.25, where v_A and v_B are the upper and lower limits of the velocity during the MRD test.

5.3.1.2 Cubic Nonlinear Damping Isolation for Sensitive Equipment

Figure 4.26 shows a study about the cubic nonlinear damping vibration isolation system of sensitive equipment, where $F(t)$, m_2, m_1, k_2, k_1, and c_1 have the same meanings as the aforementioned sensitive equipment control system. The input velocity of the cubic nonlinear damping force f_{cubic} is the relative motion velocity between the vibration-sensitive device and the placement platform (or foundation), thus $v = \dot{x}_2 - \dot{x}_1$, and the dynamic equation is shown in Eq. (5.14).

$$\begin{cases} \mathrm{m}_1 \ddot{x}_1 + \mathrm{k}_1 x_1 + \mathrm{c}_1 \dot{x}_1 - \mathrm{k}_2(x_2 - x_1) = F(t) - f_{cubic}(t) \\ \mathrm{m}_2 \ddot{x}_2 + \mathrm{k}_2(x_2 - x_1) = f_{cubic}(t) \end{cases} \tag{5.14}$$

Assume that the state variables are $z_1 = x_1$, $z_2 = x_2$, $z_3 = \dot{x}_1$, and $z_4 = \dot{x}_2$ and the state vector is $z(t) = [z_1, z_2, z_3, z_4]^{\mathrm{T}}$. Equation (5.14) can be converted into a state space function similar to Eq. (5.10). Nevertheless, the corresponding matrix

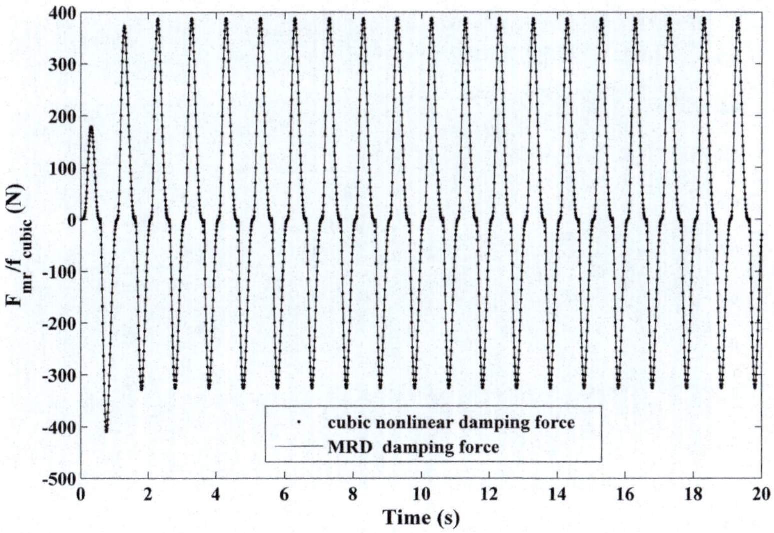

Fig. 5.22 Cubic nonlinear damping force as a function of the MRD force (power equipment)

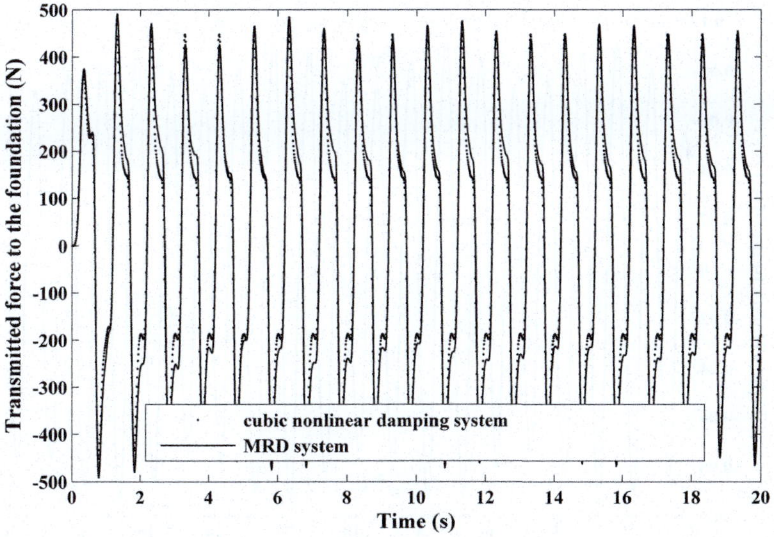

Fig. 5.23 Transmitted force from the equipment to the foundation for comparison (cubic nonlinear damping)

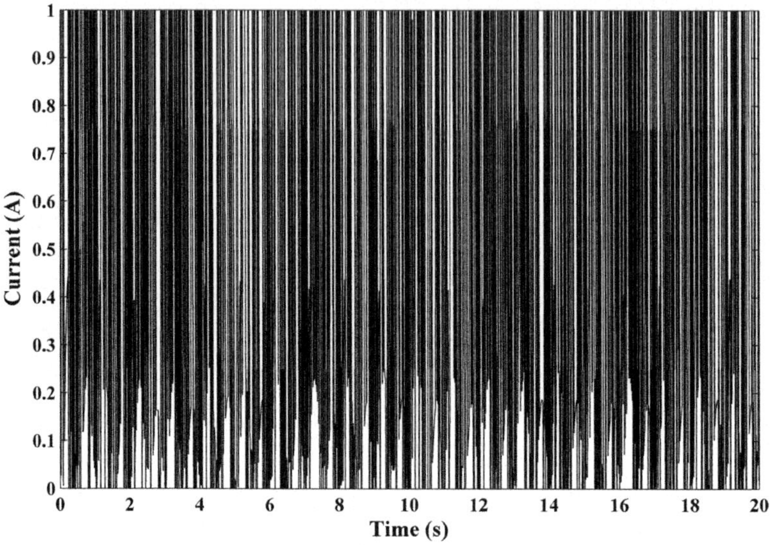

Fig. 5.24 Current supplied to the MRD (power equipment/cubic nonlinear damping)

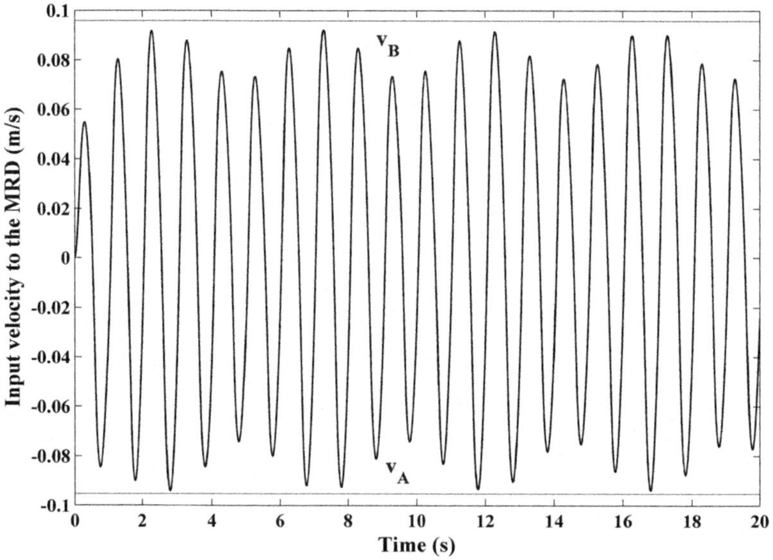

Fig. 5.25 Velocity supplied to the MRD (power equipment/cubic nonlinear damping)

coefficients must be changed to $A = \begin{bmatrix} 0 & 0 & 1 & 0 \\ 0 & 0 & 0 & 1 \\ -\frac{k_1+k_2}{m_1} & \frac{k_2}{m_1} & -\frac{c_1}{m_1} & 0 \\ \frac{k_2}{m_2} & -\frac{k_2}{m_2} & 0 & 0 \end{bmatrix}$, $B = \begin{bmatrix} 0 & 0 \\ 0 & 0 \\ \frac{1}{m_1} & -\frac{1}{m_1} \\ 0 & \frac{1}{m_2} \end{bmatrix}$,

$C = \begin{bmatrix} 1\,0\,0\,0 \\ 0\,1\,0\,0 \\ 0\,0\,1\,0 \\ 0\,0\,0\,1 \end{bmatrix}$, $D = \begin{bmatrix} 0\,0 \\ 0\,0 \\ 0\,0 \\ 0\,0 \end{bmatrix}$, and f_{cubic} is expressed as mentioned above.

Figure 5.19 shows the cubic nonlinear damping calculation strategy for sensitive equipment. The vibration isolation characteristics based on the MRD equivalent to achieve the cubic nonlinear damping will be replaced by MRD, as shown in Fig. 5.27. The MRD input velocity is the relative moving velocity between m_2 and m_1, that is $v = \dot{x}_2 - \dot{x}_1$, the dynamic equation formed by which is shown in Eq. (5.15).

$$\begin{cases} m_1 \ddot{x}_1 + k_1 x_1 + c_1 \dot{x}_1 - k_2(x_2 - x_1) = F(t) - F_{mr}(t) \\ m_2 \ddot{x}_2 + k_2(x_2 - x_1) = F_{mr}(t) \end{cases} \quad (5.15)$$

where $F_{mr}(t)$ is the output damping force of the MRD (Fig. 5.26).

Assume that the state variables are $z_1 = x_1$, $z_2 = x_2$, $z_3 = \dot{x}_1$, $z_4 = \dot{x}_2$ and the state vector is $z(t) = [z_1, z_2, z_3, z_4]^T$. Equation (5.15) can be transformed into the following state space equation:

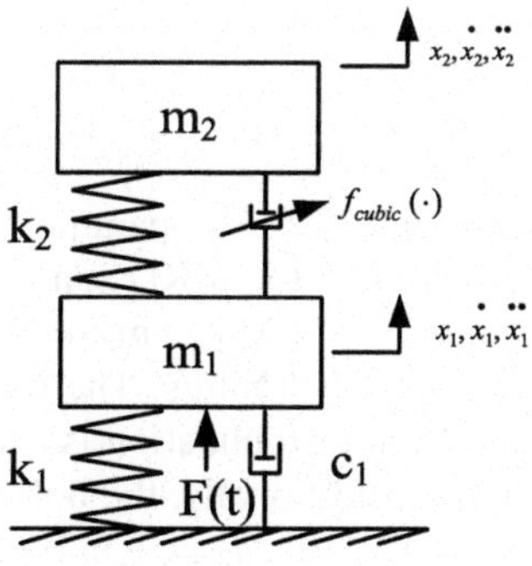

Fig. 5.26 Vibration isolation of the sensitive equipment using cubic nonlinear damping

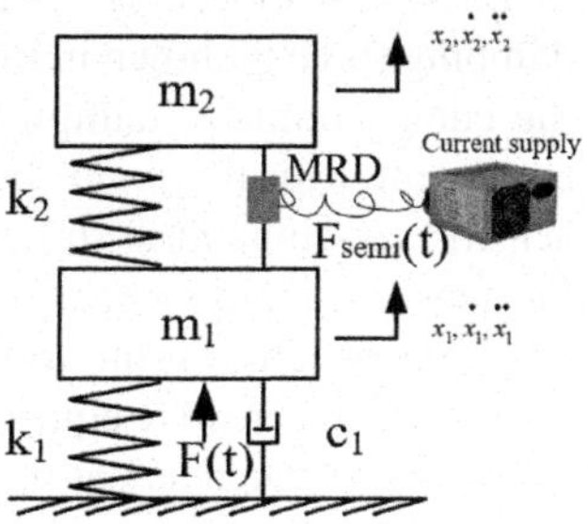

Fig. 5.27 Semi-active control of sensitive equipment using an MRD

$$\begin{cases} \dot{z}(t) = \mathrm{A}z(t) + \mathrm{B}F_{input} \\ y(t) = \mathrm{C}z(t) + \mathrm{D}F_{input} \end{cases} \tag{5.16}$$

where $F_{input} = \begin{bmatrix} F(t) \\ F_{mr}(t) \end{bmatrix}$, $\mathrm{A} = \begin{bmatrix} 0 & 0 & 1 & 0 \\ 0 & 0 & 0 & 1 \\ -\frac{\mathrm{k_1+k_2}}{\mathrm{m_1}} & \frac{\mathrm{k_2}}{\mathrm{m_1}} & -\frac{\mathrm{c_1}}{\mathrm{m_1}} & 0 \\ \frac{\mathrm{k_2}}{\mathrm{m_2}} & -\frac{\mathrm{k_2}}{\mathrm{m_2}} & 0 & 0 \end{bmatrix}$, and $\mathrm{B} = \begin{bmatrix} 0 & 0 \\ 0 & 0 \\ \frac{1}{\mathrm{m_1}} & -\frac{1}{\mathrm{m_1}} \\ 0 & \frac{1}{\mathrm{m_2}} \end{bmatrix}$.

The actual output force $F_{mr}(t)$ of the MRD is calculated according to the proposed 6-segment cubic polynomial model, and the output variable $y(t)$ is:

$\begin{bmatrix} x_1 \; x_2 \; \dot{x}_1 \; \dot{x}_2 \; \ddot{x}_2 - \ddot{x}_1 \end{bmatrix}^{\mathrm{T}}$, $\mathrm{C} = \begin{bmatrix} 1 & 0 & 0 & 0 \\ 0 & 1 & 0 & 0 \\ 0 & 0 & 1 & 0 \\ 0 & 0 & 0 & 1 \\ \frac{\mathrm{k_2}}{\mathrm{m_2}} + \frac{\mathrm{k_1+k_2}}{\mathrm{m_1}} & -\left(\frac{\mathrm{k_2}}{\mathrm{m_2}} + \frac{\mathrm{k_2}}{\mathrm{m_1}}\right) & \frac{\mathrm{c_1}}{\mathrm{m_1}} & 0 \end{bmatrix}$, and $\mathrm{D} = \begin{bmatrix} 0 & 0 \\ 0 & 0 \\ 0 & 0 \\ 0 & 0 \\ -\frac{1}{\mathrm{m_1}} & \frac{1}{\mathrm{m_1}} + \frac{1}{\mathrm{m_2}} \end{bmatrix}$.

The semi-active control strategy based on the MRD is the same as in Fig. 4.21. The actual output calculation principle of the MRD is similar to that of the MRD semi-active control for power equipment, and it is also based on the open-loop control strategy of Sect. 5.2.5, $F_{desired}$ is the cubic nonlinear damping force f_{cubic} calculated using Eq. (5.14).

Example 5.2 The parameters of the cubic nonlinear damping vibration isolation system and the MRD semi-active control system for sensitive equipment were set to: $\mathrm{m_1} = 1200$ kg, $\mathrm{m_2} = 100$ kg, $\mathrm{k_1} = 1 \times 10^6$ N/m, $\mathrm{k_2} = 1.5 \times 10^4$ N/m, $\mathrm{c_1} = 1.6 \times 10^4$ N·m/s. The external interference force $F(t)$ acting on the placement platform or foundation is a sine wave with an amplitude of 5 N and a frequency of 1 Hz. Additionally, the polynomial coefficients of the cubic nonlinear damping are $d_3 = 1 \times 10^5$, $d_2 = 1 \times 10^3$, and $d_1 = 1 \times 10^1$.

Based on the cubic nonlinear damping force, the equivalent tracking MRD damping force is shown in Fig. 5.28. The MRD damping force can accurately follow the cubic nonlinear damping force, and the MRD damping force curve is highly consistent with the controlled object. In both cases, the vibration velocity of the sensitive equipment is shown in Fig. 5.29, indicating that the vibration velocity of the sensitive equipment under the MRD system coincides with the cubic nonlinear damping system. Furthermore, the MRD control current is shown in Fig. 5.30. For the vibration-sensitive equipment's micro-vibration control system, the MRD's input velocity was taken as within the interval $[v_\mathrm{A}, \mathrm{v_B}]$ throughout the entire chapter.

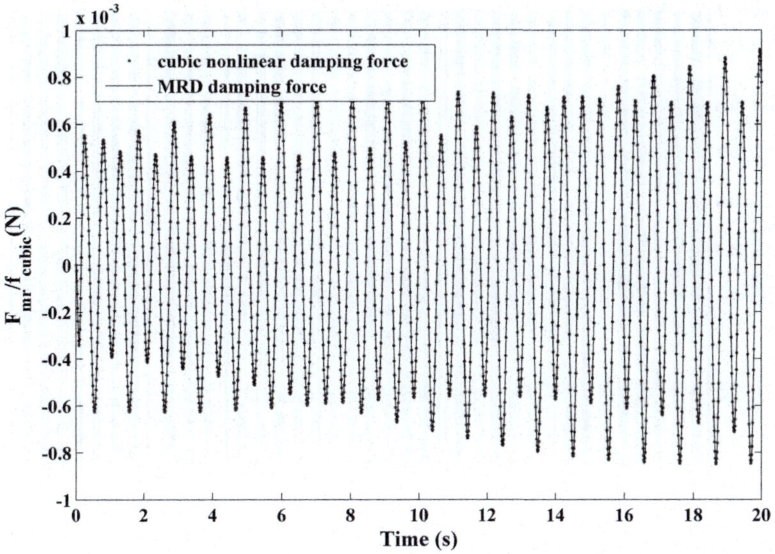

Fig. 5.28 Cubic nonlinear damping force as a function of the MRD force (sensitive equipment)

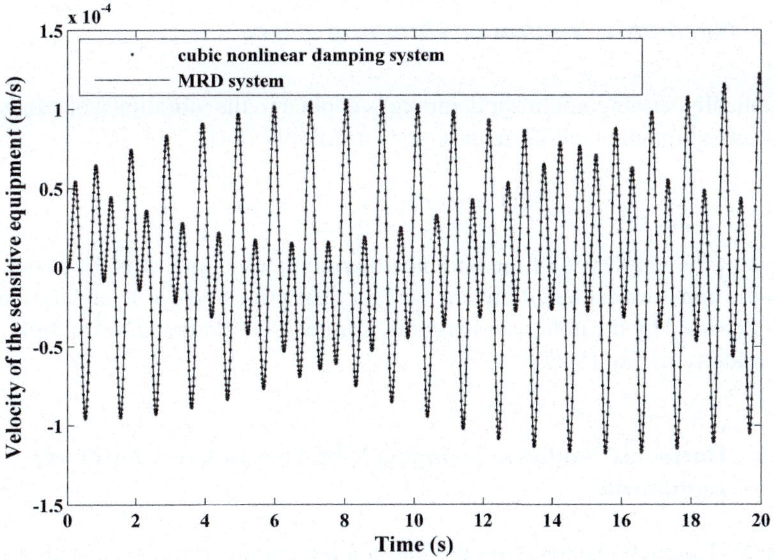

Fig. 5.29 Velocity of sensitive equipment for comparison (cubic nonlinear damping)

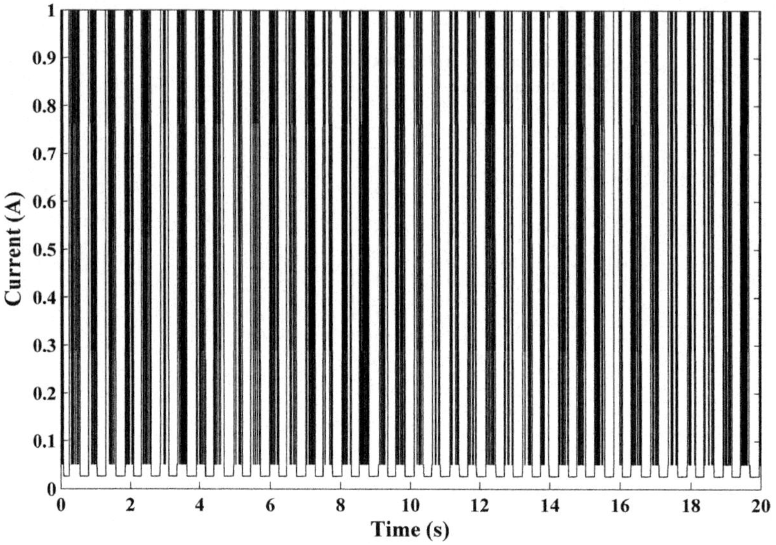

Fig. 5.30 Current supplied to the MRD (sensitive equipment/cubic nonlinear damping)

5.3.2 Harmonic Nonlinear Damping Force

Harmonically varying nonlinear damping is applied to the vibration isolation system of typical equipment, and its expression is as follows [14]:

$$\xi(t) = \alpha \cos(\omega_{\zeta} t + \beta) + \zeta_{DC} \tag{5.17}$$

where α is the amplitude of the nonlinear harmonic damping, β is the phase angle, ζ_{DC} is the non-zero center amplitude used for restrictions $\zeta(t) > 0$, and the damping force generated by the nonlinear harmonic damping is calculated by multiplying the input velocity by Eq. (5.17).

5.3.2.1 Harmonic Nonlinear Damping Vibration Isolation for Power Equipment

Figure 5.31 depicts the study on the harmonic damping vibration isolation system of power equipment, where $F(t)$, m_2, m_1, k_2, k_1, and c_1 are as previously defined, and f_h is the nonlinear harmonic damping force. The input velocity is the relative velocity between m_2 and m_1, that is $v = \dot{x}_2 - \dot{x}_1$, the dynamic equation of the system is shown in Eq. (5.18).

$$\begin{cases} \mathrm{m}_1 \ddot{x}_1 + \mathrm{k}_1 x_1 + \mathrm{c}_1 \dot{x}_1 - \mathrm{k}_2(x_2 - x_1) = f_h(t) \\ \mathrm{m}_2 \ddot{x}_2 + \mathrm{k}_2(x_2 - x_1) = F(t) - f_h(t) \end{cases} \tag{5.18}$$

where the nonlinear harmonic damping force can be expressed as $f_h = \zeta(t)(\dot{x}_2 - \dot{x}_1)$, the state variables are $z_1 = x_1$, $z_2 = x_2$, $z_3 = \dot{x}_1$, $z_4 = \dot{x}_2$, and the state vector is $z(t) = [z_1, z_2, z_3, z_4]^{\mathrm{T}}$. Equation (5.18) can be changed into the following state space equation:

$$\begin{cases} \dot{z}(t) = \mathrm{A}z(t) + \mathrm{B}F_{input} \\ y(t) = \mathrm{C}z(t) + \mathrm{D}F_{input} \end{cases} \tag{5.19}$$

where $F_{input} = \begin{bmatrix} F(t) \\ f_h(t) \end{bmatrix}$, $\mathrm{A} = \begin{bmatrix} 0 & 0 & 1 & 0 \\ 0 & 0 & 0 & 1 \\ -\frac{\mathrm{k}_1+\mathrm{k}_2}{\mathrm{m}_1} & \frac{\mathrm{k}_2}{\mathrm{m}_1} & -\frac{\mathrm{c}_1}{\mathrm{m}_1} & 0 \\ \frac{\mathrm{k}_2}{\mathrm{m}_2} & -\frac{\mathrm{k}_2}{\mathrm{m}_2} & 0 & 0 \end{bmatrix}$, $\mathrm{B} = \begin{bmatrix} 0 & 0 \\ 0 & 0 \\ 0 & \frac{1}{\mathrm{m}_1} \\ \frac{1}{\mathrm{m}_2} & -\frac{1}{\mathrm{m}_2} \end{bmatrix}$, $\mathrm{C} = \begin{bmatrix} 1 & 0 & 0 & 0 \\ 0 & 1 & 0 & 0 \\ 0 & 0 & 1 & 0 \\ 0 & 0 & 0 & 1 \end{bmatrix}$, $\mathrm{D} = \begin{bmatrix} 0 & 0 \\ 0 & 0 \\ 0 & 0 \\ 0 & 0 \end{bmatrix}$, and f_h is the state space form defined as:

$$f_h = \zeta(t)(z_4 - z_3) \tag{5.20}$$

Substituting Eq. (5.20) into Eq. (5.19) gives the calculation strategy shown in Fig. 5.32.

The principle for the semi-active control based on MRD equivalent realization, that is, to replace the nonlinear harmonic damping with MRD, is the same as that indicated in Fig. 5.20. The dynamic and state space equations are the same as Eqs.

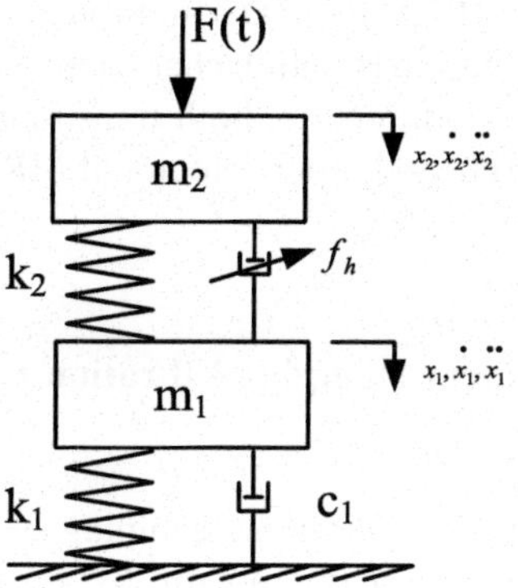

Fig. 5.31 Vibration isolation of power equipment using harmonically varying nonlinear damping

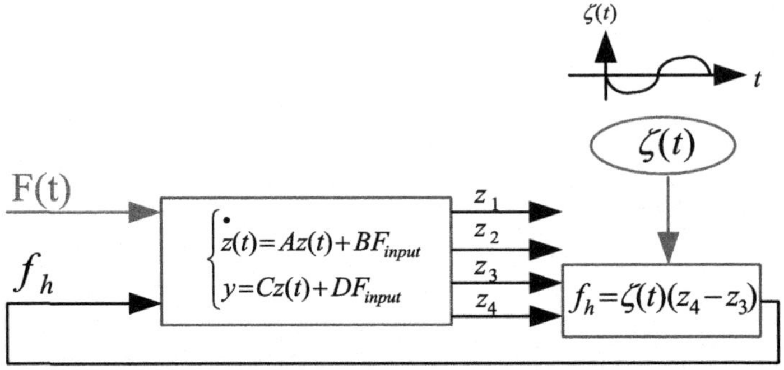

Fig. 5.32 Schematic diagram of vibration isolation with harmonically varying nonlinear damping

(5.12) and (5.13). The semi-active control strategy based on MRD is the same as in Fig. 5.21. The actual MRD output calculation is still based on the open-loop control strategy in Sect. 5.2.5, with $F_{desired}$ now being the nonlinear harmonic damping force f_h.

Example 5.3 The power equipment parameter configuration using nonlinear harmonic damping vibration isolation system and MRD semi-active control system is the same as that in Example 5.1. The excitation load $F(t)$ generated by the power equipment is in the form of a sine wave, the amplitude is 100 N and the frequency is 1 Hz. The coefficients of nonlinear harmonic damping are $\alpha = 600$, $\beta = 300$, $\zeta_{DC} = 650$, and $\omega_\zeta = 2\pi$. The change in the nonlinear harmonic damping with time is shown in Fig. 5.33.

Based on the nonlinear harmonic damping force, the equivalently tracked MRD damping force is shown in Fig. 5.34. The semi-active control based on MRD can effectively track the nonlinear harmonic damping force, where the induced MRD damping force and the tracked object are highly consistent. The comparison between the force transmitted to the foundation by the power equipment under the two control conditions is shown in Fig. 5.35. The MRD's high-precision nonlinear damping force tracking ensures that the vibration isolation characteristics of the nonlinear harmonic damping are well applied. Additionally, both transmitted forces results are highly consistent. The control current and input velocity of MRD are respectively shown in Figs. 5.36 and 5.37.

5.3.2.2 Harmonic Nonlinear Damping Vibration Isolation of Sensitive Equipment

This section studies the nonlinear harmonic damping system of sensitive equipment, as shown in Fig. 5.38. Generally, $F(t)$, m_2, m_1, k_2, k_1, and c_1 have the are as stated before, and f_h is the damping force generated by nonlinear harmonic damping. The input velocity is the relative moving velocity between the sensitive equipment and

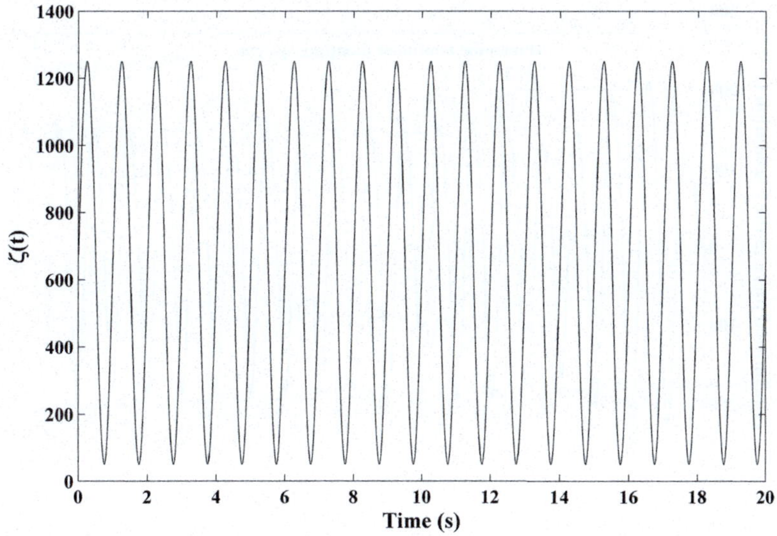

Fig. 5.33 Harmonically varying nonlinear damping varied with time (power equipment)

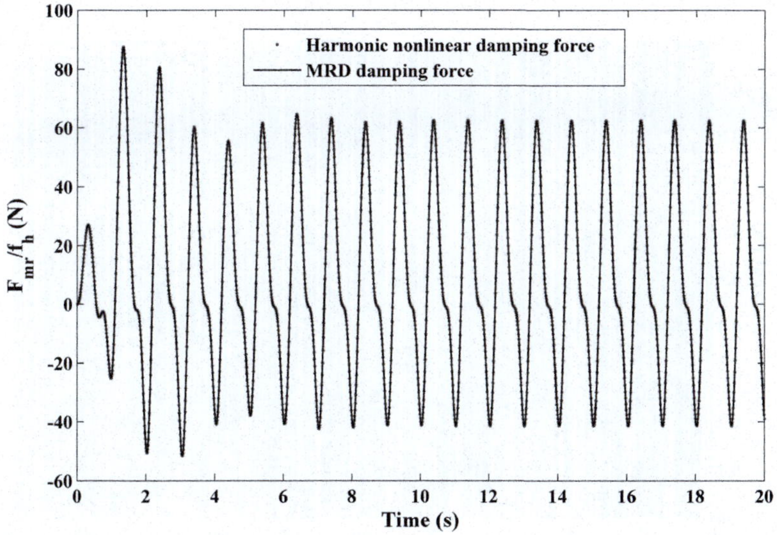

Fig. 5.34 Harmonically varying nonlinear damping force as a function of the MRD force (power equipment)

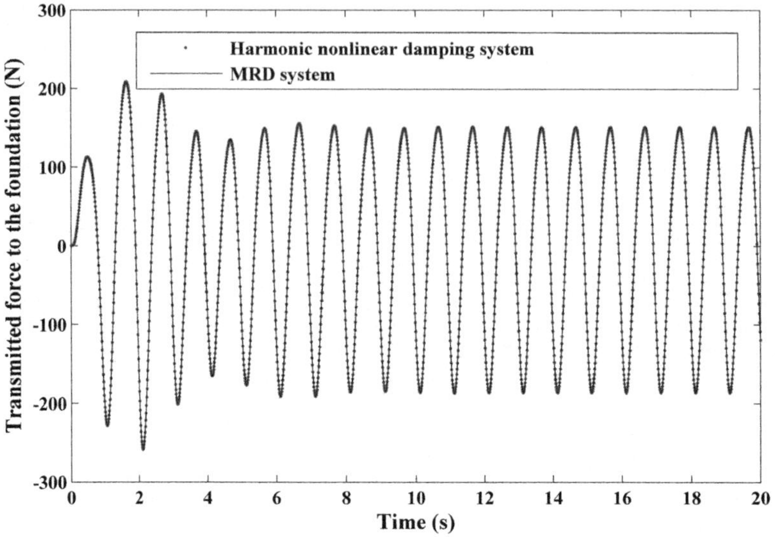

Fig. 5.35 Transmitted force from the equipment to the foundation for comparison (harmonically varying nonlinear damping)

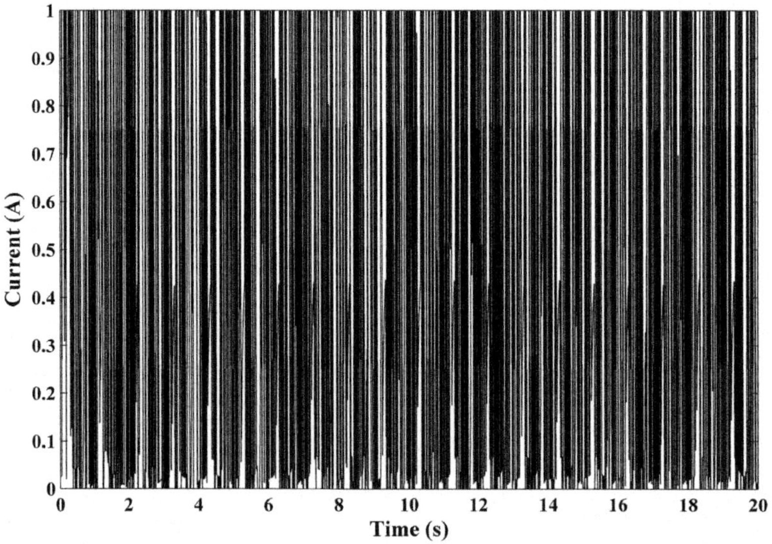

Fig. 5.36 Current supplied to the MRD (power equipment/harmonically varying nonlinear damping)

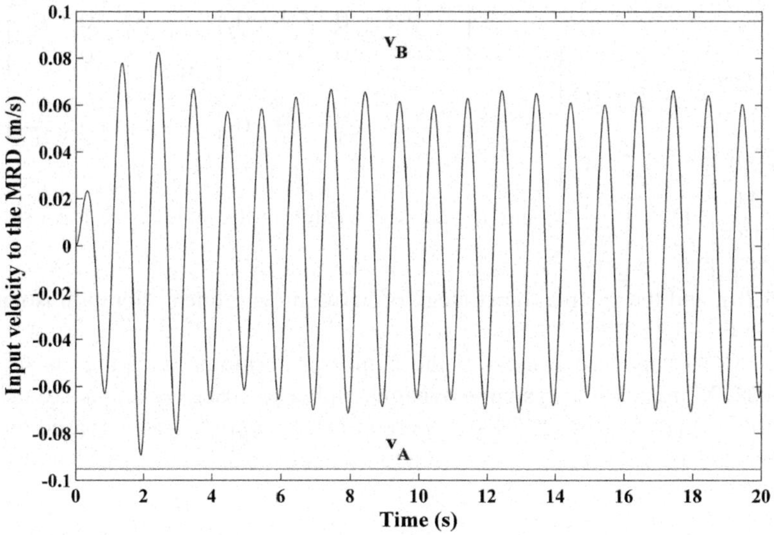

Fig. 5.37 Velocity supplied to the MRD (power equipment/harmonically varying nonlinear damping)

the placement platform, hence $v = \dot{x}_2 - \dot{x}_1$, and the resulting dynamic equation is shown in Eq. (5.21).

$$\begin{cases} \mathrm{m}_1 \ddot{x}_1 + \mathrm{k}_1 x_1 + \mathrm{c}_1 \dot{x}_1 - \mathrm{k}_2(x_2 - x_1) = F(t) - f_h(t) \\ \mathrm{m}_2 \ddot{x}_2 + \mathrm{k}_2(x_2 - x_1) = f_h(t) \end{cases} \tag{5.21}$$

where the nonlinear harmonic damping force is $f_h = \zeta(t)(\dot{x}_2 - \dot{x}_1)$, the state variables are $z_1 = x_1$, $z_2 = x_2$, $z_3 = \dot{x}_1$, and $z_4 = \dot{x}_2$, and the state vector is $z(t) = [z_1, z_2, z_3, z_4]^{\mathrm{T}}$. Equation (5.21) can be transformed into the following state space equation:

$$\begin{cases} \dot{z}(t) = \mathrm{A}z(t) + \mathrm{B}F_{input} \\ y(t) = \mathrm{C}z(t) + \mathrm{D}F_{input} \end{cases} \tag{5.22}$$

where $F_{input} = \begin{bmatrix} F(t) \\ f_h(t) \end{bmatrix}$, $A = \begin{bmatrix} 0 & 0 & 1 & 0 \\ 0 & 0 & 0 & 1 \\ -\frac{k_1+k_2}{m_1} & \frac{k_2}{m_1} & -\frac{c_1}{m_1} & 0 \\ \frac{k_2}{m_2} & -\frac{k_2}{m_2} & 0 & 0 \end{bmatrix}$, $B = \begin{bmatrix} 0 & 0 \\ 0 & 0 \\ -\frac{1}{m_1} & \frac{1}{m_1} \\ 0 & \frac{1}{m_2} \end{bmatrix}$, $C = \begin{bmatrix} 1\,0\,0\,0 \\ 0\,1\,0\,0 \\ 0\,0\,1\,0 \\ 0\,0\,0\,1 \end{bmatrix}$, $D = \begin{bmatrix} 0\,0 \\ 0\,0 \\ 0\,0 \\ 0\,0 \end{bmatrix}$, and f_h can be rewritten in the state space form shown in Eq. (5.20), and the calculation strategy of harmonic nonlinear vibration isolation is given in Fig. 5.32.

The MRD was used instead of the nonlinear harmonic damping, as shown in Fig. 5.27. Moreover, the system's dynamic and state space equations are the same as those in Eqs. (5.15) and (5.16). The semi-active control strategy based on MRD is shown in Fig. 5.21. $F_{desired}$ was used to calculate the MRD output damping force with the nonlinear harmonic damping force f_h.

Example 5.4 The parameters of the nonlinear harmonic damping vibration isolation system and the MRD semi-active control system for sensitive equipment are the same as those in Example 5.2. The external interference force $F(t)$ acting on the placement platform or foundation is in the form of a sine wave, the amplitude is 5 N, and the frequency is 1 Hz. The coefficients of nonlinear harmonic damping are $\alpha = 5, \beta = 5, \zeta_{DC} = 6$, and $\omega_\zeta = 2\pi$. Figure 5.39 shows the change in the nonlinear harmonic damping with time.

Based on the nonlinear harmonic damping force, the equivalently tracked MRD damping force is shown in Fig. 5.40. The MRD can accurately follow the nonlinear harmonic damping force, and the induced MRD damping force is highly consistent with the controlled object. The comparison between the vibration velocity of the sensitive equipment under both control conditions is shown in Fig. 5.41. The MRD's high-precision nonlinear damping force tracking ensures that the nonlinear harmonic damping characteristics can be perfectly reproduced in the vibration isolation system for equipment isolation. The MRD control current is shown in Fig. 5.42.

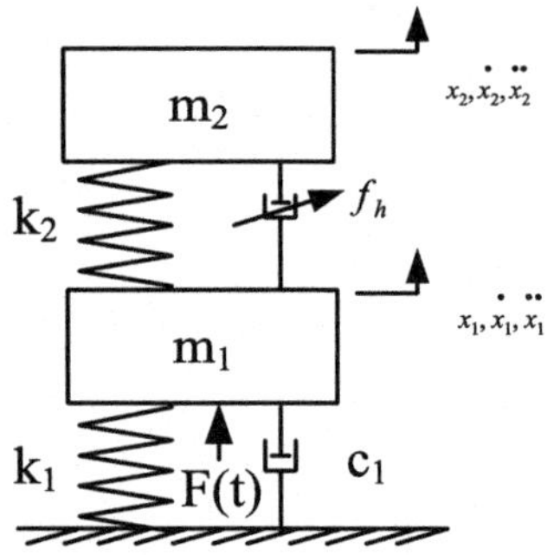

Fig. 5.38 Vibration isolation of sensitive equipment using harmonically varying nonlinear damping

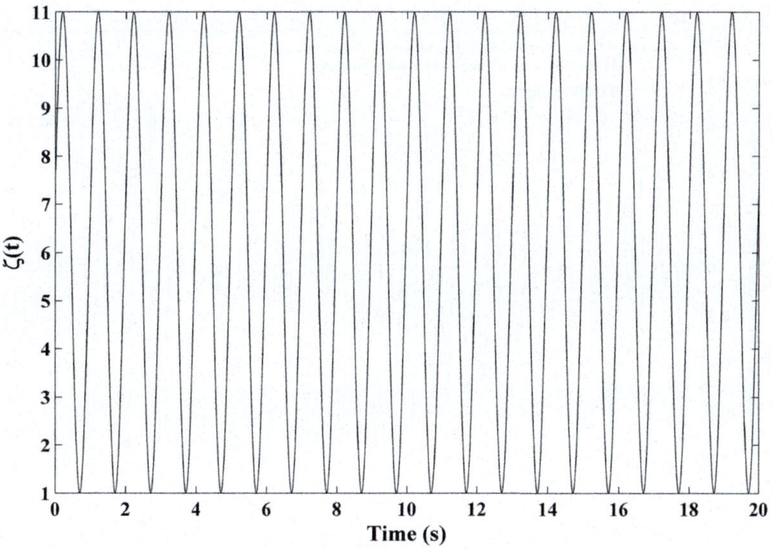

Fig. 5.39 Nonlinear harmonic damping with time (sensitive equipment)

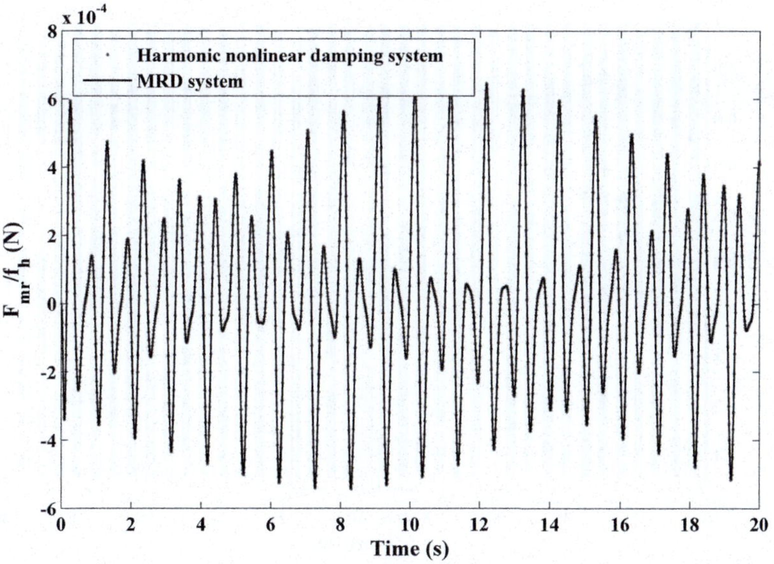

Fig. 5.40 Harmonically varying nonlinear damping force as a function of the MRD force (sensitive equipment)

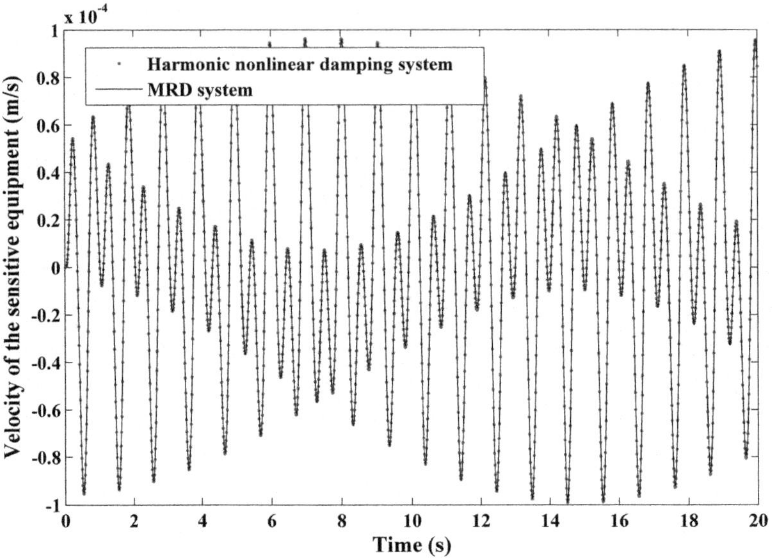

Fig. 5.41 Velocity of sensitive equipment (harmonically varying nonlinear damping)

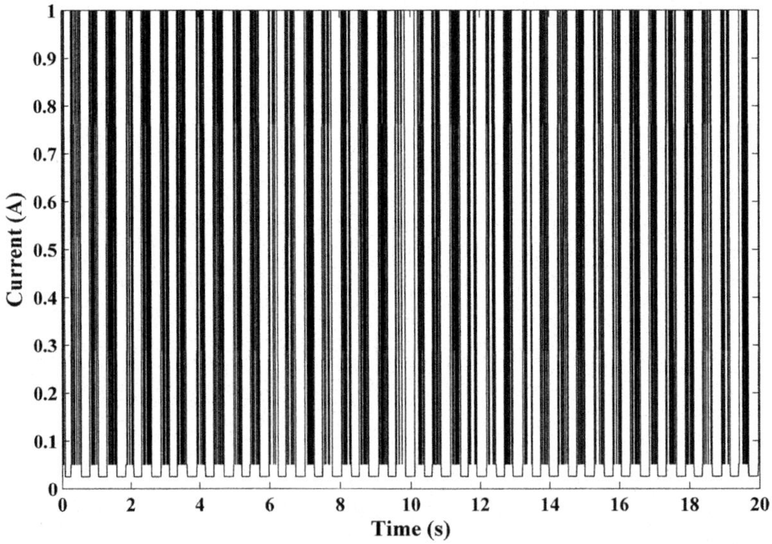

Fig. 5.42 Current supplied to the MRD (sensitive equipment/harmonically varying nonlinear damping)

5.4 MRD-Based Tracking Semi-Active Variable Damping Control Force

Many scholars have researched theoretical semi-active control strategies in recent years and proposed many effective semi-active control algorithms worldwide. The most common ones are semi-active variable damping control strategies. Liu [18] proposed several practical semi-active variable damping control algorithms, as shown in Table 5.2. As theoretical strategies, they have greatly enriched the semi-active control. The way to effectively induce these semi-active control forces in actual engineering is significant, and combining theories with practical applications is critical. This study investigates the theoretical semi-active variable damping control force based on the proposed MRD model.

5.4.1 *Semi-Active Variable Damping Control for Power Equipment*

Research on the vibration isolation system based on the introduced semi-active variable damping control strategy is performed for power equipment, and it is shown

Table 5.2 Some theoretical semi-active variable damping control strategies proposed by Liu [18]

Type	Expressions
Continuous skyhook	$c_{sa} = \begin{cases} \max[c_{\min}, \min(\frac{c_{sky} v_2}{v}, c_{\max})], v_2 v \geq 0 \\ c_{\min}, v_2 v < 0 \end{cases}$
Continuous balance	$c_{sa} = \begin{cases} \max[c_{\min}, \min(\frac{-kd}{v}, c_{\max})], dv \leq 0 \\ c_{\min}, dv > 0 \end{cases}$
On–off skyhook	$c_{sa} = \begin{cases} c_{\max}, v_2 v \geq 0 \\ c_{\min}, v_2 v < 0 \end{cases}$
On–off balance	$c_{sa} = \begin{cases} c_{\max}, dv \leq 0 \\ c_{\min}, dv > 0 \end{cases}$

Note c_{sa} is the semi-active variable damping, $c_{\max}$ and $c_{\min}$ are the upper and lower damping limits, since the minimum value of actual damping cannot reach zero, $c_{\min}$ generally takes a smaller value, c_{sky} is the skyhook damping coefficient, v_2 is the vibration velocity of the isolated object (in this study, it is the vibration velocity of the equipment), v is the input velocity to the c_{sa}, d is the input displacement to the c_{sa}, k is the stiffness coefficient. For both types of equipment in this study, $v_2 = \dot{x}_2$, $v = \dot{x}_2 - \dot{x}_1$, $d = x_2 - x_1$, and $k = k_2$. The four semi-active control strategies in Table 5.2 are shown in Fig. 5.43

Fig. 5.43 Four semi-active variable damping control strategies

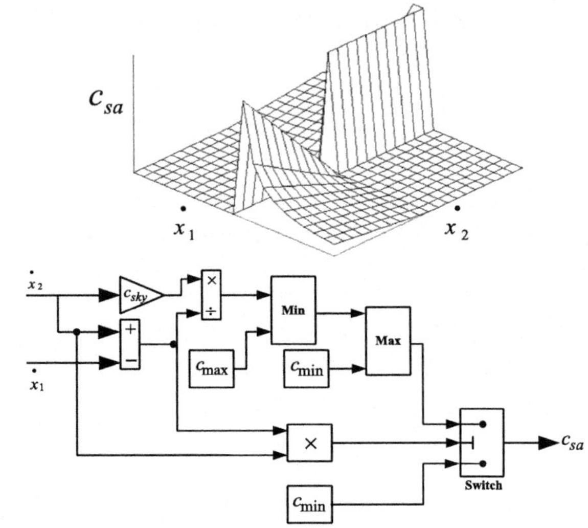

(a) Continuous skyhook

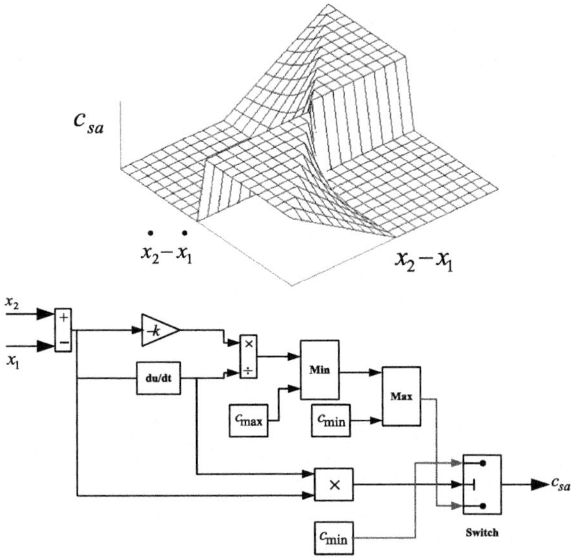

(b) Continuous balance

in Fig. 5.44, where $F(t)$, m_2, m_1, k_2, k_1, and c_1 have the same meanings as previously mentioned, and f_{sa1} is the semi-active variable damping control force generated in Eq. (5.23a). The input velocity is the relative velocity between m_2 and m_1, $v = \dot{x}_2 - \dot{x}_1$, and the resulting dynamic function is given in Eq. (5.24).

Fig. 5.43 (continued)

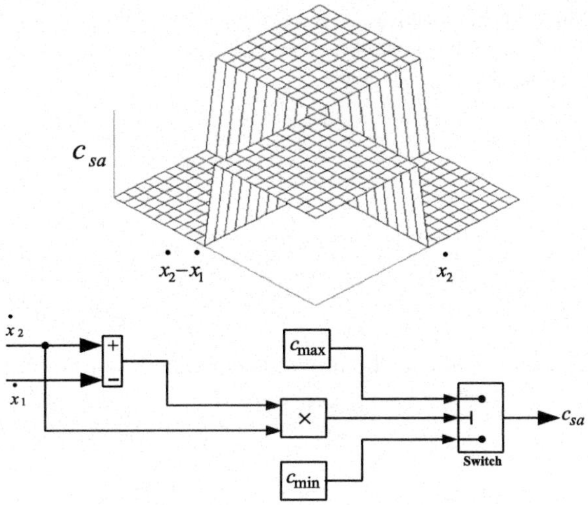

(c) On-off skyhook

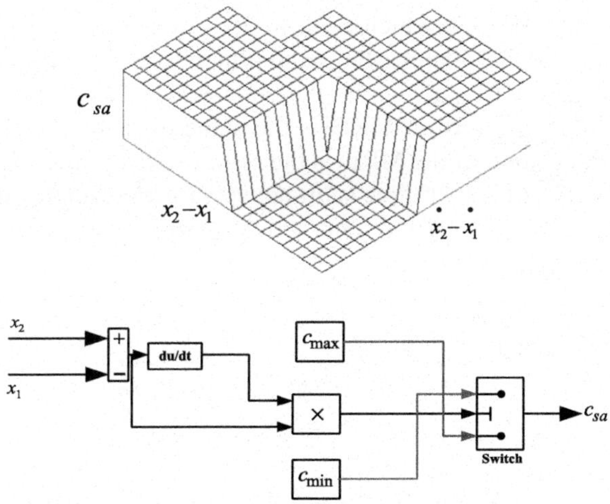

(d) On-off balance

$$\begin{cases} m_1 \ddot{x}_1 + k_1 x_1 + c_1 \dot{x}_1 - k_2(x_2 - x_1) = f_{sa1}(t) \\ m_2 \ddot{x}_2 + k_2(x_2 - x_1) = F(t) - f_{sa1}(t) \end{cases} \tag{5.23}$$

where the semi-active variable damping control force is expressed as $f_{sa1} = c_{sa}(\dot{x}_2 - \dot{x}_1)$, the state variables are $z_1 = x_1$, $z_2 = x_2$, $z_3 = \dot{x}_1$, and $z_4 = \dot{x}_2$, and the state vector is $z(t) = [z_1, z_2, z_3, z_4]^{\mathrm{T}}$. Equation (5.24) can be converted into a state space equation that is similar to Eq. (5.19). $F_{input} = \begin{bmatrix} F(t) \\ f_{sa1}(t) \end{bmatrix}$, A, B, C, and

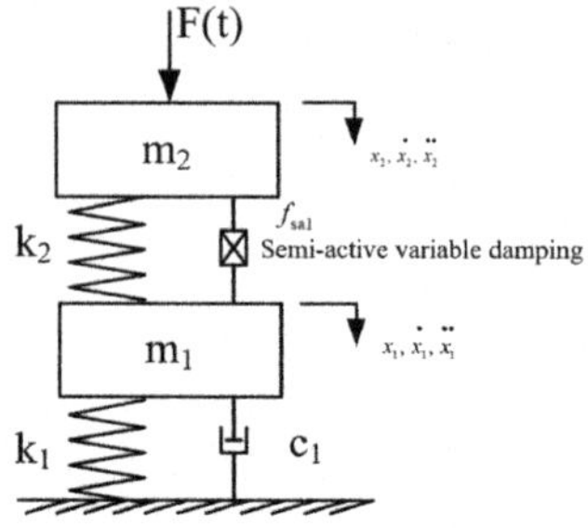

Fig. 5.44 Semi-active variable damping control system for power equipment

D are the same as the previous ones, and f_{sa1} is written in the state space form as:

$$f_{sa1} = c_{sa}(z_4 - z_3) \tag{5.24}$$

Substituting Eq. (5.25) into Eq. (5.24) gives the calculation strategy shown in Fig. 5.45.

The investigations on the equivalent realization of semi-active control were carried out on MRD, in which the semi-active variable damping control strategy was replaced by an MRD that has the same principle as that given in Fig. 5.20. The dynamic and state space equations are the same as Eqs. (5.12) and (5.13). The semi-active control strategy based on MRD is the same as in Fig. 5.21. $F_{desired}$ in the actual output calculation of the MRD is the semi-active variable damping control force f_{sa1}.

For the other three semi-active variable damping control strategies, such as Eqs. (5.23b)–(5.23d), the semi-active control forces generated are recorded as f_{sa2}, f_{sa3}, and f_{sa4}.

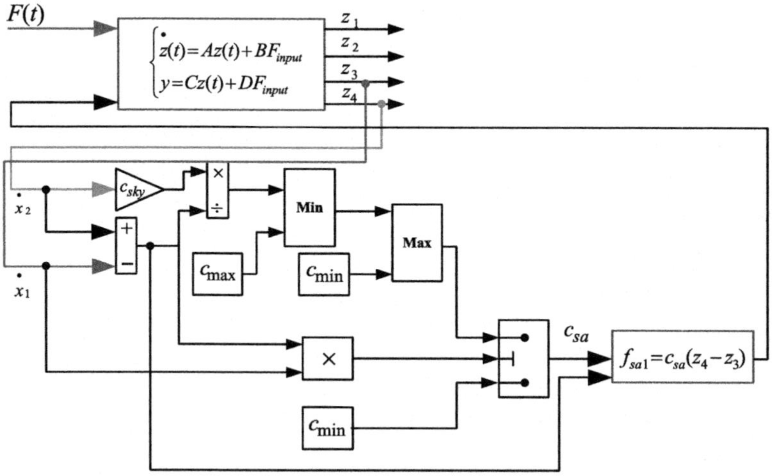

Fig. 5.45 Schematic diagram of semi-active variable damping control strategy

Example 5.5 The parameters of the vibration isolation system using theoretical semi-active variable damping control strategy and the MRD semi-active control system for power equipment are the same as those in Example 5.3. The excitation load $F(t)$ generated by the power equipment is a sine wave with an amplitude of 100 N and a frequency of 1 Hz. The coefficients in Eq. (5.23a) are $c_{\max} = 2000$, $c_{\min} = 0.01$, $c_{sky} = 500$.

The equivalent tracking for MRD damping force is shown in Fig. 5.46 based on the semi-active variable damping control force f_{sa1}. The semi-active control system implemented by the MRD can effectively track the theoretical semi-active variable damping control force, and the two results are highly consistent. The comparison between the forces transmitted to the foundation by the power equipment under the two control conditions is shown in Fig. 5.47. The conclusion indicates that the MRD can perfectly implement the theoretical semi-active variable damping control strategy. The control current and input velocity of the MRD are shown in Fig. 5.48 and Fig. 5.49, respectively. In order to avoid redundancy, the other three strategies only give the tracking of semi-active variable damping control force here, shown in Fig. 5.50. The forces transmitted to the foundation from the power equipment, MRD control current, and input velocity are not listed individually.

The MRD semi-active variable damping control force was induced based on the other three semi-active variable damping control algorithms in Eqs. (5.23b)–(5.23d), as shown in Fig. 5.50, where the coefficients in the continuous balance, on–off skyhook, and on–off balance algorithm are set to $c_{\max} = 1000$ and $c_{\min} = 0.01$.

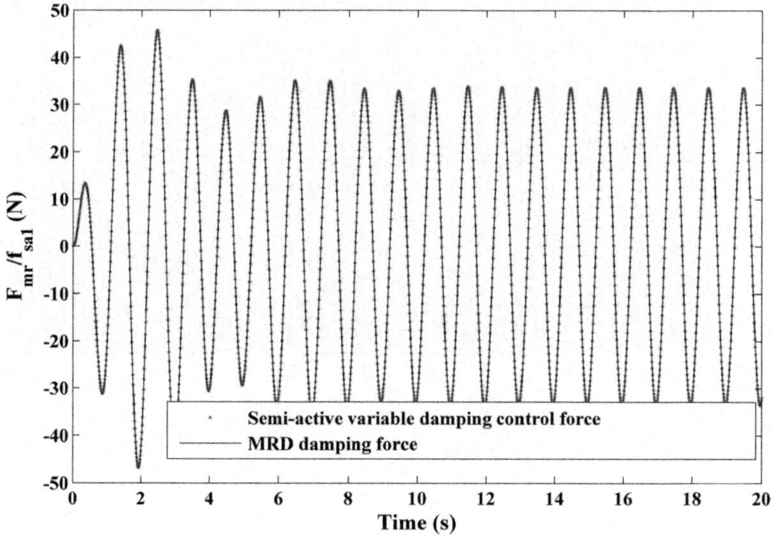

Fig. 5.46 Semi-active variable damping control force f_{sa1} as a function of the MRD force (power equipment)

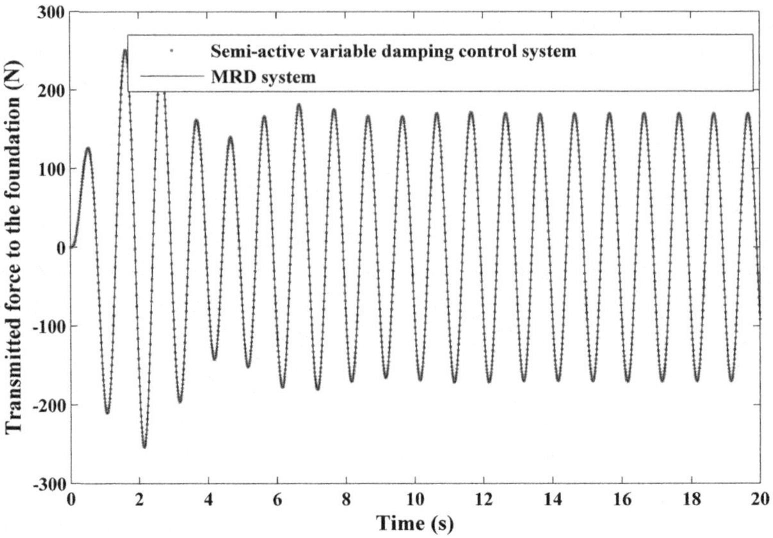

Fig. 5.47 Transmitted force from the equipment to the foundation (semi-active variable damping control)

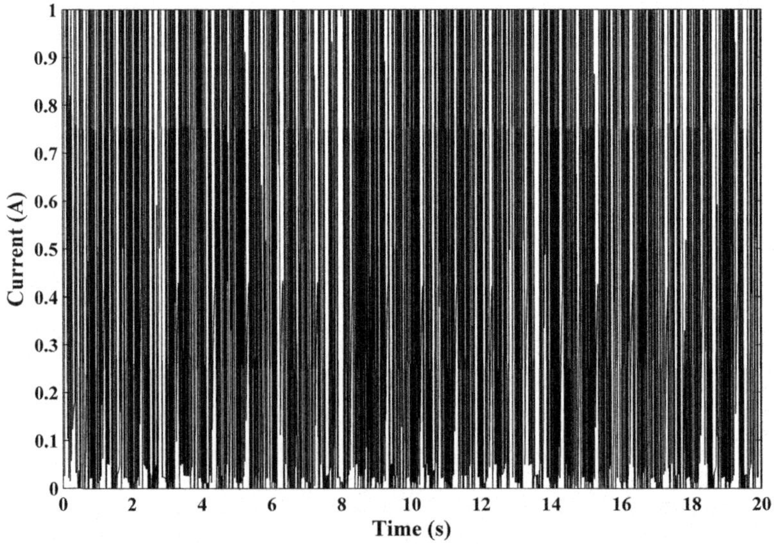

Fig. 5.48 Current supplied to the MRD (power equipment/semi-active variable damping control)

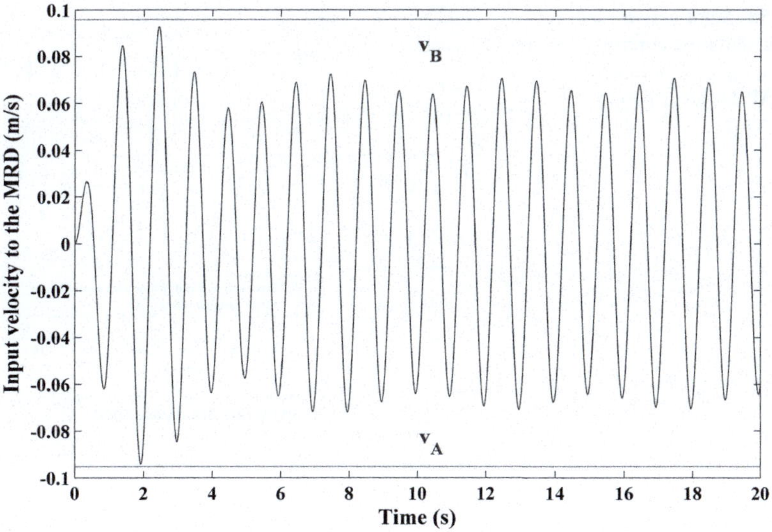

Fig. 5.49 Velocity supplied to the MRD (power equipment/semi-active variable damping control)

5.4.2 *Semi-Active Variable Damping Control for Sensitive Equipment*

This section investigates the vibration isolation system of sensitive equipment based on the semi-active variable damping control strategy, as shown in Fig. 5.51, where $F(t)$, m_2, m_1, k_2, k_1, and c_1 have the same meanings as the vibration isolation system of the aforementioned sensitive equipment, and f_{sa1} is the semi-active variable damping control force generated by Eq. (5.23a). The input velocity is the relative movement velocity between the equipment and the placement platform (or foundation), $v = \dot{x}_2 - \dot{x}_1$, and the resulting dynamic equation is given in Eq. (5.26).

$$\begin{cases} m_1 \ddot{x}_1 + k_1 x_1 + c_1 \dot{x}_1 - k_2(x_2 - x_1) = F(t) - f_{sa1}(t) \\ m_2 \ddot{x}_2 + k_2(x_2 - x_1) = f_{sa1}(t) \end{cases} \quad (5.25)$$

where the semi-active variable damping control force can be expressed as $f_{sa1} = c_{sa}(\dot{x}_2 - \dot{x}_1)$, and the set state variables are $z_1 = x_1$, $z_2 = x_2$, $z_3 = \dot{x}_1$, and $z_4 = \dot{x}_2$, and the state vector is $z(t) = [z_1, z_2, z_3, z_4]^{\mathrm{T}}$. Equation (5.26) can be transformed into a state space equation of the same form as Eq. (5.22), where the expressions of F_{input}, A, B, C, and D are the same as in Eq. (5.22). The state space form of the semi-active variable damping control force can be expressed by Eq. (5.25), and the calculation strategy of semi-active variable damping control based on the continuous skyhook is given in Fig. 5.45.

Fig. 5.50 Semi-active variable damping control force f_{sa2}–f_{sa4} as a function of the MRD force (power equipment)

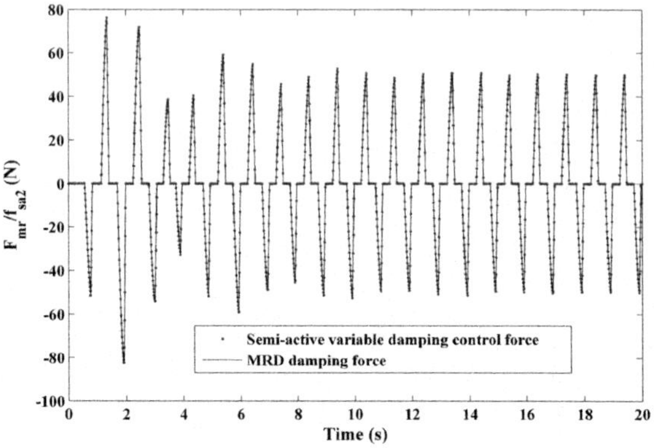

(a) Continuous balance

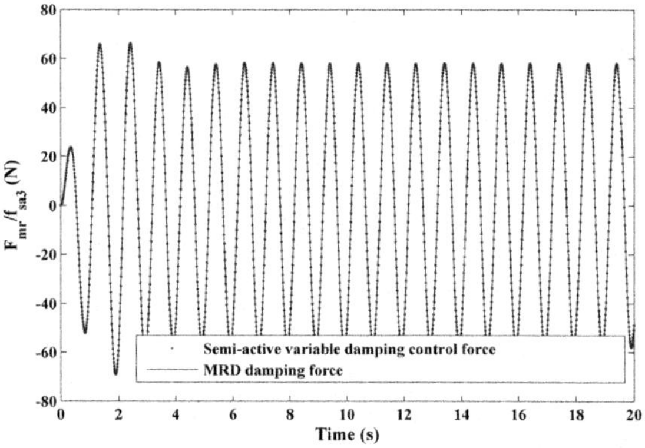

(b) On-off skyhook

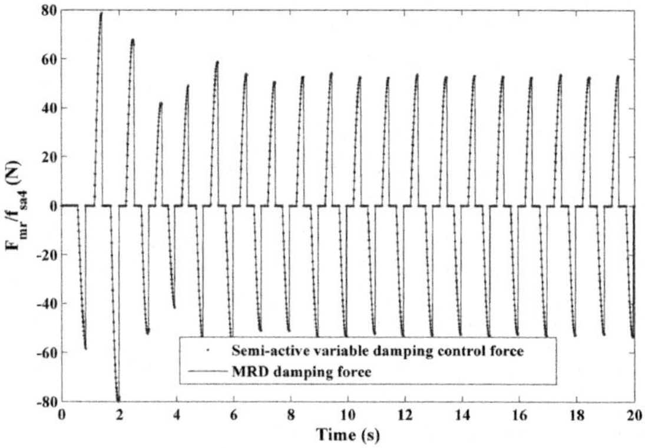

(c) On-off balance

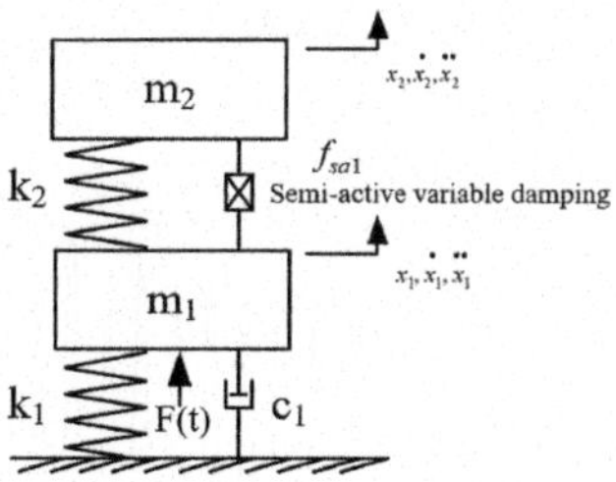

Fig. 5.51 Semi-active variable damping control for sensitive equipment

The calculation strategy for replacing the semi-active variable damping control with MRD is the same as that in Fig. 5.27. The dynamic and state space equations are the same as Eqs. (5.15) and (5.16). The calculation strategy of semi-active control based on MRD is shown in Fig. 5.21. The $F_{desired}$ in the actual MRD output calculation is the semi-active variable damping control force f_{sa1} in Fig. 5.51. The other three semi-active variable damping control strategies are based on Eqs. (5.23b) to (5.23d), and the generated semi-active control forces are recorded as f_{sa2}, f_{sa3}, and f_{sa4}.

Example 5.6 The parameters of the vibration isolation system and the MRD semi-active control system based on semi-active variable damping control are the same as that in Example 5.4. The excitation load $F(t)$ is in the form of a sine wave, and the amplitude is 5 N, and the frequency is 1 Hz. The coefficients in Eq. (5.23a) are $c_{\max} = 10$, $c_{\min} = 0.01$, and $c_{sky} = 100$.

Based on the semi-active variable damping control force, the equivalently tracked MRD damping force is shown in Fig. 5.52. The MRD can still effectively track the semi-active variable damping control force for sensitive equipment systems, and the two goals are highly consistent. The comparison between the vibration velocity of the sensitive equipment under the two control states is shown in Fig. 5.53. The MRD can perfectly reproduce the semi-active variable damping control characteristics. During the tracking f_{sa1} calculation, the control current of MRD is shown in Fig. 5.54. The tracking for semi-active variable damping control force using the other three strategies is shown in Fig. 5.55.

The MRD semi-active variable damping control force was controlled based on the other three semi-active variable damping control algorithms in Eqs. (5.23b)–(5.23d), as shown in Fig. 5.55, where the coefficients in the continuous balance, on–off skyhook, and on–off balance algorithm are set to $c_{\max} = 10$ and $c_{\min} = 0.01$.

5.5 Equivalently Achieved Optimal Active Control Strategy Using MRD Semi-Active Control

The cubic nonlinear damping force, nonlinear harmonic damping force, and theoretical semi-active variable damping control force were tracked using the proposed

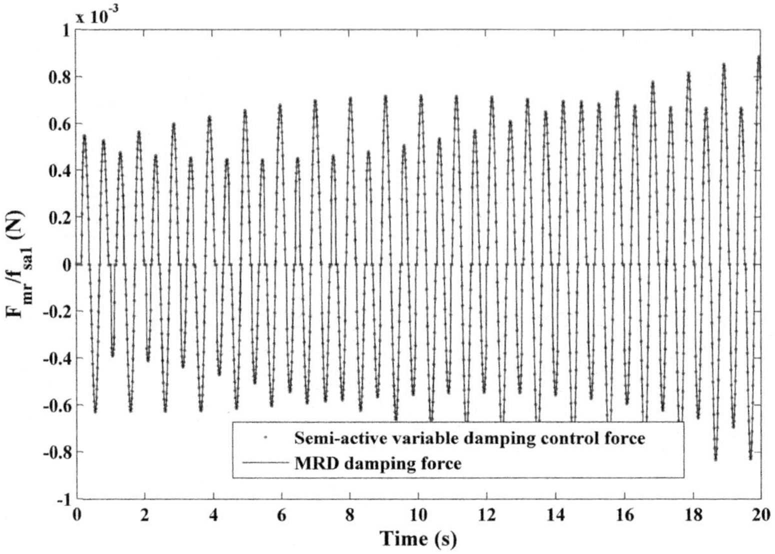

Fig. 5.52 Semi-active variable damping control force f_{sa1} as a function of the MRD force (sensitive equipment)

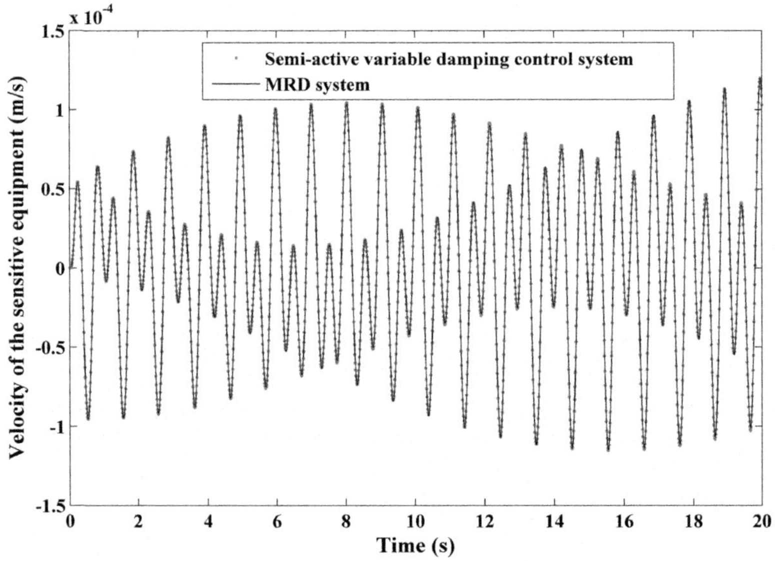

Fig. 5.53 Velocity of sensitive equipment (semi-active variable damping control)

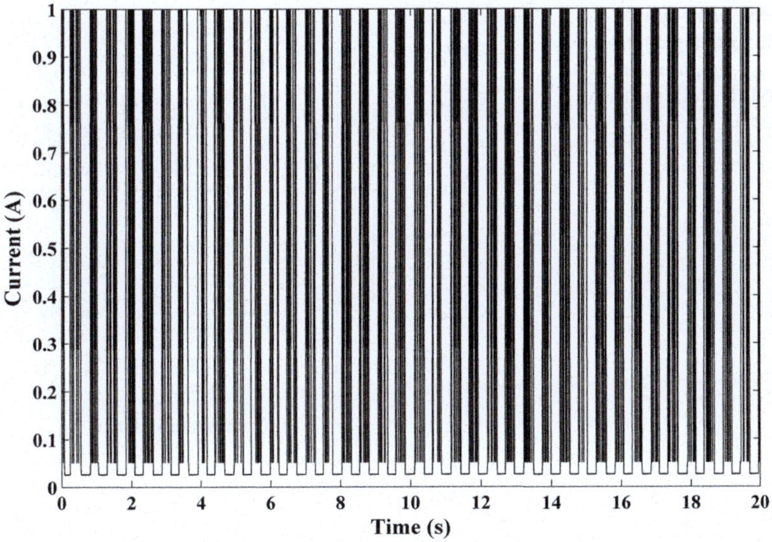

Fig. 5.54 Current supplied to the MRD (sensitive equipment/semi-active variable damping control)

MRD model and open-loop control strategy. The results demonstrated that MRD can perfectly replicate and reproduce the isolation characteristics, which has significant implications for the practical engineering application of nonlinear damping forces with certain vibration isolation advantages.

Chapter three presents various optimal active control strategies for two types of equipment. The excellent control effect of the active control strategy has been sought by vibration control [20], but its application in practical engineering is difficult due to its complex sensor/actuator design system, high energy consumption, and existing robustness.

In order to solve this problem, this research uses an optimal active control strategy to determine the optimal control force that serves as a reference for tracking utilizing the proposed MRD damping force model and open-loop control strategy. In actual engineering, the MRD semi-active control system replaces the complex active control system to achieve the same effect as an optimal active control.

5.5.1 Equivalently Achieved PSO-H$_\infty$ Optimal Active Control for Power Equipment

The PSO-H$_\infty$ optimal active control was carried out for power equipment relying on Sects. 4.4 and 4.4.2, and the optimal active control force is obtained, which is the output force F_a of the actuator under the optimal active control strategy. This

Fig. 5.55 Semi-active variable damping control force f_{sa2}–f_{sa4} as a function of the MRD force (sensitive equipment)

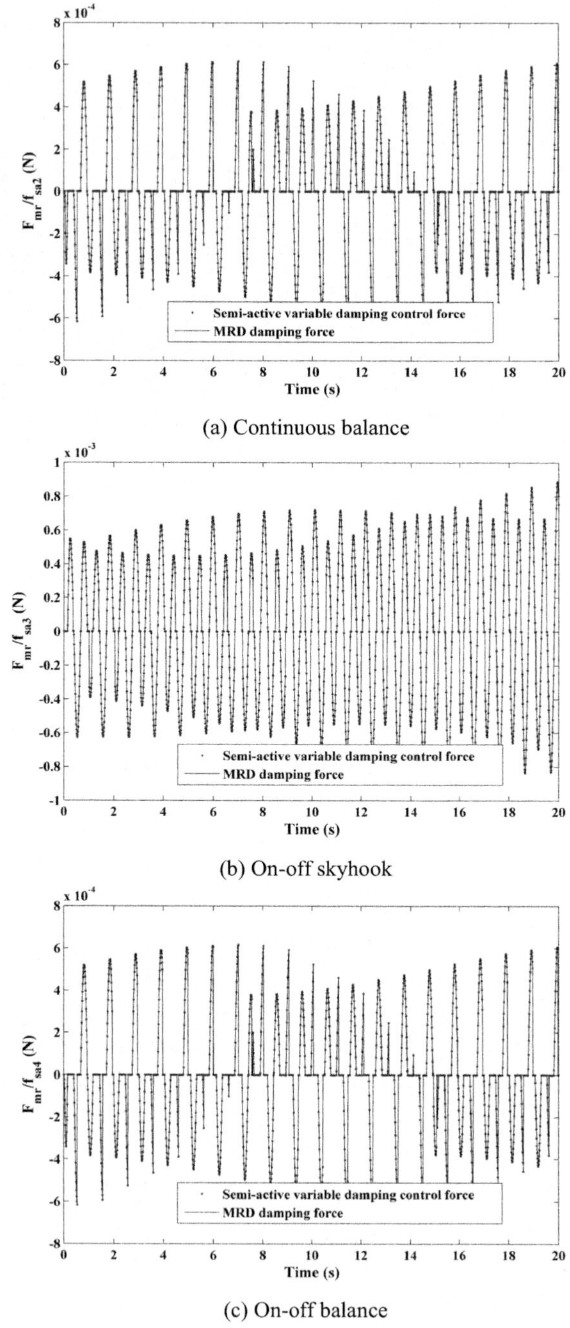

force is used as $F_{desired}$ in the MRD semi-active control, and the equivalent realization calculation is performed. Figure 5.56 reveals the semi-active control of power equipment after MRD replaced the active control system, where the damping force generated by MRD is $F_{mr}(t)$, the other parameters of the vibration isolation system have the same meanings as before, and the dynamic equation at this time is shown in Eq. (5.27).

$$\begin{cases} \mathrm{m}_1 \ddot{x}_1 + \mathrm{k}_1 x_1 + \mathrm{c}_1 \dot{x}_1 - \mathrm{c}_2(\dot{x}_2 - \dot{x}_1) - \mathrm{k}_2(x_2 - x_1) = F_{mr}(t) \\ \mathrm{m}_2 \ddot{x}_2 + \mathrm{c}_2(\dot{x}_2 - \dot{x}_1) + \mathrm{k}_2(x_2 - x_1) = F(t) - F_{mr}(t) \end{cases} \tag{5.26}$$

The state variables are:$z_1 = x_1$,$z_2 = x_2$,$z_3 = \dot{x}_1$,$z_4 = \dot{x}_2$ and the state vector is $z(t) = [z_1, z_2, z_3, z_4]^{\mathrm{T}}$, and Eq. (5.27) can be transformed into the following state space equation:

$$\begin{cases} \dot{z}(t) = \mathrm{A}z(t) + \mathrm{B}F_{input}(t) \\ y(t) = \mathrm{C}z(t) + \mathrm{D}F_{input}(t) \end{cases} \tag{5.27}$$

where $F_{input} = \begin{bmatrix} F(t) \\ F_{mr}(t) \end{bmatrix}$, $\mathrm{A} = \begin{bmatrix} 0 & 0 & 1 & 0 \\ 0 & 0 & 0 & 1 \\ -\frac{\mathrm{k}_1+\mathrm{k}_2}{\mathrm{m}_1} & \frac{\mathrm{k}_2}{\mathrm{m}_1} & -\frac{\mathrm{c}_1+\mathrm{c}_2}{\mathrm{m}_1} & \frac{\mathrm{c}_2}{\mathrm{m}_1} \\ \frac{\mathrm{k}_2}{\mathrm{m}_2} & -\frac{\mathrm{k}_2}{\mathrm{m}_2} & \frac{\mathrm{c}_2}{\mathrm{m}_2} & -\frac{\mathrm{c}_2}{\mathrm{m}_2} \end{bmatrix}$, $\mathrm{B} = \begin{bmatrix} 0 & 0 \\ 0 & 0 \\ 0 & \frac{1}{\mathrm{m}_1} \\ \frac{1}{\mathrm{m}_2} & -\frac{1}{\mathrm{m}_2} \end{bmatrix}$,

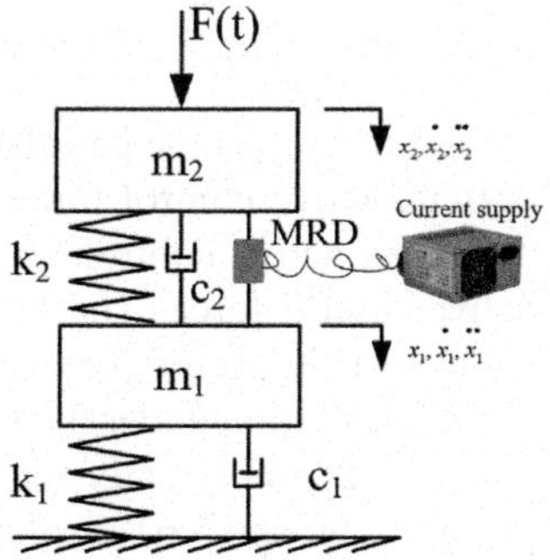

Fig. 5.56 MRD substitutes the active control system (power equipment)

$$C = \begin{bmatrix} 1 & 0 & 0 & 0 \\ 0 & 1 & 0 & 0 \\ 0 & 0 & 1 & 0 \\ 0 & 0 & 0 & 1 \\ \frac{k_2}{m_2} + \frac{k_1+k_2}{m_1} & -\left(\frac{k_2}{m_2} + \frac{k_2}{m_1}\right) & \frac{c_1+c_2}{m_1} + \frac{c_2}{m_2} & -\left(\frac{c_2}{m_1} + \frac{c_2}{m_2}\right) \end{bmatrix}, \quad D = \begin{bmatrix} 0 & 0 \\ 0 & 0 \\ 0 & 0 \\ 0 & 0 \\ \frac{1}{m_2} & -\left(\frac{1}{m_1} + \frac{1}{m_2}\right) \end{bmatrix},$$

The output variable $y(t) = \left[x_1 \; x_2 \; \dot{x}_1 \; \dot{x}_2 \; \ddot{x}_2 - \ddot{x}_1 \right]^T$. The MRD based semi-active control can be calculated according to Fig. 5.21. The actual output force $F_{mr}(t)$ of MRD is determined by the optimal active control force F_a generated by PSO-H_∞ active control.

Example 5.7 The parameter settings of the vibration isolation system of the power equipment and the PSO algorithm are configured identically to Example 4.7. The vibration load produced by the power equipment is a sine wave with an amplitude of 300 N and a frequency of 1 Hz. The constraints of the gain search range of the H_∞ state feedback controller are set to $\|K\|_\infty \leq 2 \times 10^3$. After calculation, the fitness convergence curve (Fig. 5.57) is obtained, the optimal controller derived is $K = \left[1.055 \; -1.932 \; 0.978 \; 2.000 \right] \times 10^3$, and the optimal active control force F_a based on the model shown in Fig. 4.26 is calculated. The open-loop control strategy described in Sect. 5.2.5 is substituted to calculate the actual output force of the MRD.

Figure 5.58 indicates the equivalent tracking MRD damping force using the optimal active control force F_a generated by the PSO-H_∞ active control strategy. The semi-active control implemented by MRD can effectively track the optimal active control force. The MRD damping force closely matches the optimal active control force. Figure 5.59 compares the force transmitted by the power equipment to the foundation under the two control operating conditions. The MRD semi-active control system perfectly achieves and replicates the effect of optimal active control. Figures 5.60 and 5.61 display the control current and input speed of MRD.

5.5.2 Equivalently Achieved PSO-H_∞ Optimal Active Control for Sensitive Equipment

Based on Sects. 4.4 and 4.4.3, the active control for sensitive equipment using PSO-H_∞ is examined to obtain the optimal active control force F_a. This force is utilized as $F_{desired}$ in the MRD semi-active control, and an equivalent realization calculation is performed. Figure 5.62 depicts the power equipment's semi-active control after

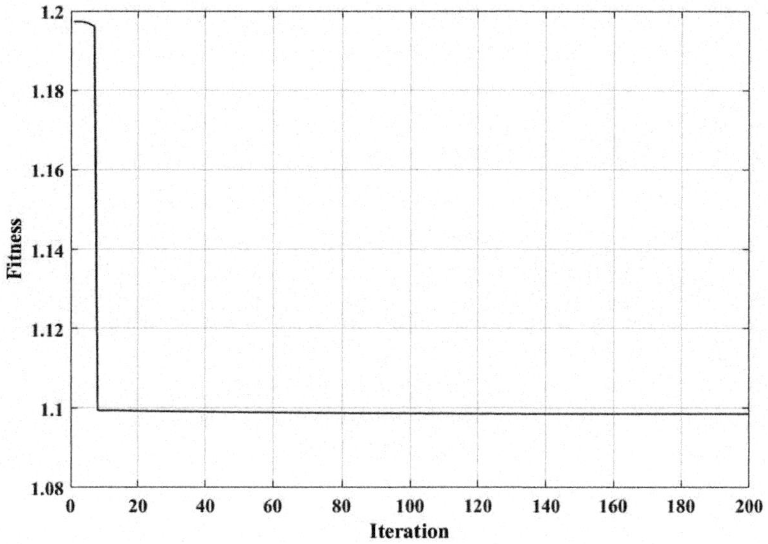

Fig. 5.57 Fitness convergence of PSO-H_∞ active control for power equipment

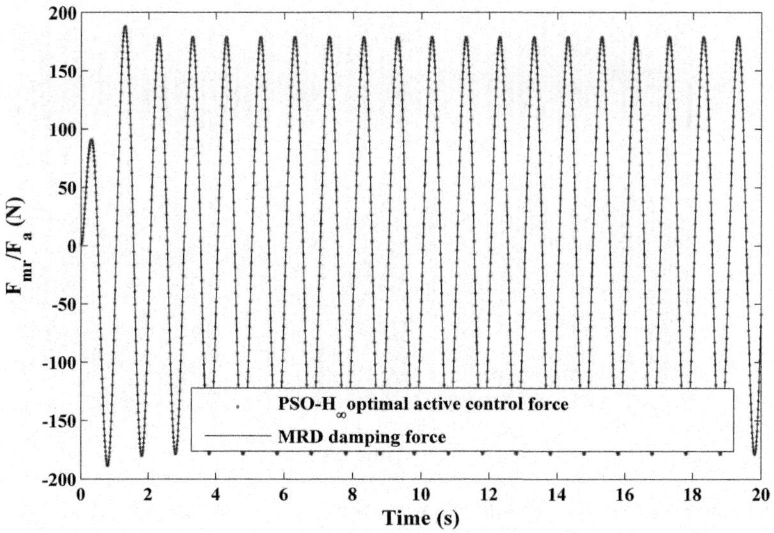

Fig. 5.58 Optimal active control force as a function of the MRD force (power equipment/PSO-H_∞)

MRD replaces the active control system. The damping force generated by MRD is $F_{mr}(t)$, and the other parameters of the vibration isolation system have the same meanings as before. The dynamic equation is shown in Eq. (5.29).

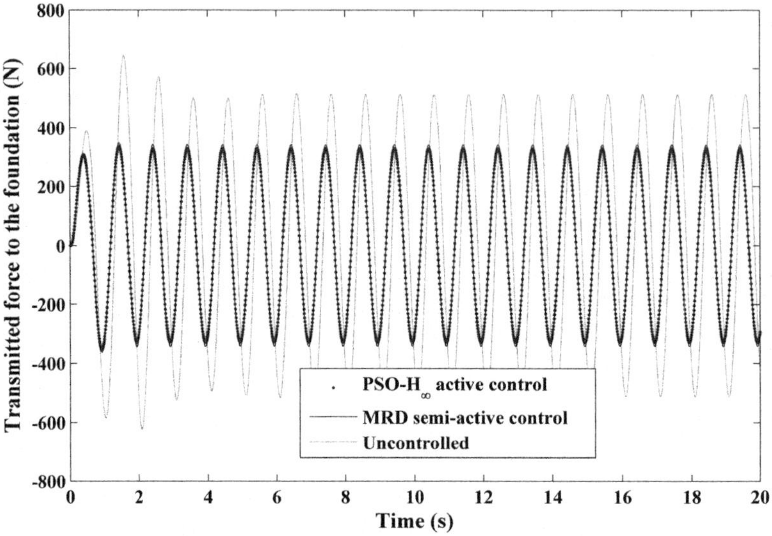

Fig. 5.59 Transmitted force from the equipment to the foundation for comparison (MRD/PSO-H_∞)

Fig. 5.60 Current supplied to the MRD (power equipment/achieving the PSO-H_∞ equivalently)

$$\begin{cases} m_1\ddot{x}_1 + k_1x_1 + c_1\dot{x}_1 - c_2(\dot{x}_2 - \dot{x}_1) - k_2(x_2 - x_1) = F(t) - F_{mr}(t) \\ m_2\ddot{x}_2 + c_2(\dot{x}_2 - \dot{x}_1) + k_2(x_2 - x_1) = F_{mr}(t) \end{cases} \tag{5.28}$$

State variables are $z_1 = x_1$, $z_2 = x_2$, $z_3 = \dot{x}_1$, $z_4 = \dot{x}_2$ and state vector is $z(t) = [z_1, z_2, z_3, z_4]^T$, and Eq. (5.29) can be transformed into the following state

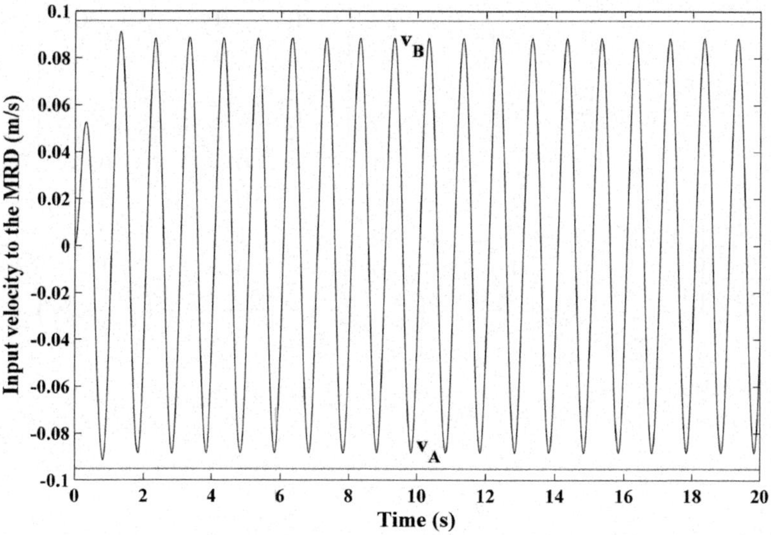

Fig. 5.61 Velocity supplied to the MRD (power equipment/ achieving the PSO-H_∞ equivalently)

Fig. 5.62 MRD substitutes the active control system (sensitive equipment)

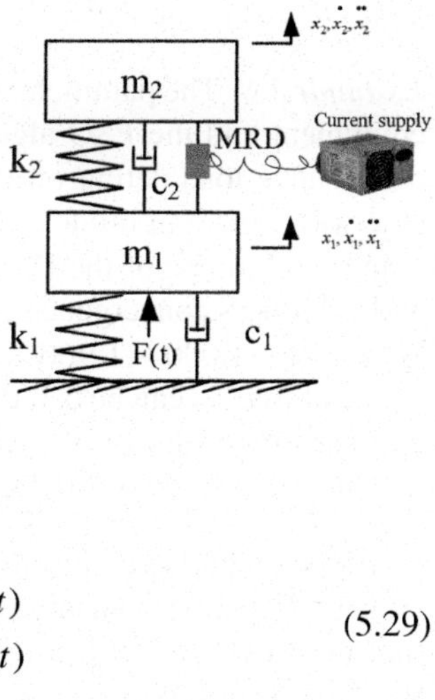

space form:

$$
\begin{cases}
\dot{z}(t) = \mathrm{A}z(t) + \mathrm{B}F_{input}(t) \\
y(t) = \mathrm{C}z(t) + \mathrm{D}F_{input}(t)
\end{cases}
\tag{5.29}
$$

where $F_{input} = \begin{bmatrix} F(t) \\ F_{mr}(t) \end{bmatrix}$, $A = \begin{bmatrix} 0 & 0 & 1 & 0 \\ 0 & 0 & 0 & 1 \\ -\frac{k_1+k_2}{m_1} & \frac{k_2}{m_1} & -\frac{c_1+c_2}{m_1} & \frac{c_2}{m_1} \\ \frac{k_2}{m_2} & -\frac{k_2}{m_2} & \frac{c_2}{m_2} & -\frac{c_2}{m_2} \end{bmatrix}$, $B = \begin{bmatrix} 0 & 0 \\ 0 & 0 \\ \frac{1}{m_1} & -\frac{1}{m_1} \\ 0 & \frac{1}{m_2} \end{bmatrix}$, $C = \begin{bmatrix} 1 & 0 & 0 & 0 \\ 0 & 1 & 0 & 0 \\ 0 & 0 & 1 & 0 \\ 0 & 0 & 0 & 1 \\ \frac{k_2}{m_2} + \frac{k_1+k_2}{m_1} & -\left(\frac{k_2}{m_2} + \frac{k_2}{m_1}\right) & \frac{c_1+c_2}{m_1} + \frac{c_2}{m_2} & -\left(\frac{c_2}{m_1} + \frac{c_2}{m_2}\right) \end{bmatrix}$, $D = \begin{bmatrix} 0 & 0 \\ 0 & 0 \\ 0 & 0 \\ 0 & 0 \\ -\frac{1}{m_1} & \frac{1}{m_1} + \frac{1}{m_2} \end{bmatrix}$,

The output variable is $y(t) = \begin{bmatrix} x_1 & x_2 & \dot{x}_1 & \dot{x}_2 & \ddot{x}_2 - \ddot{x}_1 \end{bmatrix}^T$. Figure 5.21 demonstrates MRD-based semi-active control. The actual output force $F_{mr}(t)$ calculation of MRD is based on the optimal active control force F_a generated by the PSO-H_∞ active controller of the sensitive equipment.

Example 5.8 The parameter settings of the vibration isolation system of the sensitive equipment and the PSO algorithm are configured identically to Example 4.8. The disturbance load acting on the sensitive equipment's platform (or foundation) is a sine wave, the amplitude is 5 N, and the frequency is 1 Hz. The constraints of the gain search range of the H_∞ state feedback controller are set to $\|K\|_\infty \le 4 \times 10^3$, and the fitness convergence curve (Fig. 5.63) is presented, and the optimal controller is $K = \begin{bmatrix} -4.000 & -1.966 & -1.036 & -4.000 \end{bmatrix} \times 10^3$. In addition, the optimal active control force F_a can be calculated based on the proposed model depicted in Fig. 4.26 and substituted into the open-loop control strategy of Sect. 5.2.5 to calculate the actual output force of MRD.

Based on the optimal active control force produced by the PSO-H_∞ active control strategy, Fig. 5.64 determines the equivalent tracking MRD damping force, indicating that the MRD can still effectively track the optimal active control force, and the tracked object is highly consistent with MRD force. Figure 5.65 depicts the vibration velocity of the sensitive equipment under the two control conditions, and for the micro-vibration control system of the sensitive equipment, the MRD semi-active control can still equivalently conduct the optimal active control effect. Figure 5.66 demonstrates the MRD control current.

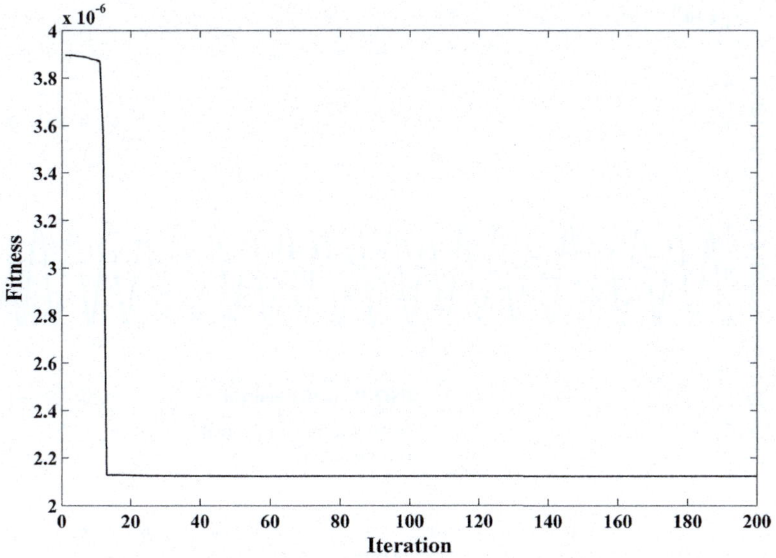

Fig. 5.63 Fitness convergence of PSO-H_∞ active control for sensitive equipment

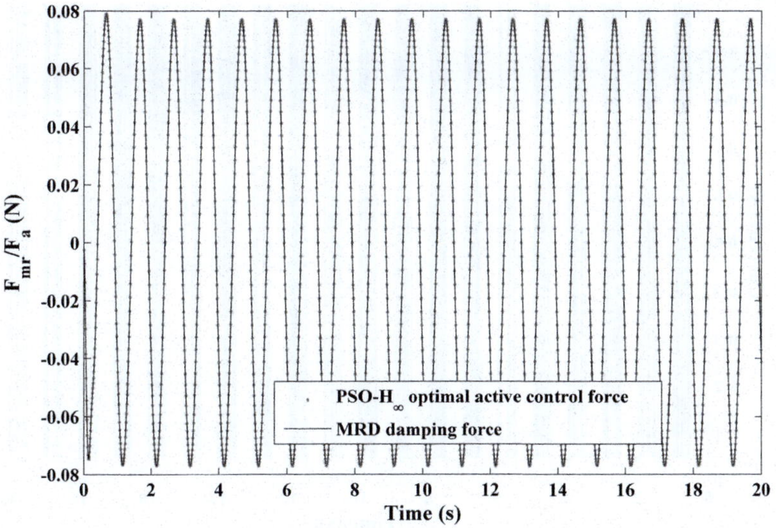

Fig. 5.64 Optimal active control force traced by MRD force (sensitive equipment/PSO-H_∞)

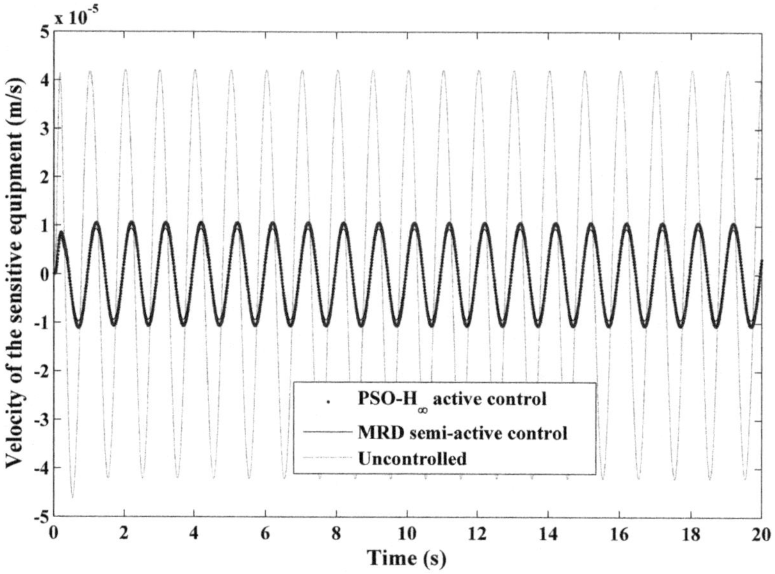

Fig. 5.65 Comparing velocity of sensitive equipment (MRD/PSO-H_∞)

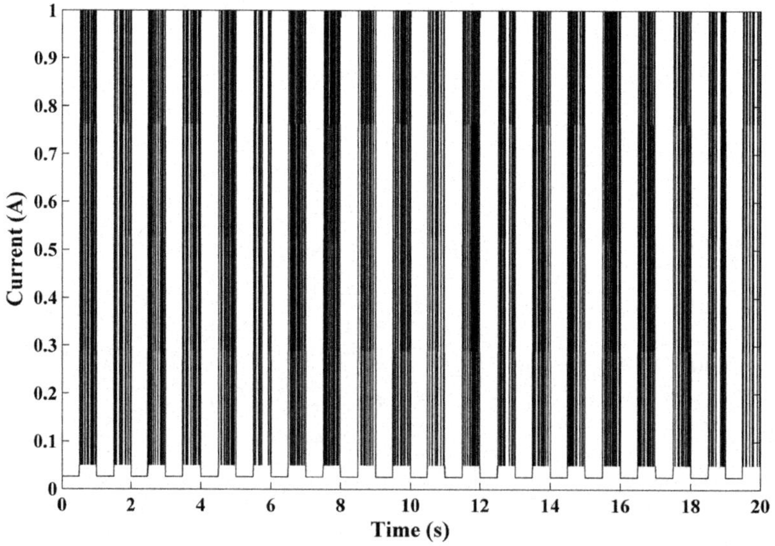

Fig. 5.66 Current supplied to the MRD (sensitive equipment/achieving the PSO-H_∞ equivalently)

5.6 Equivalently Achieved Multi-Objective Control Output Active Control Strategy Using MRD Semi-Active Control

Section 5.5 constructs the MRD semi-active control system to perform optimal active control characteristics equivalently and conducts semi-active control research on power and sensitive equipment with the PSO-H_∞ active controller as the objective. The optimal active control force F_a is employed as the input $F_{desired}$ of the MRD, and the result confirms that the proposed method is effective. In reality, the other optimal active control strategies introduced in Chap. 4 are efficient and have the same principles.

Chapter 4.7 presents a multi-objective control output active control strategy using the MOPSO algorithm. This strategy eliminates the deficiencies of traditional active control that only considers a single control output and the multi-objective characteristics of the active control system. Hence, this section develops the equivalent realization of the multi-objective control output active control strategy relying on the MRD semi-active control system.

5.6.1 Equivalently Achieved Multi-Objective Control Output Active Control for Power Equipment

Based on the study of the multi-objective control output active control of power equipment in Sects. 4.7 and 4.7.1, the optimal active control force F_a that meets the multi-objective control characteristics was derived, and the definition and calculation strategy of multi-objective control output active control was described in Sect. 4.7.1. The optimal active control force is utilized as $F_{desired}$ in MRD semi-active control, and the equivalent realization computation was conducted. Figure 5.56 displays the semi-active control strategy after MRD replaced the active control system, and the dynamic equation of the MRD semi-active control system is identical Eq. (5.27).

Example 5.9 The parameter settings of the vibration isolation system of the power equipment and the MOPSO algorithm are configured identically to Example 4.13. The vibration load generated by the power equipment is a sine wave, the amplitude is 300 N, and the frequency is 1 Hz. The constraints of the gain search range of the H_∞ state feedback controller are set to $\|K\|_\infty \leq 2 \times 10^3$, and the multi-objective optimization Pareto frontier (Fig. 5.67) is obtained. The optimal controller is $K = \begin{bmatrix} -0.463 & -1.988 & 1.651 & 1.894 \end{bmatrix} \times 10^3$, and the optimal active control force F_a based on the model illustrated in Fig. 5.26 is computed and substituted into the open-loop control strategy in Sect. 5.2.5 to determine the MRD's actual output force.

Figure 5.68 reveals the MRD damping force for equivalent tracking using the optimal active control force F_a generated by the multi-objective control output active control strategy. The semi-active control system implemented by MRD can

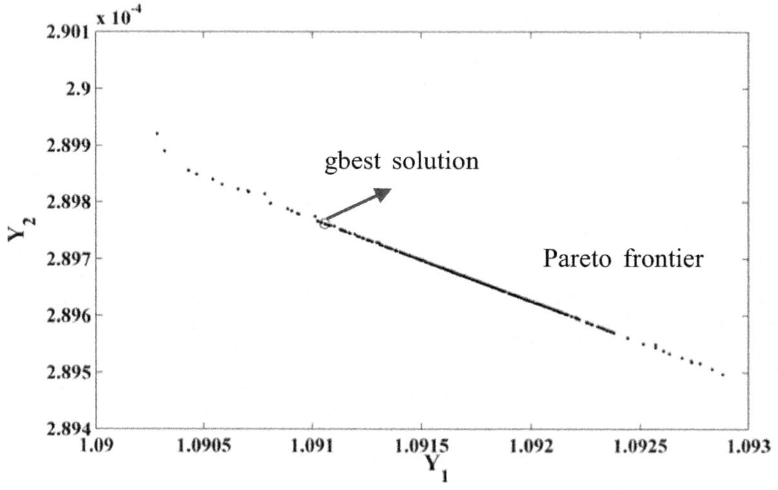

Fig. 5.67 Pareto frontier of multi-objective control output active control for power equipment

effectively track the optimal active control force, and the achieved MRD damping force is highly consistent with the tracked object. Under the two control conditions, Fig. 5.69 compares the force transmitted by the power equipment to the foundation and the vibration velocity of the equipment. The MRD semi-active control system can perform the multi-objective control output active control characteristics perfectly. Figures 5.70 and 5.71 indicates the control current and input velocity of MRD.

5.6.2 *Equivalently Achieved Multi-Objective Control Output Active Control for Sensitive Equipment*

Based on the study of the multi-objective control output active control of sensitive equipment in Sects. 4.7 and 4.7.2, the optimal active control force F_a that satisfies the multi-objective control characteristics was determined, and the definition and calculation strategy of multi-objective control output active control was described in Sect. 4.7.2. The optimal active control force was used as $F_{desired}$ in MRD semi-active control, and the equivalent realization calculation was performed. Figure 5.62 depicts the semi-active control strategy after MRD replaced the active control system, and the dynamic equation of the MRD semi-active control system is identical to Eq. (5.29).

Example 5.10 The parameter setting of the vibration isolation system of the sensitive equipment and the MOPSO algorithm are configured identically to Example 4.14. The disturbance load acting on the sensitive equipment's platform (or foundation) is a sine wave with an amplitude of 5 N and a frequency of 1 Hz. The constraints of the gain search range of the H_∞ state feedback controller are set to $\|K\|_\infty \leq 4 \times 10^3$,

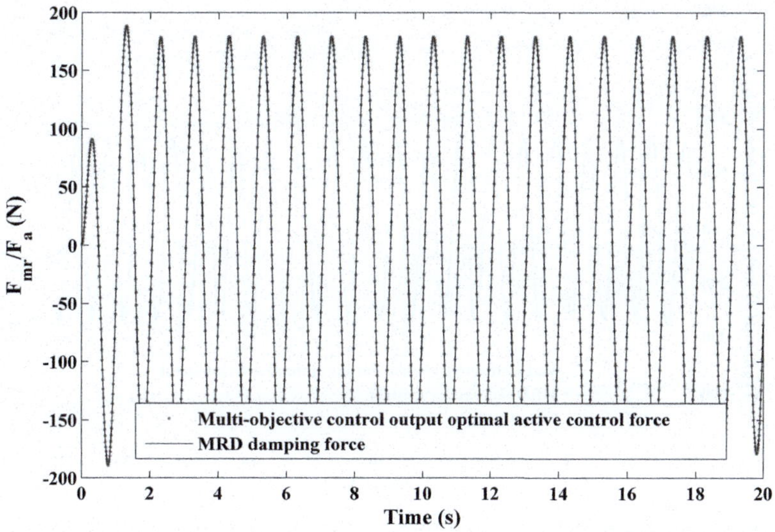

Fig. 5.68 Optimal active control force as a function of the MRD force (power equipment/multi-objective control output)

and the Pareto frontier of multi-objective optimization (Fig. 5.72) is derived, and the optimal controller is $K = \left[-3.907\ -3.255\ 1.461\ -3.834\right] \times 10^3$. The optimal active control force based on the model indicated in Fig. 4.26 is calculated and substituted into the open-loop control strategy described in Sect. 5.2.5 to determine the actual output force of the MRD.

Figure 5.73 illustrates the equivalent tracking MRD damping force relying on the optimal active control force F_a generated by the multi-objective control output active control strategy. The MRD semi-active control system implemented effectively tracks the optimal active control force, and the MRD damping force curve is highly consistent with the tracked object. Figure 5.74 depicts the sensitive equipment vibrational velocity and the relative displacement between the equipment and the placement platform for the two control conditions. The MRD's high-precision active control force tracking ensures that the active control effect of the sensitive equipment's multi-objective control output can be perfectly achieved. Figure 5.75 displays the MRD control current.

5.7 Summary

A 6-segment cubic polynomial model that can accurately predict the MRD output is proposed utilizing the MRD mechanical test data. This model overcomes the accuracy issues associated with low-order polynomial fitting and eliminates the possibility of

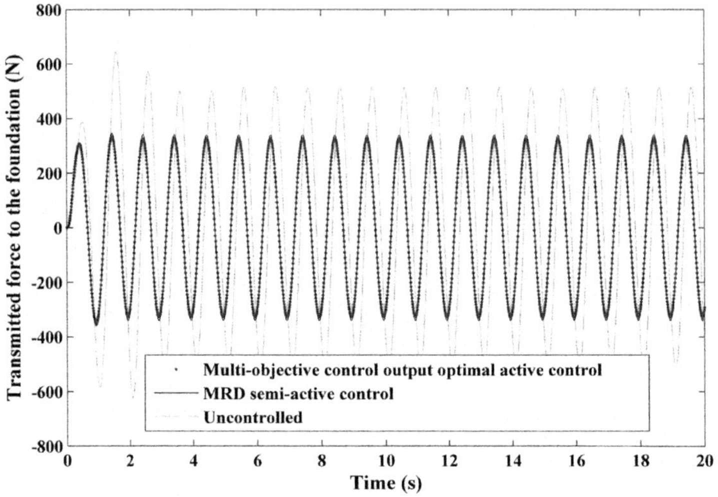

(a) Comparison of transmitted force to the power equipment foundation

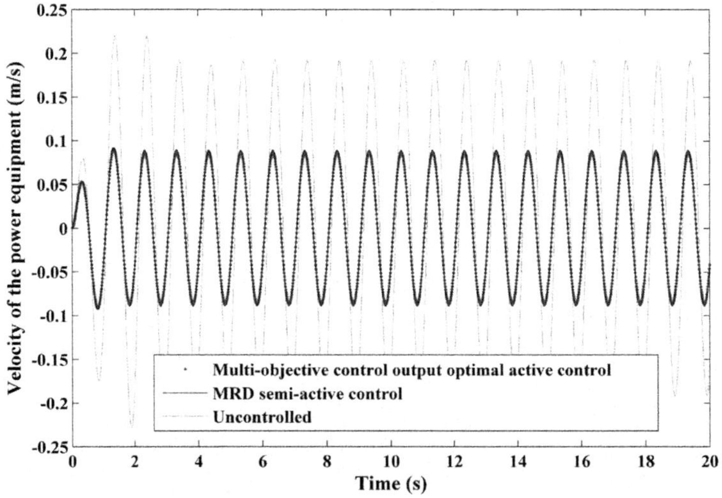

(b) Comparison of vibration velocity of power equipment

Fig. 5.69 Comparing vibration responses of power equipment

the Runge phenomenon in high-order fitting. The predicted MRD output is highly consistent with the test data, and the error is minimal. In addition, the model does not require the fitting polynomial coefficients to be linear with the control current, but it considers the cubic polynomial that conforms more closely to the fitting law. This study confirms the necessity of this consideration. Based on this model, an MRD

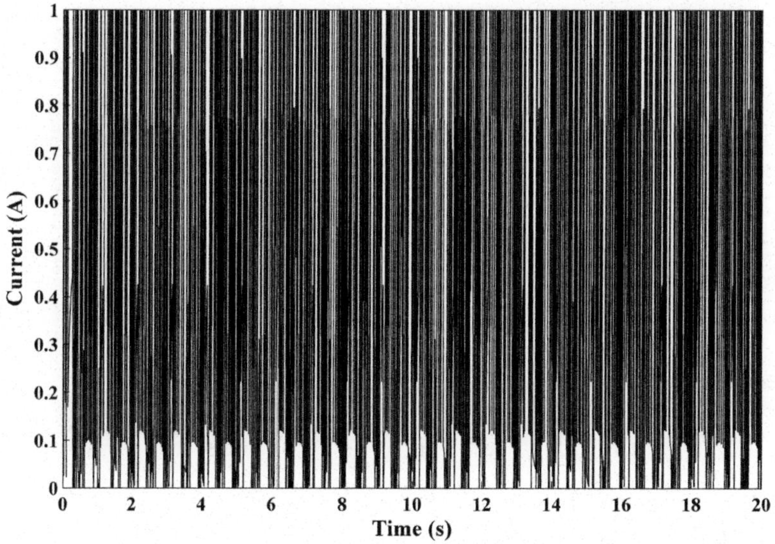

Fig. 5.70 Current supplied to the MRD (power equipment/achieving the multi-objective control output equivalently)

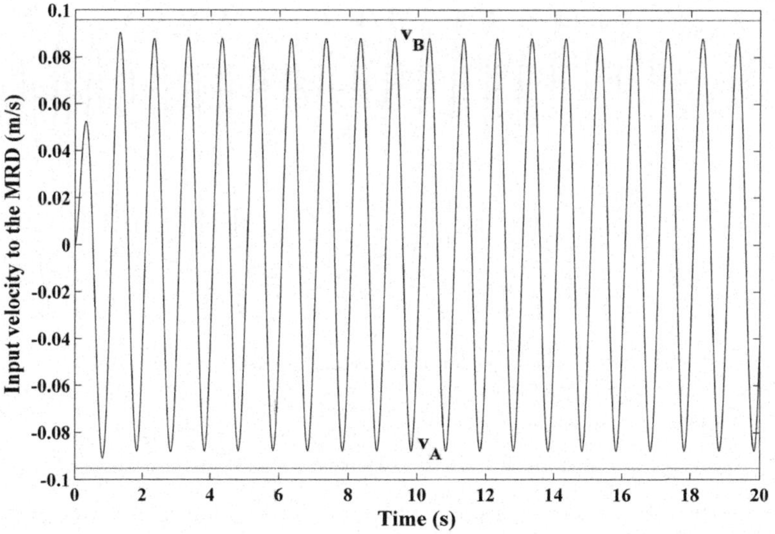

Fig. 5.71 Velocity supplied to the MRD (power equipment/achieving the multi-objective control output equivalently)

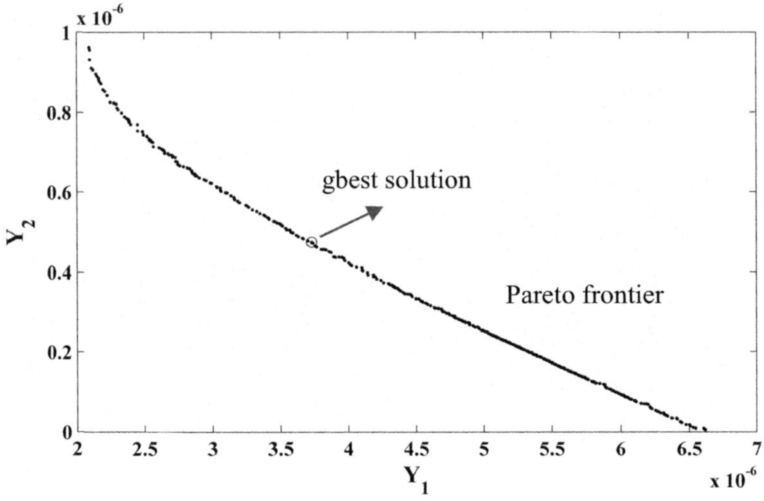

Fig. 5.72 Pareto frontier of multi-objective control output active control for sensitive equipment

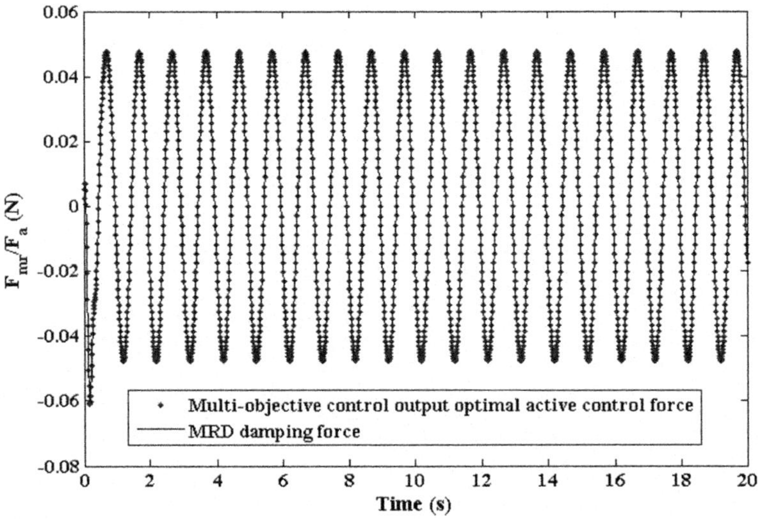

Fig. 5.73 Optimal active control force as a function of the MRD force (sensitive equipment/multi-objective control output)

open-loop control strategy is introduced, which uses the inverse solution of the ideal input force to determine the MRD control current.

Nonlinear damping is inevitably presented in the vibration isolation system, and some nonlinear damping has superior vibration isolation characteristics than linear damping. This study conducted tracking research on the nonlinear damping force

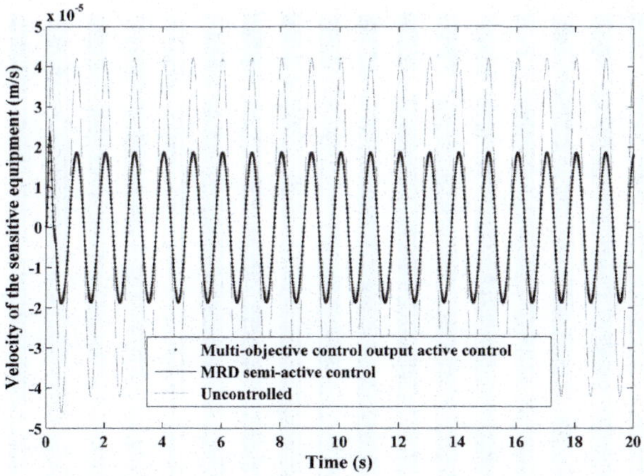

(a) Comparison of sensitive equipment vibrational velocity

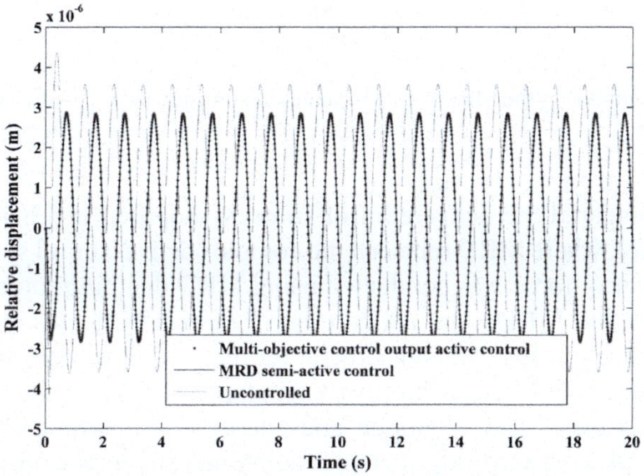

(b) Comparing the relative displacement of sensitive equipment and the placement platform

Fig. 5.74 Vibration responses of sensitive equipment for comparison

relying on the proposed MRD model and open-loop control strategy. The results indicated that the MRD can precisely track the nonlinear damping force, while the target control characteristics are nearly identical to the nonlinear damping effect. Accordingly, it is possible to employ MRD to equivalently achieve specific effects, such as frequency-domain vibration isolation characteristics of nonlinear damping. Recently, research on semi-active control strategies has increased in scope. An important topic is how to perfectly apply the theoretical method to actual engineering in practical engineering. Based on the proposed MRD model and open-loop control strategy, this study successfully tracked the theoretical semi-active variable damping control force, and the target control characteristics were equivalently achieved.

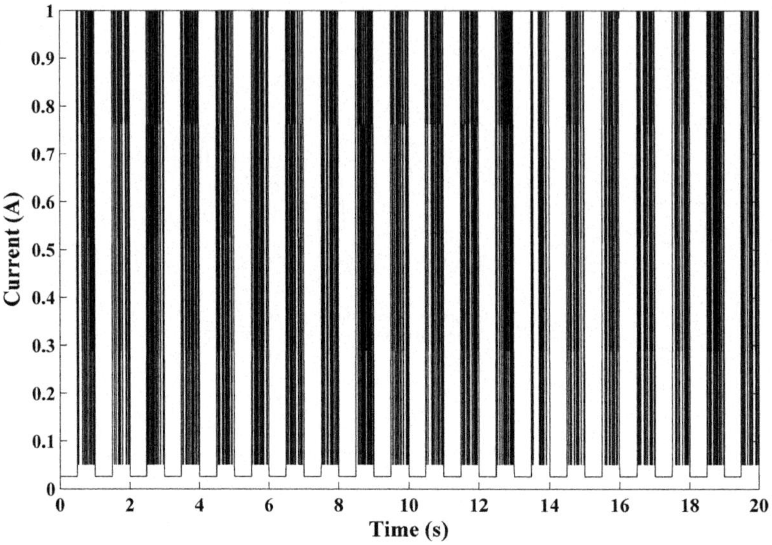

Fig. 5.75 Current supplied to the MRD (sensitive equipment/achieving the multi-objective control output equivalently)

This study compared the active control strategy proposed in Chap. 3 to MRD semi-active control, where MRD replaced the complex active control system, and the optimal active control force was conducted using MRD damping force. The results demonstrated that the MRD semi-active control system perfectly recognizes the active control characteristics, which is crucial for the vibration control technologies that favor the active control effect but fear the complex design of the active control system, poor robustness, and economic and energy consumption. Promoting the realized system will serve as a bridge between active and practical semi-active control and provide convenient conditions for the future realizing semi-active control based on the optimal active control guidance, which has significant practical implications.

References

1. Semi-active vibration control based on MR damper [D]. Tsinghua University (2009)
2. Stanway R, Sproston JL, El-Wahed AK (1996) Applications of electro-rheological fluids in vibration control: a survey [J]. Smart Mater Struct 5(4):464
3. Pang L, Kamath GM, Wereley NM (1998) Analysis and testing of a linear stroke magnetorheological damper[C]. In: Proceedings on 39th AIAA/ASME/ASCE/AHS/ASC structures, structural dynamics, and materials conference and exhibit and AIAA/ASME/AHS adaptive structures forum, 2841–2856
4. Spencer BF Jr, Dyke SJ, Sain MK et al (1997) Phenomenological model for magnetorheological dampers[J]. J Eng Mech 123(3):230–238
5. Dyke SJ, Spencer BF Jr, Sain MK et al (1996) Modeling and control of magnetorheological dampers for seismic response reduction[J]. Smart Mater Struct 5(5):565

6. Choi SB, Lee SK, Park YP, et al (2001) A hysteresis model for the field-dependent damping force of a magnetorheological damper [J]. J Sound Vibration 245(2):375.383
7. Zhou TM, Chen EW, Lu YM, Liu ZS, Chen WW (2014) Modified polynomial model and its verification for a MR damper [J]. J Vibration Shock 7(7):221–226
8. Mallik AK, Kher V, Puri M et al (1999) On the modelling of non-linear elastomeric vibration isolators [J]. J Sound Vib 219(2):239–253
9. Peng ZK, Lang ZQ (2008) The effects of nonlinearity on the output frequency response of a passive engine mount [J]. J Sound Vib 318(1):313–328
10. Jing XJ, Lang ZQ, Billings SA et al (2008) Frequency domain analysis for suppression of output vibration from periodic disturbance using nonlinearities [J]. J Sound Vib 314(3):536–557
11. Lang ZQ, Billings SA, Tomlinson GR et al (2006) Analytical description of the effects of system nonlinearities on output frequency responses: a case study [J]. J Sound Vib 295(3):584–601
12. Lang ZQ, Billings SA, Yue R et al (2007) Output frequency response function of nonlinear Volterra systems[J]. Automatica 43(5):805–816
13. Ho C, Lang ZQ, Sapiński B et al (2013) Vibration isolation using nonlinear damping implemented by a feedback-controlled MR damper [J]. Smart Mater Struct 22(10):1–12
14. Iba D (2011) Vibration control using harmonically-varying damping [J]. J Syst Des Dyn 5(5):727–736
15. Hirohata S, Iba D (2012) Frequency response analysis of multi-degree-of-freedom system with harmonically varying damping[C]. SPIE smart structures and materials+ nondestructive evaluation and health monitoring. international society for optics and photonics, 83453R-15
16. Iba D, Masuda A, Sone A (2008) Resonance by parametric excitation of variable damping [J]. J Syst Des Dyn 2(1):331–341
17. Liu Y, Waters TP, Brennan MJ (2005) A comparison of semi-active damping control strategies for vibration isolation of harmonic disturbances [J]. J Sound Vib 280(1):21–39
18. Liu Y (2004) Semi-active damping control for vibration isolation of base disturbances [D]. University of Southampton
19. Cafferty S, Worden K, Tomlinson G (1995) Characterization of automotive shock absorbers using random excitation [J]. Proc Inst Mech Eng Part D: J Autom Eng 209(4):239–248
20. Ou JP (2003) Structural vibration control, active, semi-active and intelligent control [M]. Science Press, Beijing

Chapter 6
Vibration Control for Equipment-Structure

Abstract In this chapter, the equipment-frame structure is used as a composite vibration system, and the TMD and ATMD control strategies are introduced to investigate the approaches to control power equipment-induced structural vibration and the vibration on sensitive equipment induced by pedestrian loads in the structure. Based on the MRD mechanical model and open-loop control strategy proposed in Chap. 5, the MRD is used to replace the active control system, and the SATMD control strategy to equivalently achieve the optimal ATMD control characteristics is developed.

In industrial engineering, the vibration of power equipment causes severe disturbance to the auxiliary structure, affects the structure's life, and even threatens the safety of life and property. The external environment's vibration, such as traffic loads, explosions, and pedestrian loads in the structure, affects the normal use of sensitive equipment. Therefore, studying the vibration control of the equipment and building structure composite systems is of great significance.

In this chapter, the power equipment-frame structure is used as a composite vibration system, and the TMD and ATMD control strategies are introduced to investigate the approaches to control power equipment-induced structural vibration. Since the sensitive equipment-frame structure is a composite vibration system, TMD and ATMD control strategies are implemented to assess the vibration control method of sensitive equipment induced by pedestrian loads in the structure. Generally, TMD does not consider the control energy input, whereas ATMD considers the control energy input based on the TMD control. Additionally, the control method adopted herein is the optimal active control strategy proposed in Chap. 3.

Based on the MRD mechanical model and open-loop control strategy proposed in Chap. 5 the MRD is used to replace the active control system, and the SATMD control strategy to equivalently achieve the optimal ATMD control characteristics is developed.

W. Huang and J. Xu, *Optimized Engineering Vibration Isolation, Absorption and Control*, Springer Tracts in Civil Engineering, https://doi.org/10.1007/978-981-99-2213-0_6

6.1 Vibration Control Strategy for Power Equipment-Structure

6.1.1 TMD/ATMD Vibration Control for Power Equipment-Structure

Currently, tuned mass damper (TMD) control technology [1–3] has been widely used in structural seismic and wind vibrations control. Since TMD does not consider the control energy input, to improve the TMD's performance, scholars have proposed an active tuned mass damper (ATMD) control technology [4–6]. The ATMD comprises a mass and an active controller (in this study, it refers to the optimal active control strategy proposed in Chap. 3) and can also be equipped with springs and dampers.

The TMD/ATMD vibration control system for the power equipment-frame structure is shown in Fig. 6.1, where $\mathrm{m_p}$ is the mass of the power equipment, $\mathrm{k_p}$ and $\mathrm{c_p}$ are the stiffness and damping of the vibration isolation system for equipment, respectively, x_p, $\dot{x}_p$ and $\ddot{x}_p$ are the displacement, velocity, and acceleration of equipment, respectively, $\mathrm{m_b}$ is the mass of the platform or foundation where the equipment is placed, $\mathrm{k_b}$ and $\mathrm{c_b}$ are the stiffness and damping of the foundation or support system, respectively, x_b, $\dot{x}_b$ and $\ddot{x}_b$ are the displacement, velocity, and acceleration of the foundation or support system, respectively, $\mathrm{m_1}$, $\mathrm{m_2}$, and $\mathrm{m_3}$ are the mass of the 1st to the 3rd layers of structure, respectively, $\mathrm{k_1}$, $\mathrm{k_2}$, and $\mathrm{k_3}$ are the stiffness of 1st to the 3rd layers of the structure, respectively, $\mathrm{c_1}$, $\mathrm{c_2}$, and $\mathrm{c_3}$ are the damping of the 1st to the 3rd layers of the structure, respectively, x_1, x_2 and x_3 are the displacement of the 1st to the 3rd layers of the structure, respectively, $\dot{x}_1$, $\dot{x}_2$ and $\dot{x}_3$ are the velocity of the 1st to the 3rd layers of the structure, respectively, $\ddot{x}_1$, $\ddot{x}_2$ and $\ddot{x}_3$ are the acceleration of the 1st to the 3rd layers of the structure, respectively, $\mathrm{m_d}$ is the mass of the TMD/ATMD, $\mathrm{k_d}$ and $\mathrm{c_d}$ are the stiffness and damping of the TMD/ATMD, respectively, $F_a(t)$ is the active control force generated by actuator when the ATMD is utilized for structural vibration control, and $F_d(t)$ is the vibration load generated by the power equipment.

For the TMD/ATMD control system of the power equipment-frame structure shown in Fig. 6.1, the dynamic equation can be expressed as follows:

$$\begin{cases} \mathrm{m_1}\ddot{x}_1 + (\mathrm{c_1}+\mathrm{c_2}+\mathrm{c_b})\dot{x}_1 - \mathrm{c_2}\dot{x}_2 - \mathrm{c_b}\dot{x}_b + (\mathrm{k_1}+\mathrm{k_2}+\mathrm{k_b})x_1 - \mathrm{k_2}x_2 - \mathrm{k_b}x_b = 0 \\ \mathrm{m_2}\ddot{x}_2 - \mathrm{c_2}\dot{x}_1 + (\mathrm{c_2}+\mathrm{c_3})\dot{x}_2 - \mathrm{c_3}\dot{x}_3 - \mathrm{k_2}x_1 + (\mathrm{k_2}+\mathrm{k_3})x_2 - \mathrm{k_3}x_3 = 0 \\ \mathrm{m_3}\ddot{x}_3 - \mathrm{c_3}\dot{x}_2 + \mathrm{c_3}\dot{x}_3 - \mathrm{k_3}x_2 + \mathrm{k_3}x_3 + \mathrm{c_d}\dot{x}_3 + \mathrm{k_d}x_3 - \mathrm{c_d}\dot{x}_d - \mathrm{k_d}x_d = -F_a(t) \\ \mathrm{m_b}\ddot{x}_b - \mathrm{c_b}\dot{x}_1 + (\mathrm{c_b}+\mathrm{c_p})\dot{x}_b - \mathrm{c_p}\dot{x}_p - \mathrm{k_b}x_1 + (\mathrm{k_b}+\mathrm{k_p})x_b - \mathrm{k_p}x_p = 0 \\ \mathrm{m_p}\ddot{x}_p - \mathrm{c_p}\dot{x}_b + \mathrm{c_p}\dot{x}_p - \mathrm{k_p}x_b + \mathrm{k_p}x_p = F_d(t) \\ \mathrm{m_d}\ddot{x}_d + \mathrm{c_d}\dot{x}_d + \mathrm{k_d}x_d - \mathrm{c_d}\dot{x}_3 - \mathrm{k_d}x_3 = F_a(t) \end{cases} \tag{6.1}$$

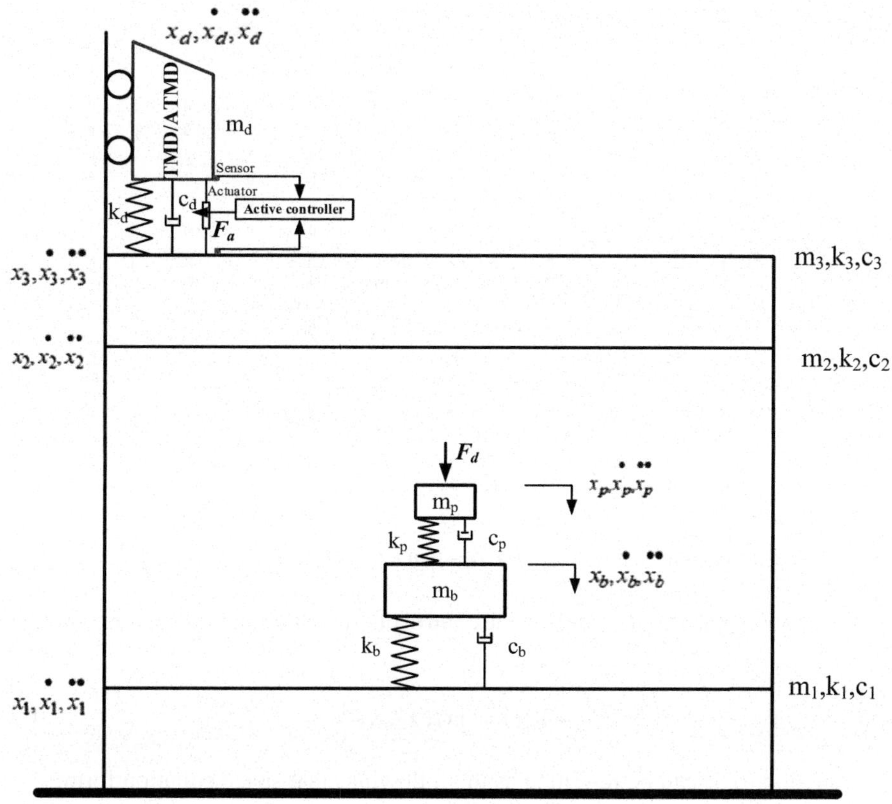

Fig. 6.1 TMD/ATMD vibration control system for power equipment-frame structure

In Eq. (6.1), when considering the active control force $F_a(t)$, it is the ATMD control system. Otherwise, it is the dynamic response of the TMD control system. For the ATMD control system, the state variables are set to $z_1 = x_1, z_2 = x_2, z_3 = x_3$, $z_4 = x_b$, $z_5 = x_p$, $z_6 = x_d$, $z_7 = \dot{x}_1$, $z_8 = \dot{x}_2$, $z_9 = \dot{x}_3$, $z_{10} = \dot{x}_b$, $z_{11} = \dot{x}_p$, and $z_{12} = \dot{x}_d$, and the state vector is obtained as:

$$z = \{z_1, z_2, z_3, z_4, z_5, z_6, z_7, z_8, z_9, z_{10}, z_{11}, z_{12}\}^{\mathrm{T}} \tag{6.2}$$

The state space form of Eq. (6.2) is described as

$$\dot{z} = \mathrm{A}z + \mathrm{B}_1 F_{\mathrm{d}}(t) + \mathrm{B}_2 F_a(t) \tag{6.3}$$

where

$$A=\begin{bmatrix}
0 & 0 & 0 & 0 & 0 & 0 & 1 & 0 & 0 & 0 & 0 & 0\\
0 & 0 & 0 & 0 & 0 & 0 & 0 & 1 & 0 & 0 & 0 & 0\\
0 & 0 & 0 & 0 & 0 & 0 & 0 & 0 & 1 & 0 & 0 & 0\\
0 & 0 & 0 & 0 & 0 & 0 & 0 & 0 & 0 & 1 & 0 & 0\\
0 & 0 & 0 & 0 & 0 & 0 & 0 & 0 & 0 & 0 & 1 & 0\\
0 & 0 & 0 & 0 & 0 & 0 & 0 & 0 & 0 & 0 & 0 & 1\\
-\frac{k_1+k_2+k_b}{m_1} & \frac{k_2}{m_1} & 0 & \frac{k_b}{m_1} & 0 & 0 & -\frac{c_1+c_2+c_b}{m_1} & \frac{c_2}{m_1} & 0 & \frac{c_b}{m_1} & 0 & 0\\
\frac{k_2}{m_2} & -\frac{k_2+k_3}{m_2} & \frac{k_3}{m_2} & 0 & 0 & 0 & \frac{c_2}{m_2} & -\frac{c_2+c_3}{m_2} & \frac{c_3}{m_2} & 0 & 0 & 0\\
0 & \frac{k_3}{m_3} & -\frac{k_3+k_d}{m_3} & 0 & 0 & \frac{k_d}{m_3} & 0 & \frac{c_3}{m_3} & -\frac{c_3+c_d}{m_3} & 0 & 0 & \frac{c_d}{m_3}\\
\frac{k_b}{m_b} & 0 & 0 & -\frac{k_b+k_p}{m_b} & \frac{k_p}{m_b} & 0 & \frac{c_b}{m_b} & 0 & 0 & -\frac{c_b+c_p}{m_b} & \frac{c_p}{m_b} & 0\\
0 & 0 & 0 & \frac{k_p}{m_p} & -\frac{k_p}{m_p} & 0 & 0 & 0 & 0 & \frac{c_p}{m_p} & -\frac{c_p}{m_p} & 0\\
0 & 0 & \frac{k_d}{m_d} & 0 & 0 & -\frac{k_d}{m_d} & 0 & 0 & \frac{c_d}{m_d} & 0 & 0 & -\frac{c_d}{m_d}
\end{bmatrix},$$

$B_1=[0,0,0,0,0,0,0,0,0,0,1/m_p,0]^T$, and $B_2=[0,0,0,0,0,0,0,0,-1/m_3,0,0,1/m_d]^T$.

Under the TMD/ATMD control, the control output of the frame structure is defined as

$$Y(t)=[x_1, x_2, x_3]^T \tag{6.4}$$

Equation (6.4) can be rewritten as the following state space equation form:

$$Y(t)=C_1 z+D_{11}F_d(t)+D_{12}F_a(t) \tag{6.5}$$

where

$$C_1=\begin{bmatrix} 1\,0\,0\,0\,0\,0\,0\,0\,0\,0\,0\,0\\ 0\,1\,0\,0\,0\,0\,0\,0\,0\,0\,0\,0\\ 0\,0\,1\,0\,0\,0\,0\,0\,0\,0\,0\,0 \end{bmatrix}$$

$D_{11}=[0,0,0]^T$, and $D_{12}=[0,0,0]^T$.

In practical engineering, it is generally difficult to observe the entire state of the state vector. Therefore, this study assumes the following observation output:

$$y(t)=\left[x_b, x_p, \dot{x}_b, \dot{x}_p, \dot{x}_d-\dot{x}_3\right]^T \tag{6.6}$$

The state space form of Eq. (6.6) is written as

$$y(t)=C_2 z+D_{21}F_d(t)+D_{22}F_a(t) \tag{6.7}$$

where

$$
C_2 = \begin{bmatrix} 0\,0\,0\,1\,0\,0\,0\,0 & 0 & 0\,0\,0 \\ 0\,0\,0\,0\,1\,0\,0\,0 & 0 & 0\,0\,0 \\ 0\,0\,0\,0\,0\,0\,0\,0 & 0 & 1\,0\,0 \\ 0\,0\,0\,0\,0\,0\,0\,0 & 0 & 0\,1\,0 \\ 0\,0\,0\,0\,0\,0\,0\,0 & -1 & 0\,0\,1 \end{bmatrix},
$$

$$
D_{21} = [0, 0, 0, 0, 0]^{T} \text{ and } D_{22} = [0, 0, 0, 0, 0]^{T}
$$

When the input noise $\varepsilon_1(t)$ in Eq. (6.3) and the measurement noise $\varepsilon_2(t)$ in Eq. (6.7) must be considered, the state space equations and observation equations can be modified as follows:

$$
\dot{z} = Az + B_1 F_d(t) + B_2 F_a(t) + \varepsilon_1(t) \tag{6.8a}
$$

$$
y(t) = C_2 z + D_{21} F_d(t) + D_{22} F_a(t) + \varepsilon_2(t) \tag{6.8b}
$$

where $\varepsilon_1(t)$ and $\varepsilon_2(t)$ are Gaussian random white noise with zero mean.

In order to eliminate the influence of the above noise to obtain better active control, this chapter uses the LQG optimal control theory, with the details shown in Sect. 3.3. The obtained optimal active control force using LQG theory can be expressed as:

$$
F_a(t) = -K\hat{y}(t) \tag{6.9}
$$

where K is the optimal feedback control gain generated by the LQR control algorithm, and $\hat{y}(t)$ is the estimation of $y(t)$.

The solution of $\hat{y}(t)$ requires constructing the following Kalman filter, whose state space form is:

$$
\begin{cases} \dot{\hat{z}}(t) = A\hat{z}(t) + B_1 F_d(t) + B_2 F_a(t) + K_e\left[y(t) - \hat{y}(t)\right] \\ \hat{y}(t) = C_3\hat{z}(t) \end{cases} \tag{6.10}
$$

where $\hat{z}(t)$ is the estimated state vector, K_e is the Kalman filter gain, and $C_3 =$ diag([ones(1, 12]) in which diag(•) generates a diagonal matrix and ones(m, n) generates an m × n unit diagonal matrix.

Figure 6.2 shows the constructed LQG control with the Kalman filter.

Based on the PSO algorithm, PSO-LQG active control of the frame structure vibration caused by the power equipment is adopted here, and the control strategy is applied to the ATMD system. The fitness function is considered as follows:

$$
\sum_{i\Delta t}^{T=N\Delta t} \sum_{l=1}^{3} |x_l(i\Delta t)| \tag{6.11}
$$

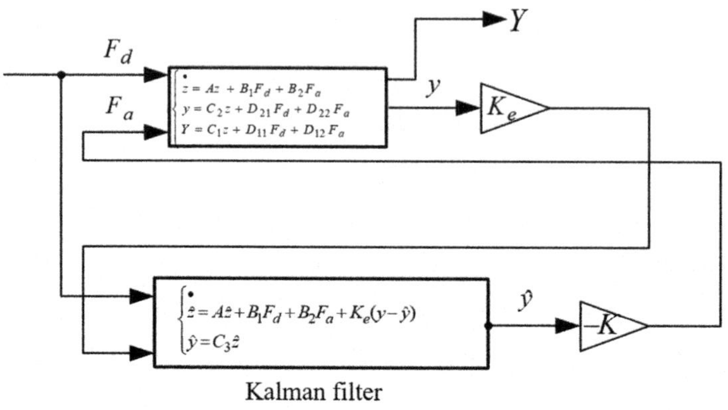

Fig. 6.2 LQG control with a Kalman filter

where the calculation of x_l is based on Y in Fig. 6.2, Δt is the discretized time interval, and $T = N\Delta t$ is the total simulation time.

Example 6.1 For the power equipment-frame structure composite vibration system shown in Fig. 6.1, the structural parameters are $m_1 = m_2 = m_3 = 4 \times 10^5$ kg, $k_1 = k_2 = k_3 = 2 \times 10^8$ N/m, $c_1 = c_2 = c_3 = 1 \times 10^6$ N·s/m. The parameters of the vibration isolation system in the power equipment are $m_p = 600$ kg, $m_b = 1200$ kg, $k_p = 1.5 \times 10^4$ N/m, $k_b = 1.0 \times 10^6$ N/m, $c_p = 1.0 \times 10^3$ N·s/m, and $c_b = 1.6 \times 10^4$ N·s/m. The vibration load generated by the power equipment was assumed to be in the form of a sine wave, and the amplitude is 2.0×10^3 N, and the frequency is 1.0 Hz. The natural frequencies of the power equipment-frame structure system are calculated as follows: 0.7901, 1.5835, 4.4311, 4.6336, and 6.4148 Hz. The TMD's natural frequency was adjusted to be the same as the first-order natural frequency of the composite system, the mass ratio of the TMD mass to the structure is 0.05, and the damping ratio of the TMD / ATMD system is 0.07. For ATMD control, the weight matrix Q and R must be optimally configured to obtain the optimal feedback controller. Therefore, the PSO algorithm is used to calculate the optimal value.

The PSO algorithm has 100 particle populations and 200 iterations, with $c_1 = 2$, $c_2 = 1$, and $\omega = 0.99^t$ (t is the tth iteration). The parameter search range was set to:

$$\begin{aligned}
&[1 \times 10^{-7}, 1 \times 10^{-7}, 1 \times 10^{-7}, 1 \times 10^{-7}, 1 \times 10^{-7}, 1 \times 10^{-7}, 1 \times 10^{-7},\\
&\quad 1 \times 10^{-7}, 1 \times 10^{-7}, 1 \times 10^{-7}, 1 \times 10^{-7}, 1 \times 10^{-7}, 1 \times 10^{-7}]\\
&\sim [1 \times 10^{7}, 1 \times 10^{7}, 1 \times 10^{7}, 1 \times 10^{7}, 1 \times 10^{7}, 1 \times 10^{7}, 1 \times 10^{7},\\
&\quad 1 \times 10^{7}, 1 \times 10^{7}, 1 \times 10^{7}, 1 \times 10^{7}, 1 \times 10^{7}, 1]
\end{aligned}$$

The fitness convergence curve was obtained as shown in Fig. 6.3, and the obtained gbest solution is as follows:

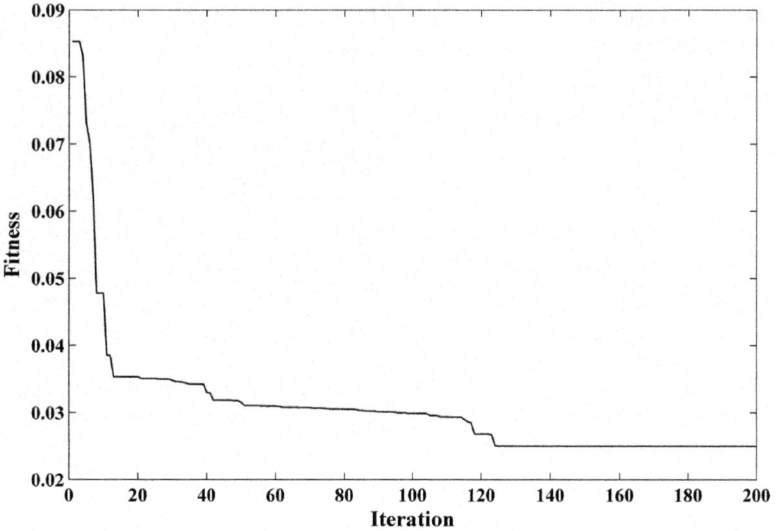

Fig. 6.3 Fitness convergence of PSO-LQG ATMD control for power equipment-frame structure

$$Q = diag\begin{bmatrix} 1.000 & 0.9505 & 0.04476 & 0.06599 & 0.9150 & 0.05840 \\ 0.003837 & 0.90095 & 0.04367 & 1.000 & 1 \times 10^{-14} & 0.06556 \end{bmatrix} \times 10^7,$$

$$R = 4.867 \times 10^{-5}$$

Based on the obtained optimal LQG active controller, the TMD/ATMD vibration control response for the power equipment-frame structure is shown in Fig. 6.4. ATMD significantly affects the vibration response control of each frame structure layer. Although the vibration response using TMD has a certain control effect, the effect is not significant. Hence, the ATMD greatly improves the TMD's control performance. Therefore, applying it to the research on vibration control of frame structures caused by power equipment is reasonable and applicable. The active control force generated by the ATMD is shown in Fig. 6.5.

6.1.2 SATMD Vibration Control for Power Equipment-Structure

In Chap. 4, a 6-segment cubic polynomial model that can accurately simulate the output force of MRD was proposed based on the experimental data of the mechanical characteristics of MRD, and the model was used to track the optimal active control force of the power equipment. An efficient semi-active control strategy was proposed, reproducing and replicating the optimal active control characteristics. This research

Fig. 6.4 Vibration control responses of the power equipment-frame structure for comparison (ATMD)

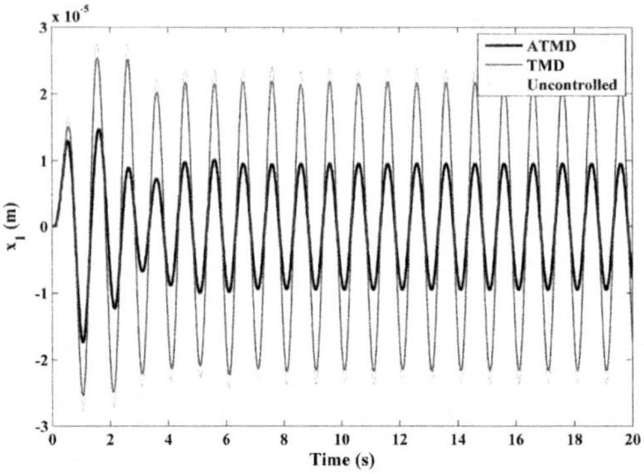

(a) Displacement of the first layer

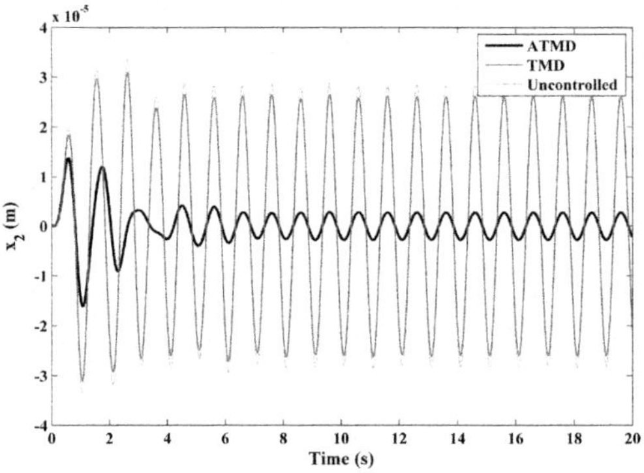

(b) Displacement of the second layer

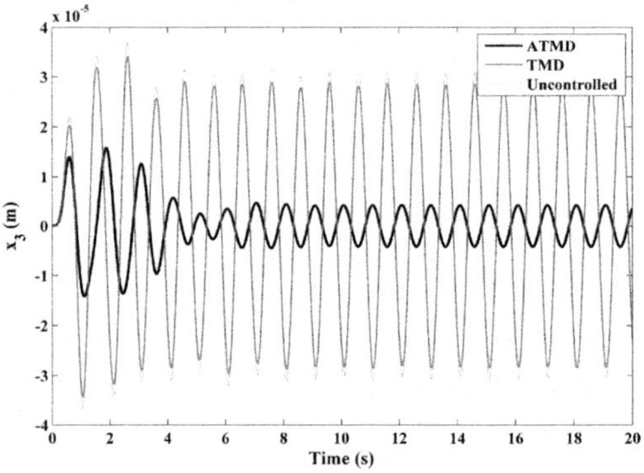

(c) Displacement of the third layer

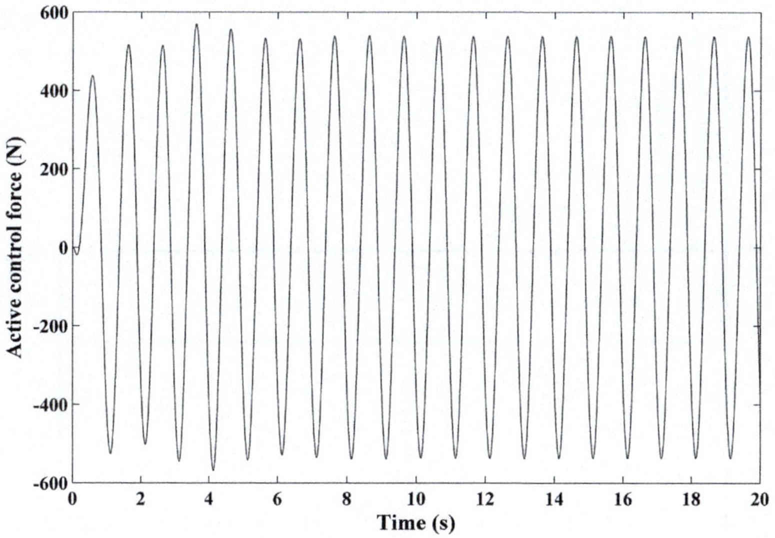

Fig. 6.5 ATMD active control force of the power equipment-frame structure

is based on the ATMD optimal active control strategy for the power equipment-frame structure described in Sect. 5.1.1. An equivalent MRD-based semi-active control implementation method to replace ATMD's complex active control system was proposed to provide a convenient and easy application in practical engineering.

The parameters of the vibration control system for the power equipment-frame structure shown in Fig. 5.6 are defined as those mentioned before. The original ATMD active control system was replaced by an MRD to form a new semi-active TMD (SATMD) control strategy, where $F_{mr}(t)$ is the damping force generated by MRD when using the SATMD for structural vibration control.

For the SATMD control system shown in Fig. 6.6, the dynamic equation can be expressed as follows:

$$\begin{cases} \mathrm{m_1}\ddot{x}_1 + (\mathrm{c_1}+\mathrm{c_2}+\mathrm{c_b})\dot{x}_1 - \mathrm{c_2}\dot{x}_2 - \mathrm{c_b}\dot{x}_b + (\mathrm{k_1}+\mathrm{k_2}+\mathrm{k_b})x_1 - \mathrm{k_2}x_2 - \mathrm{k_b}x_b = 0 \\ \mathrm{m_2}\ddot{x}_2 - \mathrm{c_2}\dot{x}_1 + (\mathrm{c_2}+\mathrm{c_3})\dot{x}_2 - \mathrm{c_3}\dot{x}_3 - \mathrm{k_2}x_1 + (\mathrm{k_2}+\mathrm{k_3})x_2 - \mathrm{k_3}x_3 = 0 \\ \mathrm{m_3}\ddot{x}_3 - \mathrm{c_3}\dot{x}_2 + \mathrm{c_3}\dot{x}_3 - \mathrm{k_3}x_2 + \mathrm{k_3}x_3 + \mathrm{c_d}\dot{x}_3 + \mathrm{k_d}x_3 - \mathrm{c_d}\dot{x}_d - \mathrm{k_d}x_d = -F_{mr}(t) \\ \mathrm{m_b}\ddot{x}_b - \mathrm{c_b}\dot{x}_1 + (\mathrm{c_b}+\mathrm{c_p})\dot{x}_b - \mathrm{c_p}\dot{x}_p - \mathrm{k_b}x_1 + (\mathrm{k_b}+\mathrm{k_p})x_b - \mathrm{k_p}x_p = 0 \\ \mathrm{m_p}\ddot{x}_p - \mathrm{c_p}\dot{x}_b + \mathrm{c_p}\dot{x}_p - \mathrm{k_p}x_b + \mathrm{k_p}x_p = F_d(t) \\ \mathrm{m_d}\ddot{x}_d + \mathrm{c_d}\dot{x}_d + \mathrm{k_d}x_d - \mathrm{c_d}\dot{x}_3 - \mathrm{k_d}x_3 = F_{mr}(t) \end{cases} \tag{6.12}$$

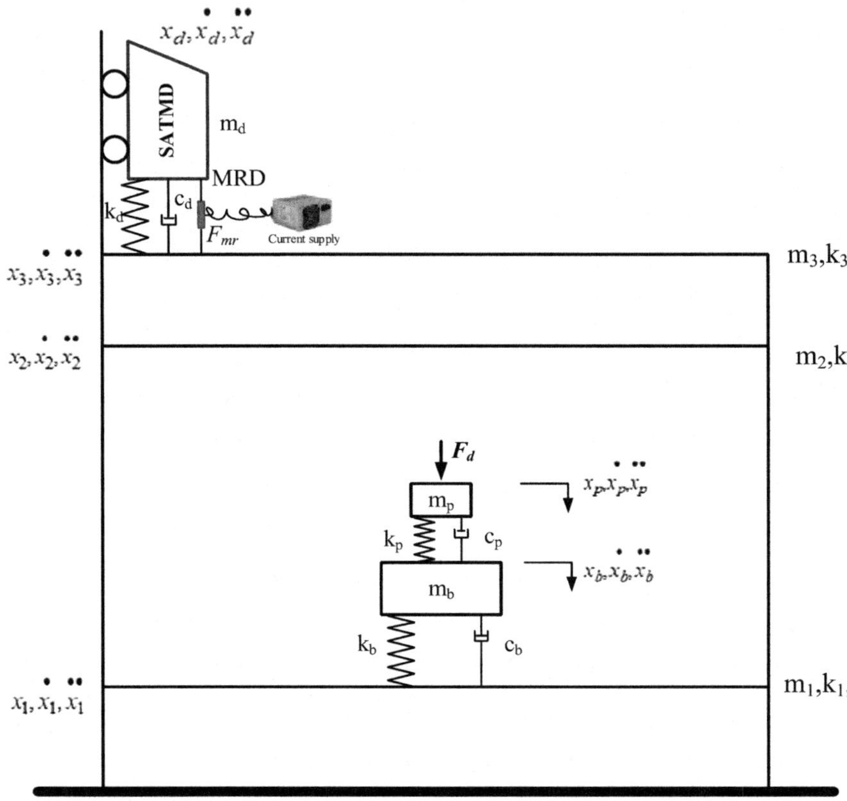

Fig. 6.6 SATMD vibration control for power equipment-frame structure

Additionally, the state variables are set to $z_1 = x_1$, $z_2 = x_2$, $z_3 = x_3$, $z_4 = x_b$, $z_5 = x_p$, $z_6 = x_d$, $z_7 = \dot{x}_1$, $z_8 = \dot{x}_2$, $z_9 = \dot{x}_3$, $z_{10} = \dot{x}_b$, $z_{11} = \dot{x}_p$, and $z_{12} = \dot{x}_d$, and the state vector is defined as $z = \{z_1, z_2, z_3, z_4, z_5, z_6, z_7, z_8, z_9, z_{10}, z_{11}, z_{12}\}^{\mathrm{T}}$. Besides, Eq. (6.12) can be expressed in the form of the following state space:

$$\dot{z} = \mathrm{A}z + \mathrm{B}F_{\mathrm{input}}(t) \tag{6.13}$$

where A has the same form as that in Eq. (6.3).

$$F_{input} = \begin{bmatrix} F_d \\ F_{mr} \end{bmatrix}, \text{ and } \mathrm{B} = \begin{bmatrix} 0\,0\,0\,0\,0\,0\,0\,0\,0 & 0 & 0 & \frac{1}{\mathrm{m_p}} & 0 \\ 0\,0\,0\,0\,0\,0\,0\,0 & \frac{-1}{\mathrm{m_3}} & 0 & 0 & \frac{1}{\mathrm{m_d}} \end{bmatrix}^T$$

Under SATMD control, the frame structure's control output vibration response is defined as:

$$Y(t) = [x_1, x_2, x_3]^{\mathrm{T}} \tag{6.14}$$

Equation (6.14) can be rewritten in the following state space form:

$$Y(t) = \mathrm{C}_1 z + \mathrm{D}_1 F_{\text{input}}(t) \tag{6.15}$$

where

$$\mathrm{C}_1 = \begin{bmatrix} 1\,0\,0\,0\,0\,0\,0\,0\,0\,0\,0\,0 \\ 0\,1\,0\,0\,0\,0\,0\,0\,0\,0\,0\,0 \\ 0\,0\,1\,0\,0\,0\,0\,0\,0\,0\,0\,0 \end{bmatrix}, \text{ and } \mathrm{D}_1 = \begin{bmatrix} 0, 0, 0 \\ 0, 0, 0 \end{bmatrix}^{\mathrm{T}}$$

Moreover, the observation output is described as follows:

$$y(t) = \left[x_b, x_p, \dot{x}_b, \dot{x}_p, \dot{x}_d - \dot{x}_3, \ddot{x}_d - \ddot{x}_3\right]^{\mathrm{T}} \tag{6.16}$$

where $\dot{x}_d - \dot{x}_3$ is the input velocity of the MRD, which is typically used to distinguish the left and right branches' characteristics during its movement.

Equation (6.16) can be expressed in the form of state space as:

$$y(t) = \mathrm{C}_2 z + \mathrm{D}_2 F_{input}(t) \tag{6.17}$$

where

$$\mathrm{C}_2 = \begin{bmatrix} 0 & 0 & 0 & 1 & 0 & 0 & 0 & 0 & 0 & 0 & 0 & 0 \\ 0 & 0 & 0 & 0 & 1 & 0 & 0 & 0 & 0 & 0 & 0 & 0 \\ 0 & 0 & 0 & 0 & 0 & 0 & 0 & 0 & 0 & 1 & 0 & 0 \\ 0 & 0 & 0 & 0 & 0 & 0 & 0 & 0 & 0 & 0 & 1 & 0 \\ 0 & 0 & 0 & 0 & 0 & 0 & 0 & 0 & -1 & 0 & 0 & 1 \\ 0 & \frac{-\mathrm{k}_3}{\mathrm{m}_3} & \frac{\mathrm{k}_\mathrm{d}}{\mathrm{m}_\mathrm{d}} + \frac{\mathrm{k}_3+\mathrm{k}_\mathrm{d}}{\mathrm{m}_3} & 0 & 0 & -\left(\frac{\mathrm{k}_\mathrm{d}}{\mathrm{m}_\mathrm{d}} + \frac{\mathrm{k}_\mathrm{d}}{\mathrm{m}_3}\right) & 0 & \frac{-\mathrm{c}_3}{\mathrm{m}_3} & \frac{\mathrm{c}_\mathrm{d}}{\mathrm{m}_\mathrm{d}} + \frac{\mathrm{c}_3+\mathrm{c}_\mathrm{d}}{\mathrm{m}_3} & 0 & 0 & -\left(\frac{\mathrm{c}_\mathrm{d}}{\mathrm{m}_\mathrm{d}} + \frac{\mathrm{c}_\mathrm{d}}{\mathrm{m}_3}\right) \end{bmatrix},$$

$$\text{and } \mathrm{D}_2 = \begin{bmatrix} 0\,0\,0\,0\,0 & 0 \\ 0\,0\,0\,0\,0 & \frac{1}{\mathrm{m}_\mathrm{d}} + \frac{1}{\mathrm{m}_3} \end{bmatrix}^{\mathrm{T}}$$

The calculation block diagram of the MRD-based SATMD control strategy for vibration control of the power equipment-frame structure is shown in Fig. 6.7.

The actual MRD output force calculation is based on the open loop control strategy in Sect. 4.2.5, where $F_{desired}$ is the optimal active control force generated by ATMD in Sect. 6.1.1.

Example 6.2 The parameter configuration of the power equipment's vibration isolation system, its vibration load, and the three-story frame structure are as defined in

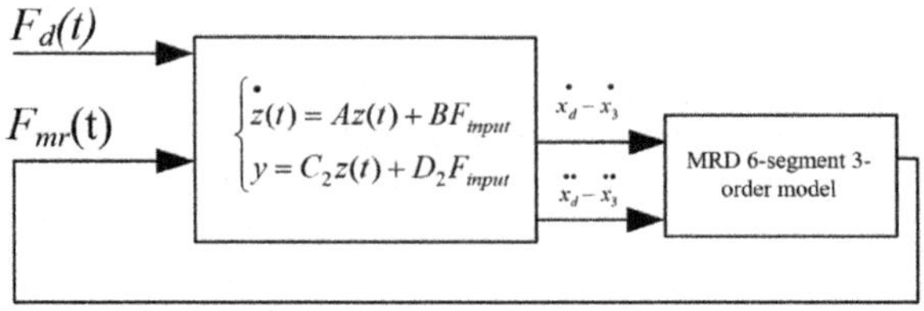

Fig. 6.7 Schematic diagram of SATMD control with a MRD

Example 6.1. Furthermore, the MRD semi-active control research for equivalent realization will be carried out based on the PSO-LQG-based ATMD control proposed in Sect. 6.1.1.

The equivalent tracking MRD damping force is shown in Fig. 6.8 based on the optimal active control force generated by ATMD, indicating that the MRD can effectively follow the ATMD active control force, and the achieved MRD damping force curve is highly consistent with the controlled object. Under both control conditions, the control response of the frame structure vibration caused by the power equipment is shown in Fig. 6.9. The MRD high-precision optimal active control force tracking ensures that the ATMD characteristics are well realized. Hence, the MRD-based SATMD control strategy can fully observe the control characteristics of ATMD. The control current and input velocity of MRD are shown in Figs. 6.10 and 6.11.

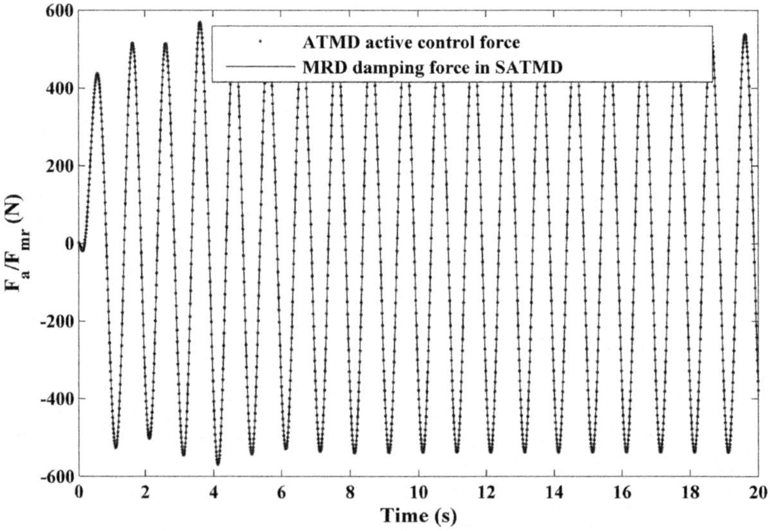

Fig. 6.8 Optimal active control force of the ATMD traced by the MRD force of SATMD (power equipment-frame structure)

Fig. 6.9 Vibration control responses of the power equipment-frame structure (SATMD)

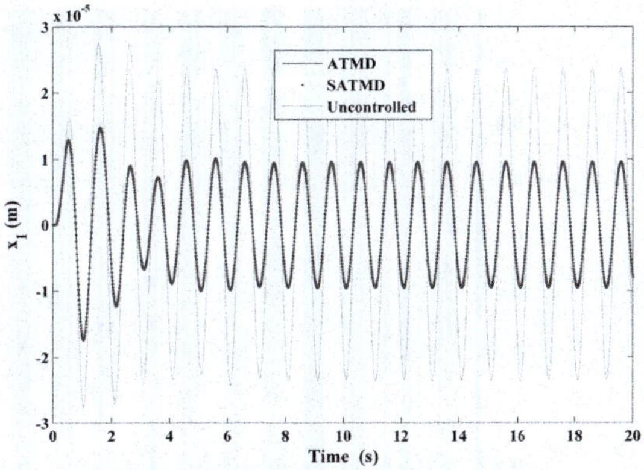

(a) Displacement of the first layer

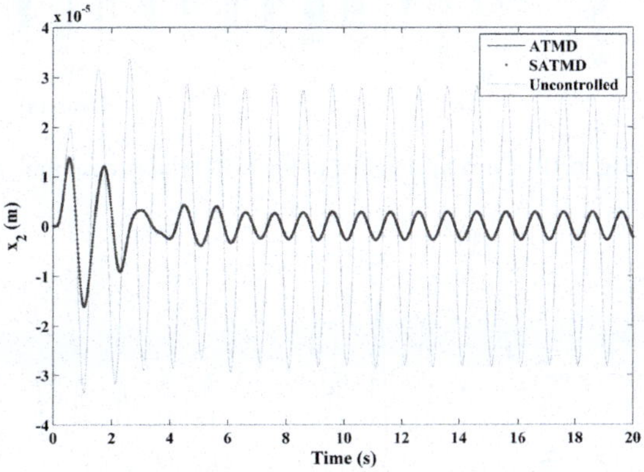

(b) Displacement of the second layer

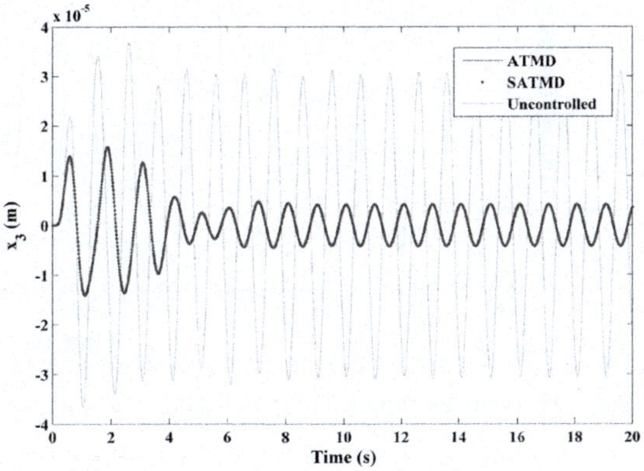

(c) Displacement of the third layer

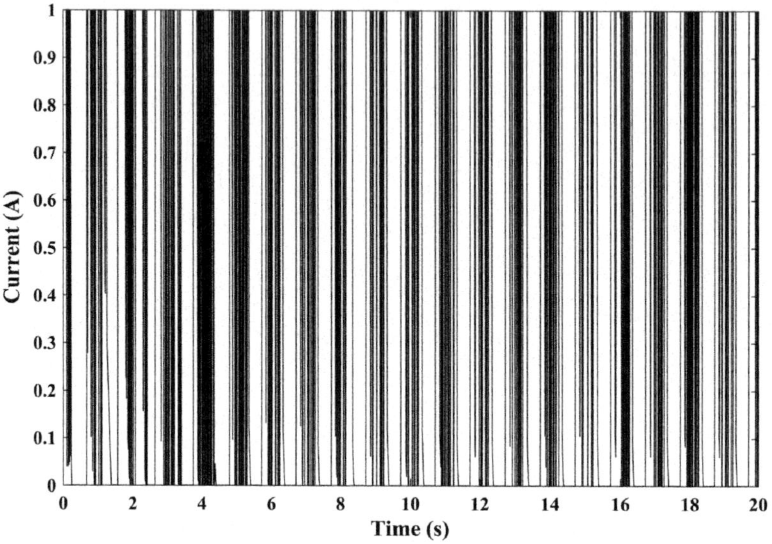

Fig. 6.10 Current supplied to the MRD (power equipment-frame structure/SATMD)

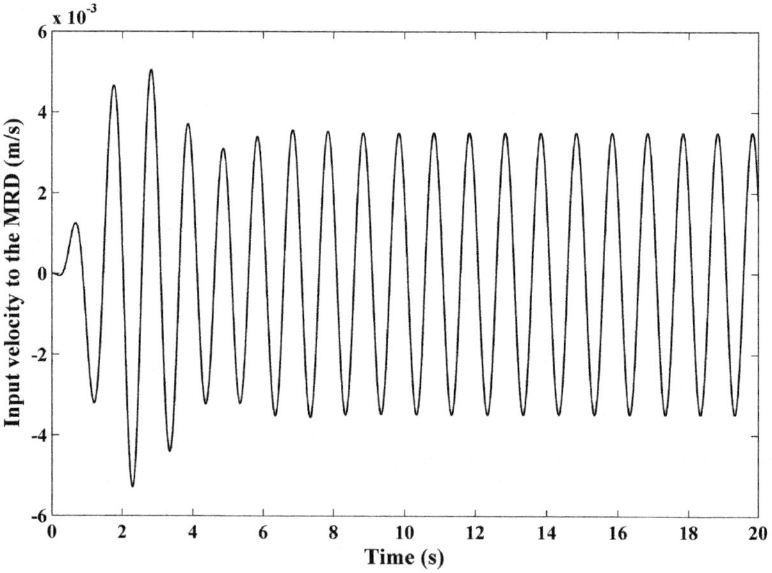

Fig. 6.11 Velocity supplied to the MRD (power equipment-frame structure/SATMD)

6.2 Vibration Control Strategy for Sensitive Equipment-Structure

6.2.1 TMD/ATMD Vibration Control for Sensitive Equipment-Structure

Figure 6.12 shows the TMD/ATMD vibration control system for sensitive equipment-frame structure, where m_p is the mass of sensitive equipment, k_p and c_p are the stiffness and damping of isolation system for the equipment, respectively, x_p, $\dot{x}_p$ and $\ddot{x}_p$ are the displacement, velocity, and acceleration of the equipment, respectively, m_b is the mass of the foundation or supporting system of the equipment, k_b and c_b are the stiffness and damping of the foundation or supporting system, respectively, the definitions of the frame structure parameters are the same as those in Sect. 6.1.1, $F_a(t)$ is the active control force generated by the actuator when the ATMD is adopted for structural vibration, and $F_d(t)$ is the vibration load acting on the frame structure. This book considers the pedestrian load of a single pedestrian in the frame structure.

Single-person continuous walking excitation is referred to the regulations in the International Association for Bridge and Structural Engineering (IABSE) [7], and the series function simulation represented by Eq. (6.18) is used here, and the function curve is shown in Fig. 6.13.

$$F_d(t) = G\left[1 + \sum_{i=1}^{3} \alpha_i \sin(2i\pi f_s t - \Phi_i)\right] \tag{6.18}$$

where G is the weight of a single pedestrian, which is 70 kg/person according to the AISC [8], α_i is the dynamic factor, $\alpha_1 = 0.4 + 0.25(f_s - 2)$, $\alpha_2 = \alpha_3 = 0.1$, f_s is the walking frequency assumed to be 1.0 Hz here, Φ_i is the phase angle, $\Phi_1 = 1$, and $\Phi_2 = \Phi_3 = \pi/2$.

For the TMD/ATMD control system of the sensitive equipment-frame structure shown in Fig. 6.12, the dynamic equation can be expressed as follows:

$$\begin{cases} m_1\ddot{x}_1 + (c_1 + c_2 + c_b)\dot{x}_1 - c_2\dot{x}_2 - c_b\dot{x}_b + (k_1 + k_2 + k_b)x_1 - k_2x_2 - k_bx_b = F_d(t) \\ m_2\ddot{x}_2 - c_2\dot{x}_1 + (c_2 + c_3)\dot{x}_2 - c_3\dot{x}_3 - k_2x_1 + (k_2 + k_3)x_2 - k_3x_3 = 0 \\ m_3\ddot{x}_3 - c_3\dot{x}_2 + c_3\dot{x}_3 - k_3x_2 + k_3x_3 + c_d\dot{x}_3 + k_dx_3 - c_d\dot{x}_d - k_dx_d = -F_a(t) \\ m_b\ddot{x}_b - c_b\dot{x}_1 + (c_b + c_p)\dot{x}_b - c_p\dot{x}_p - k_bx_1 + (k_b + k_p)x_b - k_px_p = 0 \\ m_p\ddot{x}_p - c_p\dot{x}_b + c_p\dot{x}_p - k_px_b + k_px_p = 0 \\ m_d\ddot{x}_d + c_d\dot{x}_d + k_dx_d - c_d\dot{x}_3 - k_dx_3 = F_a(t) \end{cases} \tag{6.19}$$

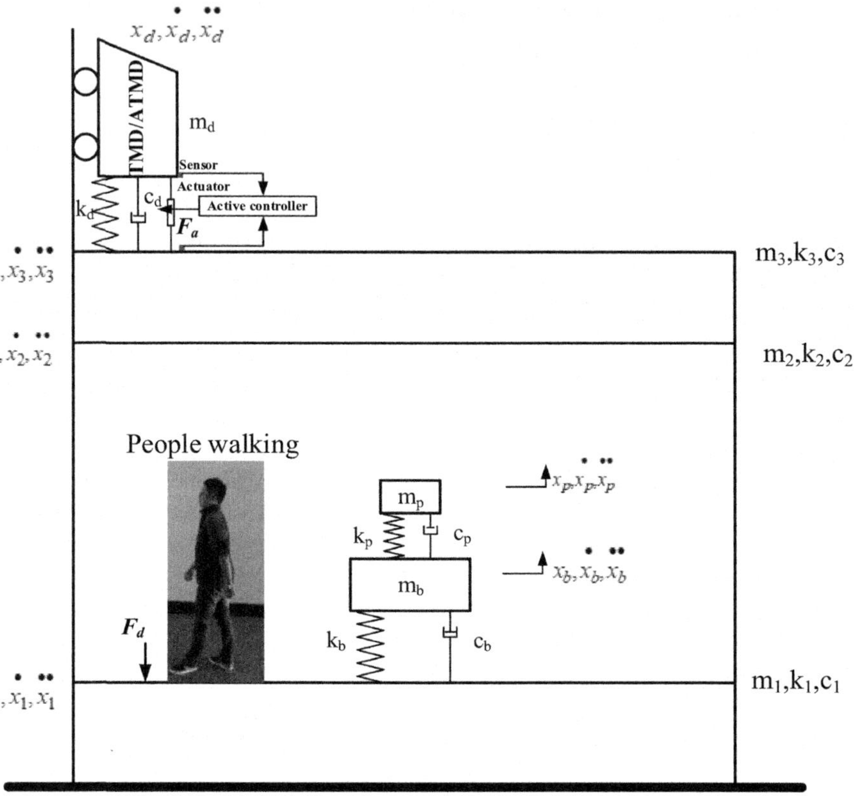

Fig. 6.12 TMD/ATMD vibration control for sensitive equipment-frame structure

In the ATMD control system represented by Eq. (6.19), the state variables are $z_1 = x_1$, $z_2 = x_2$, $z_3 = x_3$, $z_4 = x_b$, $z_5 = x_p$, $z_6 = x_d$, $z_7 = \dot{x}_1$, $z_8 = \dot{x}_2$, $z_9 = \dot{x}_3$, $z_{10} = \dot{x}_b$, $z_{11} = \dot{x}_p$, and $z_{12} = \dot{x}_d$, and the state vector is $z = \{z_1, z_2, z_3, z_4, z_5, z_6, z_7, z_8, z_9, z_{10}, z_{11}, z_{12}\}^{\mathrm{T}}$. Additionally, the state space form is written as follows:

$$\dot{z} = \mathrm{A}z + \mathrm{B}_1 F_{\mathrm{d}}(t) + \mathrm{B}_2 F_a(t) \tag{6.20}$$

where

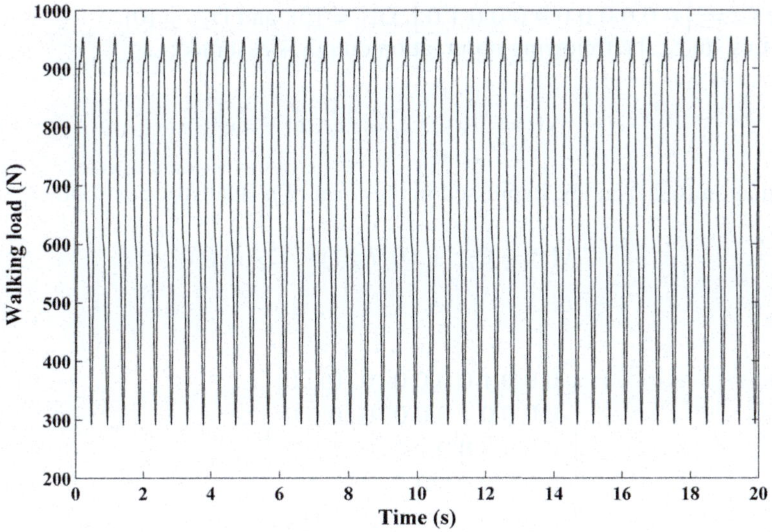

Fig. 6.13 Vibration load by the single person walking

$$A=\begin{bmatrix}
0 & 0 & 0 & 0 & 0 & 0 & 1 & 0 & 0 & 0 & 0 & 0\\
0 & 0 & 0 & 0 & 0 & 0 & 0 & 1 & 0 & 0 & 0 & 0\\
0 & 0 & 0 & 0 & 0 & 0 & 0 & 0 & 1 & 0 & 0 & 0\\
0 & 0 & 0 & 0 & 0 & 0 & 0 & 0 & 0 & 1 & 0 & 0\\
0 & 0 & 0 & 0 & 0 & 0 & 0 & 0 & 0 & 0 & 1 & 0\\
0 & 0 & 0 & 0 & 0 & 0 & 0 & 0 & 0 & 0 & 0 & 1\\
-\frac{k_1+k_2+k_b}{m_1} & \frac{k_2}{m_1} & 0 & \frac{k_b}{m_1} & 0 & 0 & -\frac{c_1+c_2+c_b}{m_1} & \frac{c_2}{m_1} & 0 & \frac{c_b}{m_1} & 0 & 0\\
\frac{k_2}{m_2} & -\frac{k_2+k_3}{m_2} & \frac{k_3}{m_2} & 0 & 0 & 0 & \frac{c_2}{m_2} & -\frac{c_2+c_3}{m_2} & \frac{c_3}{m_2} & 0 & 0 & 0\\
0 & \frac{k_3}{m_3} & -\frac{k_3+k_d}{m_3} & 0 & 0 & \frac{k_d}{m_3} & 0 & \frac{c_3}{m_3} & -\frac{c_3+c_d}{m_3} & 0 & 0 & \frac{c_d}{m_3}\\
\frac{k_b}{m_b} & 0 & 0 & -\frac{k_b+k_p}{m_b} & \frac{k_p}{m_b} & 0 & \frac{c_b}{m_b} & 0 & 0 & -\frac{c_b+c_p}{m_b} & \frac{c_p}{m_b} & 0\\
0 & 0 & 0 & \frac{k_p}{m_p} & -\frac{k_p}{m_p} & 0 & 0 & 0 & 0 & \frac{c_p}{m_p} & -\frac{c_p}{m_p} & 0\\
0 & 0 & \frac{k_d}{m_d} & 0 & 0 & -\frac{k_d}{m_d} & 0 & 0 & \frac{c_d}{m_d} & 0 & 0 & -\frac{c_d}{m_d}
\end{bmatrix},$$

$B_1=[0,0,0,0,0,0,1/m_1,0,0,0,0,0]^T$, and $B_2=[0,0,0,0,0,0,0,0,-1/m_3,0,0,1/m_d]^T$.

Under the TMD/ATMD control, the vibration control output response of the sensitive equipment-frame structure is defined as:

$$Y(t)=\left\{\dot{x}_p\right\} \tag{6.21}$$

Equation (6.21) can be rewritten in the state space form as follows:

$$Y(t)=\mathrm{C}_1 z+\mathrm{D}_{11}F_{\mathrm{d}}(t)+\mathrm{D}_{12}F_a(t) \tag{6.22}$$

where $C_1 = \begin{bmatrix} 0\,0\,0\,0\,0\,0\,0\,0\,0\,0\,0\,1\,0 \end{bmatrix}$, $D_{11} = [0]$, and $D_{12} = [0]$.

In this study, the following observation output is assumed:

$$y(t) = \left[x_b, x_p, \dot{x}_b, \dot{x}_p, \dot{x}_d - \dot{x}_3\right]^T \tag{6.23}$$

Equation (6.23) can be expressed in the form of a state space as follows:

$$y(t) = C_2 z + D_{21} F_d(t) + D_{22} F_a(t) \tag{6.24}$$

where

$$C_2 = \begin{bmatrix} 0\,0\,0\,1\,0\,0\,0\,0 & 0 & 0\,0\,0 \\ 0\,0\,0\,0\,1\,0\,0\,0 & 0 & 0\,0\,0 \\ 0\,0\,0\,0\,0\,0\,0\,0 & 0 & 1\,0\,0 \\ 0\,0\,0\,0\,0\,0\,0\,0 & 0 & 0\,1\,0 \\ 0\,0\,0\,0\,0\,0\,0\,0 & -1 & 0\,0\,1 \end{bmatrix},$$

$$D_{21} = [0, 0, 0, 0, 0]^T, \text{ and } D_{22} = [0, 0, 0, 0, 0]^T.$$

When the input noise $\varepsilon_1(t)$ for Eq. (6.20) and measurement noise $\varepsilon_2(t)$ for Eq. (6.24) must be considered, the state space equations and observation equations can be modified to:

$$\dot{z} = Az + B_1 F_d(t) + B_2 F_a(t) + \varepsilon_1(t) \tag{6.25a}$$

$$y(t) = C_2 z + D_{21} F_d(t) + D_{22} F_a(t) + \varepsilon_2(t) \tag{6.25b}$$

where $\varepsilon_1(t)$ and $\varepsilon_2(t)$ are the Gaussian random white noise with zero mean.

This study uses the LQG optimal control theory to eliminate the noise and obtain better active control.

$$F_a(t) = -K\hat{y}(t) \tag{6.26}$$

where K is the optimal feedback control gain generated by the LQR control algorithm, $\hat{y}(t)$ is the estimation of $y(t)$, which is the solution required to construct a Kalman filter, and the state space form is the same as Eq. (6.11).

Figure 6.2 indicates the calculation diagram.

The ATMD control strategy using the PSO-LQG active control method is studied based on the PSO algorithm. The fitness function is considered as follows:

$$\sum_{i\Delta t}^{T=N\Delta t} \left| \dot{x}_p - \dot{x}_{p,desired} \right| \tag{6.27}$$

where $\overset{\bullet}{x}_{p,desired} = 0$.

Example 6.3 For the sensitive equipment-frame structure composite vibration system shown in Fig. 6.12, the structural parameters are $m_1 = m_2 = m_3 = 4 \times 10^5$ kg, $k_1 = k_2 = k_3 = 2 \times 10^8$ N/m, $c_1 = c_2 = c_3 = 1 \times 10^6$ N s/m.

The parameters of the vibration isolation system of sensitive equipment are $m_p = 100$ kg, $m_b = 1200$ kg, $k_p = 1.5 \times 10^4$ N/m, $k_b = 1.0 \times 10^6$ N/m, $c_p = 1.0 \times 10^3$ N s/m, and $c_b = 1.6 \times 10^4$ N s/m. The vibration excitation on the sensitive equipment was considered a single pedestrian load, as shown in Fig. 6.13. The natural frequencies of the sensitive equipment-structure system were calculated as 1.5835, 1.9418, 4.4316, 4.6178, and 6.4149 Hz. The natural frequency of the TMD is adjusted to be the same as the first-order natural frequency of the composite system, the mass ratio of the TMD to the structure is 0.05, and the damping ratio of the TMD/ATMD system is 0.07. For ATMD control, the weight matrix Q and R must be optimally configured to obtain the optimal feedback controller. Therefore, the PSO algorithm is used herein to calculate the optimal value.

The parameters of the PSO algorithm are 100 particle populations and 200 iterations, with $c_1 = 2$, $c_2 = 1$, and $\omega = 0.99^t$ (t is the tth iteration). The search range is:

$$
\begin{aligned}
&[1 \times 10^{-7}, 1 \times 10^{-7}, 1 \times 10^{-7}, 1 \times 10^{-7}, 1 \times 10^{-7}, 1 \times 10^{-7}, 1 \times 10^{-7},\\
&\quad 1 \times 10^{-7}, 1 \times 10^{-7}, 1 \times 10^{-7}, 1 \times 10^{-7}, 1 \times 10^{-7}, 1 \times 10^{-7}]\\
&\sim [1 \times 10^{7}, 1 \times 10^{7}, 1 \times 10^{7}, 1 \times 10^{7}, 1 \times 10^{7}, 1 \times 10^{7}, 1 \times 10^{7},\\
&\quad 1 \times 10^{7}, 1 \times 10^{7}, 1 \times 10^{7}, 1 \times 10^{7}, 1 \times 10^{7}, 1].
\end{aligned}
$$

The obtained fitness convergence curve is shown in Fig. 6.14, and the obtained gbest solution is:

$$
Q = diag\begin{bmatrix} 1.0094\ 9.1158\ 0.8579\ 7.0394\ 5.6843\ 1.4496 \\ 5.8881\ 6.0191\ 3.0465\ \ 0.7714\ 3.8158\ 1.3811 \end{bmatrix} \times 10^6,
$$

$$
R = 0.9942.
$$

Based on the optimal LQG active control, the TMD/ATMD vibration control response of the sensitive equipment-frame structure is shown in Fig. 6.15, indicating that ATMD has a noticeable control effect on sensitive equipment. In contrast, although the TMD has certain control, the effect is insignificant, and the ATMD control performance is much better. Therefore, applying it to the vibration control of sensitive equipment-frame structures is reasonable and feasible. Figure 6.16 shows the ATMD-generated active control force.

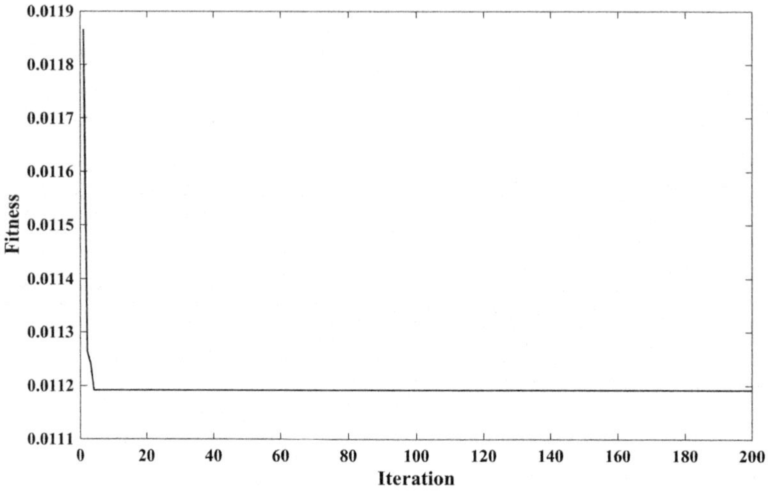

Fig. 6.14 Fitness convergence of PSO-LQG ATMD control for sensitive equipment-frame structure

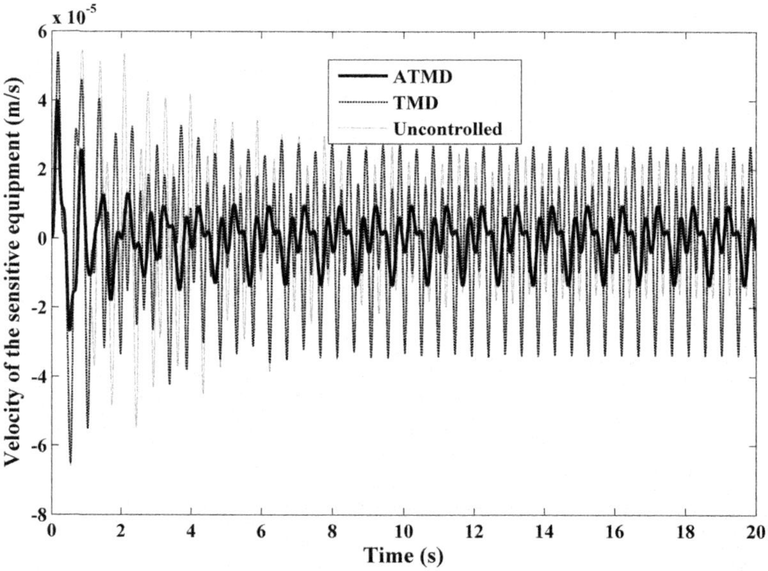

Fig. 6.15 Vibration control responses of the sensitive equipment-frame structure (ATMD)

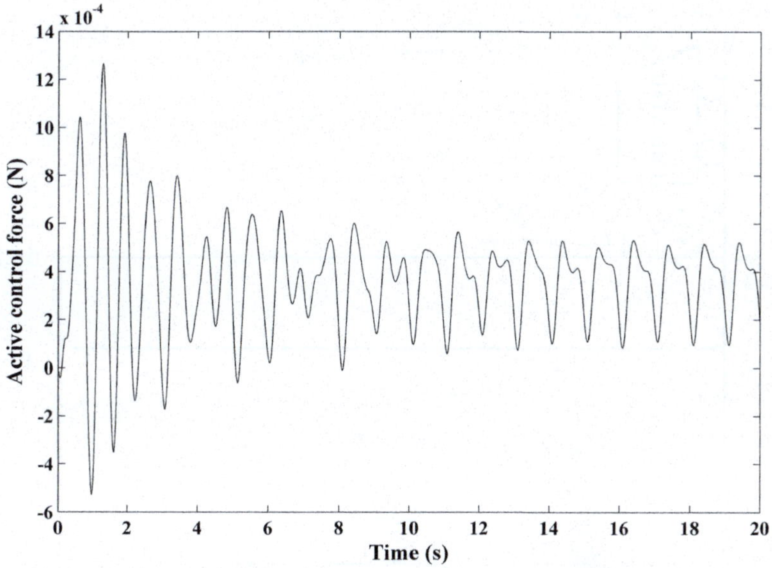

Fig. 6.16 ATMD active control force for sensitive equipment-frame structure

6.2.2 *SATMD Vibration Control for Sensitive Equipment-Structure*

The parameters of the vibration control system in the sensitive equipment-frame hstructure shown in Fig. 6.17 have the same definitions as those in Sect. 6.2.1. The original ATMD active control system is replaced by MRD to form a SATMD control strategy, where $F_{mr}(t)$ is the damping force generated by the MRD when the SATMD is used for structural vibration control.

For the SATMD control system shown in Fig. 6.17, the dynamic equation can be expressed as follows:

$$\begin{cases} \mathrm{m_1}\ddot{x}_1 + (\mathrm{c_1}+\mathrm{c_2}+\mathrm{c_b})\dot{x}_1 - \mathrm{c_2}\dot{x}_2 - \mathrm{c_b}\dot{x}_b + (\mathrm{k_1}+\mathrm{k_2}+\mathrm{k_b})x_1 - \mathrm{k_2}x_2 - \mathrm{k_b}x_b = F_d(t) \\ \mathrm{m_2}\ddot{x}_2 - \mathrm{c_2}\dot{x}_1 + (\mathrm{c_2}+\mathrm{c_3})\dot{x}_2 - \mathrm{c_3}\dot{x}_3 - \mathrm{k_2}x_1 + (\mathrm{k_2}+\mathrm{k_3})x_2 - \mathrm{k_3}x_3 = 0 \\ \mathrm{m_3}\ddot{x}_3 - \mathrm{c_3}\dot{x}_2 + \mathrm{c_3}\dot{x}_3 - \mathrm{k_3}x_2 + \mathrm{k_3}x_3 + \mathrm{c_d}\dot{x}_3 + \mathrm{k_d}x_3 - \mathrm{c_d}\dot{x}_d - \mathrm{k_d}x_d = -F_{mr}(t) \\ \mathrm{m_b}\ddot{x}_b - \mathrm{c_b}\dot{x}_1 + (\mathrm{c_b}+\mathrm{c_p})\dot{x}_b - \mathrm{c_p}\dot{x}_p - \mathrm{k_b}x_1 + (\mathrm{k_b}+\mathrm{k_p})x_b - \mathrm{k_p}x_p = 0 \\ \mathrm{m_p}\ddot{x}_p - \mathrm{c_p}\dot{x}_b + \mathrm{c_p}\dot{x}_p - \mathrm{k_p}x_b + \mathrm{k_p}x_p = 0 \\ \mathrm{m_d}\ddot{x}_d + \mathrm{c_d}\dot{x}_d + \mathrm{k_d}x_d - \mathrm{c_d}\dot{x}_3 - \mathrm{k_d}x_3 = F_{mr}(t) \end{cases} \tag{6.28}$$

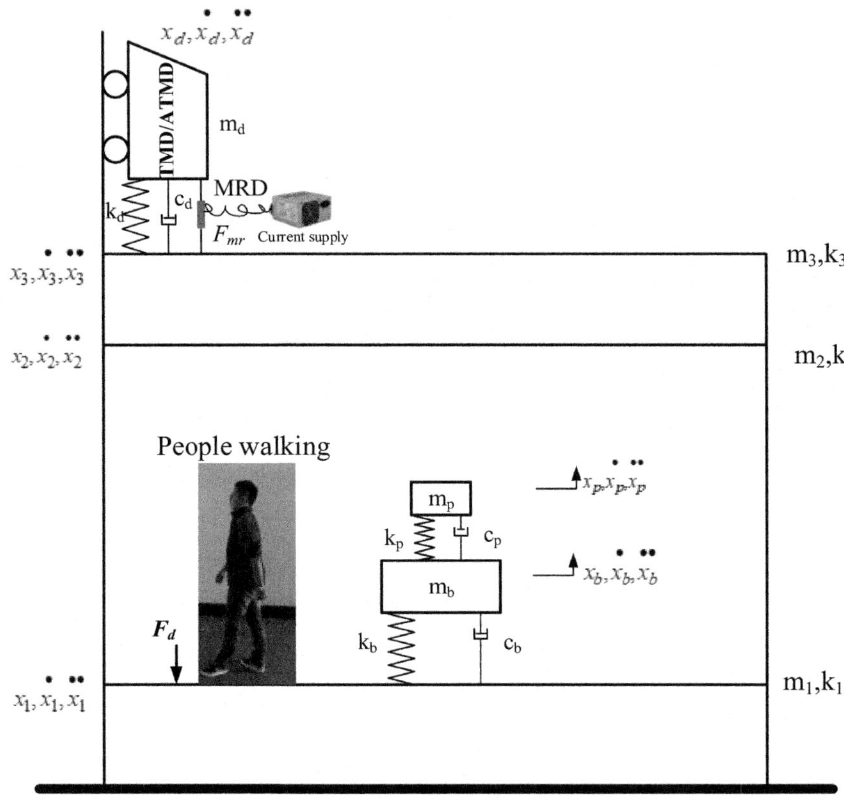

Fig. 6.17 SATMD vibration control system for sensitive equipment-frame structure

Similarly, the state variables are $z_1 = x_1$, $z_2 = x_2$, $z_3 = x_3$, $z_4 = x_b$, $z_5 = x_p$, $z_6 = x_d$, $z_7 = \dot{x}_1$, $z_8 = \dot{x}_2$, $z_9 = \dot{x}_3$, $z_{10} = \dot{x}_b$, $z_{11} = \dot{x}_p$, and $z_{12} = \dot{x}_d$, and the state vector is $z = \{z_1, z_2, z_3, z_4, z_5, z_6, z_7, z_8, z_9, z_{10}, z_{11}, z_{12}\}^{\mathrm{T}}$. Equation (6.28) can be expressed in the following state space form:

$$\dot{z} = \mathrm{A}z + \mathrm{B}F_{\mathrm{input}}(t) \tag{6.29}$$

where the form of A is the same as that in Eq. (6.13),

$$F_{input} = \begin{bmatrix} F_d \\ F_{mr} \end{bmatrix}, \text{ and } \mathrm{B} = \begin{bmatrix} 0\,0\,0\,0\,0\,0 & \frac{1}{m_1} & 0 & 0 & 0\,0 & 0 \\ 0\,0\,0\,0\,0\,0 & 0 & 0 & \frac{-1}{m_3} & 0\,0 & \frac{1}{m_d} \end{bmatrix}^{\mathrm{T}}.$$

Under the SATMD control, the vibration control output response of the sensitive equipment-frame structure is:

$$Y(t) = \left\{ \dot{x}_p \right\} \tag{6.30}$$

Equation (6.30) can be rewritten in the following state space equation form:

$$Y(t) = \mathrm{C}_1 z + \mathrm{D}_1 F_{\mathrm{input}}(t) \tag{6.31}$$

where $\mathrm{C}_1 = \begin{bmatrix} 0\,0\,0\,0\,0\,0\,0\,0\,0\,0\,0\,1\,0 \end{bmatrix}$, and $\mathrm{D}_1 = \begin{bmatrix} 0\,0 \end{bmatrix}$.

Furthermore, the following observation output is defined as:

$$y(t) = \left[x_b, x_p, \dot{x}_b, \dot{x}_p, \dot{x}_d - \dot{x}_3, \ddot{x}_d - \ddot{x}_3 \right]^{\mathrm{T}} \tag{6.32}$$

where $\dot{x}_d - \dot{x}_3$ is the input velocity to the MRD, and $\ddot{x}_d - \ddot{x}_3$ is the input acceleration to MRD, which is used to determine the characteristics of the left and right branches during its movement.

Equation (6.32) can be expressed in the following state space form:

$$y(t) = \mathrm{C}_2 z + \mathrm{D}_2 F_{input}(t) \tag{6.33}$$

where

$$C_2 = \begin{bmatrix} 0\ 0 & 0 & 1\,0 & 0 & 0\ 0 & 0 & 0\,0 & 0 \\ 0\ 0 & 0 & 0\,1 & 0 & 0\ 0 & 0 & 0\,0 & 0 \\ 0\ 0 & 0 & 0\,0 & 0 & 0\ 0 & 0 & 1\,0 & 0 \\ 0\ 0 & 0 & 0\,0 & 0 & 0\ 0 & 0 & 0\,1 & 0 \\ 0\ 0 & 0 & 0\,0 & 0 & 0\ 0 & -1 & 0\,0 & 1 \\ 0\ \frac{-\mathrm{k}_3}{\mathrm{m}_3} & \frac{\mathrm{k}_d}{\mathrm{m}_d} + \frac{\mathrm{k}_3+\mathrm{k}_d}{\mathrm{m}_3} & 0\,0 & -\left(\frac{\mathrm{k}_d}{\mathrm{m}_d} + \frac{\mathrm{k}_d}{\mathrm{m}_3}\right) & 0\ \frac{-\mathrm{c}_3}{\mathrm{m}_3} & \frac{\mathrm{c}_d}{\mathrm{m}_d} + \frac{\mathrm{c}_3+\mathrm{c}_d}{\mathrm{m}_3} & 0\,0 & -\left(\frac{\mathrm{c}_d}{\mathrm{m}_d} + \frac{\mathrm{c}_d}{\mathrm{m}_3}\right) \end{bmatrix},$$

and $\mathrm{D}_2 = \begin{bmatrix} 0\,0\,0\,0\,0 & 0 \\ 0\,0\,0\,0\,0 & \frac{1}{\mathrm{m_d}} + \frac{1}{\mathrm{m}_3} \end{bmatrix}^T$.

Figure 6.7 shows the sensitive equipment-frame structure's MRD-based SATMD vibration control strategy. The actual output force calculation of the MRD is based on the open-loop control strategy in Sect. 4.2.5, with $F_{desired}$ being the optimal active control force generated by the ATMD in Sect. 6.2.1.

Example 6.4 The vibration isolation system of the sensitive equipment, the vibration load generated by the single-person walking, and the parameter settings of the three-story frame structure are the same as those described in Example 6.3. The equivalent implementation of MRD-based SATMD control will be carried out based on the ATMD control in Sect. 6.2.1.

Figure 6.18 depicts the equivalent tracking MRD damping force based on the optimal active control force generated by ATMD. MRD can accurately follow the ATMD optimal active control force, and the achieved MRD damping force is highly

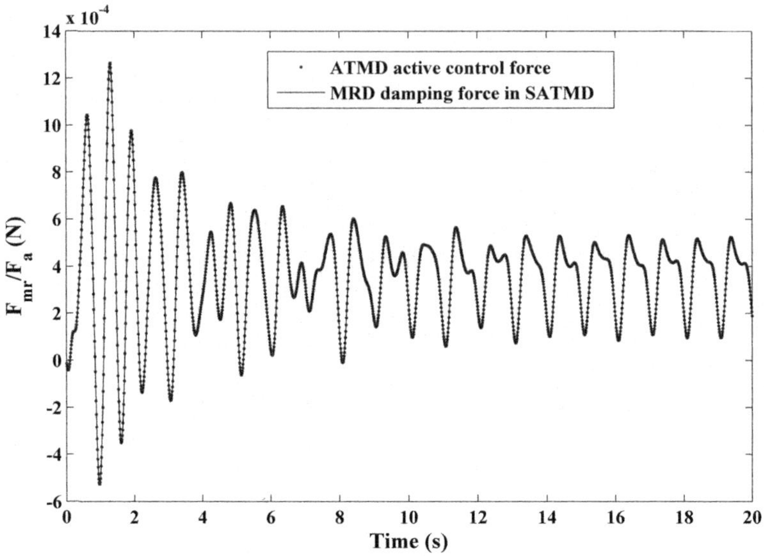

Fig. 6.18 Optimal active control force of the ATMD tracking the SATMD's MRD force (sensitive equipment-frame structure)

consistent with the controlled object. Under both control conditions, the comparison of the controlled velocity response for the walking-induced vibration of the frame structure is shown in Fig. 6.19. The MRD-based SATMD control strategy can perfectly implement the ATMD control characteristics. The control current and input velocity of the MRD are shown in Figs. 6.20 and 6.21.

6.3 Vibration and Seismic Control Strategy for Sensitive Equipment-Isolated Frame Structure

6.3.1 Vibration and Seismic Investigation for Isolated Frame Structure

On the basis of the study in Sect. 6.2, the vibration and seismic control for isolated frame structure and sensitive equipment will be carried out here. Firstly, the frame structure studied before is taken into account here again, but the sensitive equipment is not installed in the structure, and the model is shown in Fig. 6.22.

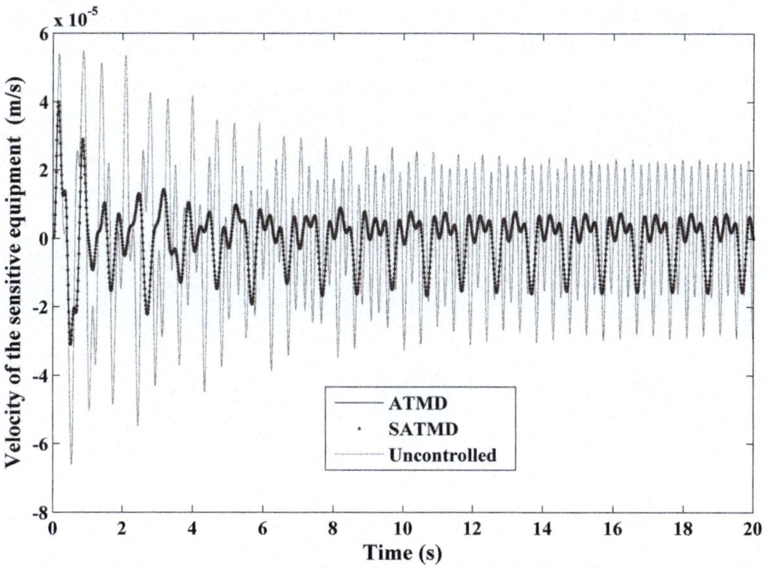

Fig. 6.19 Vibration control responses of the sensitive equipment-frame structure (SATMD)

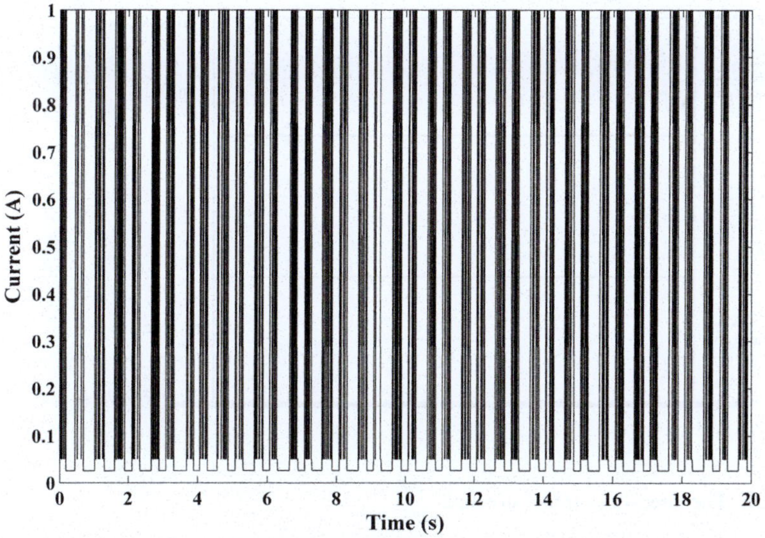

Fig. 6.20 Current supplied to the MRD (sensitive equipment-frame structure/SATMD)

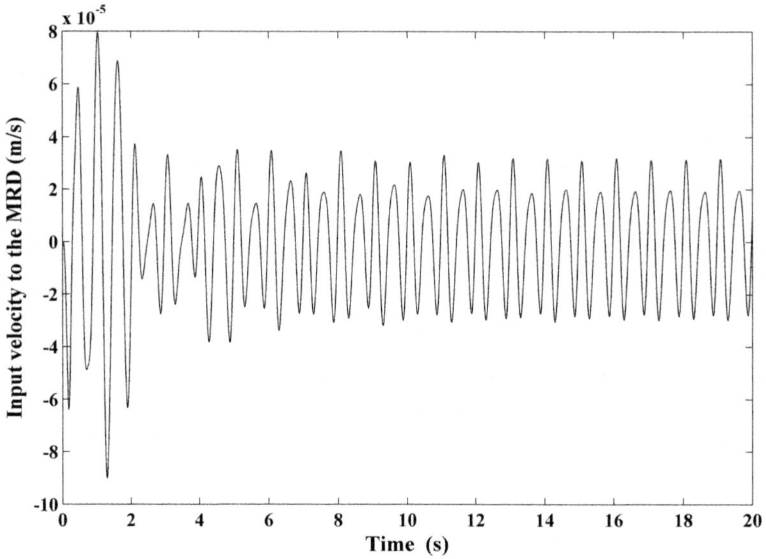

Fig. 6.21 Velocity supplied to the MRD (sensitive equipment-frame structure/SATMD)

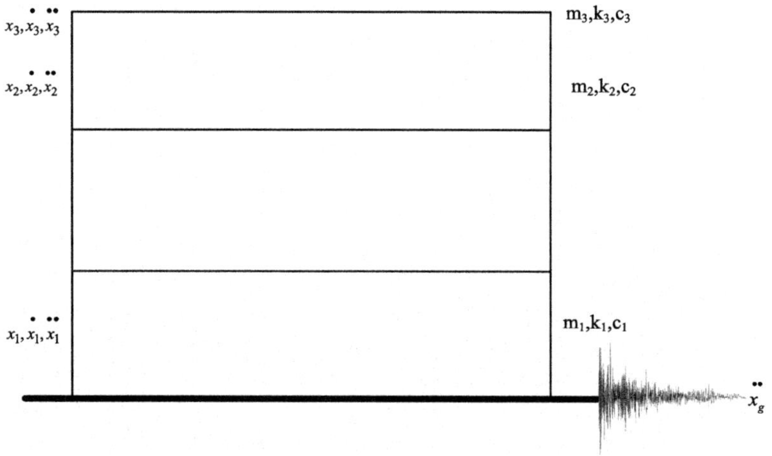

Fig. 6.22 The three-story frame structure

The dynamic equation can be expressed as follows:

$$\begin{cases} m_1\ddot{x}_1 + (c_1 + c_2)\dot{x}_1 - c_2\dot{x}_2 + (k_1 + k_2)x_1 - k_2x_2 = -m_1\ddot{x}_g \\ m_2\ddot{x}_2 - c_2\dot{x}_1 + (c_2 + c_3)\dot{x}_2 - c_3\dot{x}_3 - k_2x_1 + (k_2 + k_3)x_2 - k_3x_3 = 0 \\ m_3\ddot{x}_3 - c_3\dot{x}_2 + c_3\dot{x}_3 - k_3x_2 + k_3x_3 = 0 \end{cases} \tag{6.34}$$

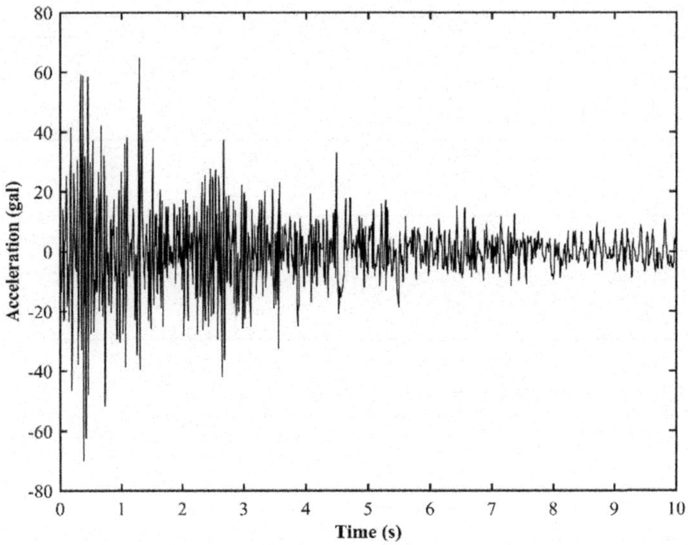

Fig. 6.23 Seismic excitation

where the $\ddot{x}_g$ is the seismic excitation shown in Fig. 6.23. The three-dimensional parameters of the frame structure are all set as $\mathrm{m}_1 = \mathrm{m}_2 = \mathrm{m}_3 = 4 \times 10^5$ kg, $\mathrm{k}_1 = \mathrm{k}_2 = \mathrm{k}_3 = 2 \times 10^8$ N/m, $\mathrm{c}_1 = \mathrm{c}_2 = \mathrm{c}_3 = 1 \times 10^6$ N s/m.

The output response of each story of the building structure excited by the seismic vibration is shown in Figure 6.24.

In recent years, with the rapid development of rail transit, the vibration disturbance of rail transit on the superstructures and adjacent structures cannot be ignored, which can affect the office work, the residence comfort, and the internal sensitive equipment [9–12]. Therefore, vibration isolation measures need to be taken for the building structure affected by transit vibration.

An isolated frame structure is proposed here in Fig. 6.25, in which the k_s is the stiffness of the isolator, and the $\overline{\mathrm{k}_1}$ is the stiffness after being connected in series with the vibration isolator's stiffness k_s, and $\overline{\mathrm{k}_1} = \frac{1}{4}\mathrm{k}_1$ is assumed here. The expression form of dynamic equation is same as the Eq. (6.34), but the k_1 should be replaced by $\overline{\mathrm{k}_1}$ here. When the input excitation is the transit vibration, the seismic excitation $\ddot{x}_g$ in the Eq. (6.34) should be replaced by $\ddot{x}_t$ which is shown in Fig. 6.26.

The vibration response comparison of each story of the isolated and non-isolation structure is presented in Fig. 6.27, where the blue dashed line represents the non-isolation structure, and the black solid line represents the isolated structure.

Due to the significant weakening of the stiffness of isolated structure compared to the non-isolation structure, it is necessary to investigate the displacement of the isolated structure excited by the seismic vibration, and the displacement comparison of each story of the isolated and non-isolation structure is shown in Fig. 6.28, the black dashed line represents the non-isolation structure and the red solid line represents

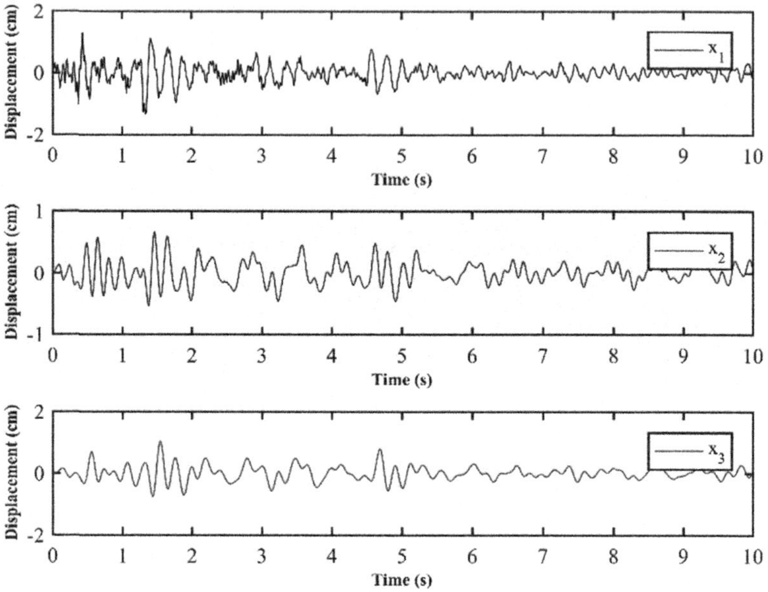

Fig. 6.24 Displacement of each story of the structure excited by seismic vibration

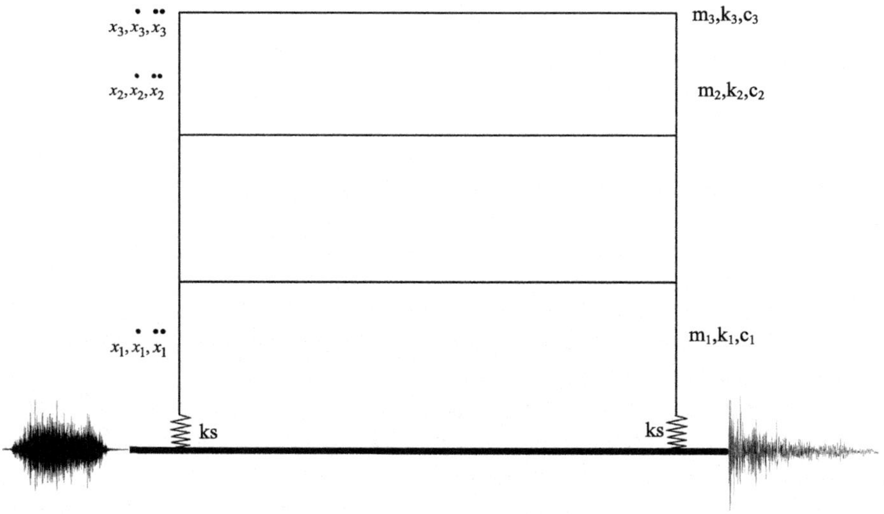

Fig. 6.25 The isolated three-story frame structure

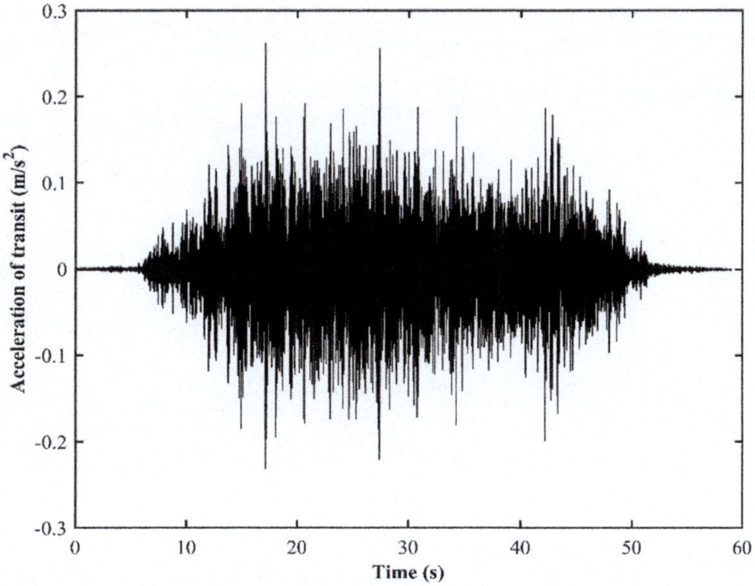

Fig. 6.26 Transit vibration excitation

the isolated structure, and the 3 subfigures from the top to the bottom represents respectively the floor 1 to floor 3.

The numerical results indicate that the maximum seismic displacement of floor 1 of the non-isolation structure is 1.33 cm, and the maximum of isolated structure is 1.69 cm, and the maximums of floor 2 are respectively 0.66 cm and 1.66 cm, and the maximums of floor 3 are respectively 1.02 cm and 1.38 cm. Seen from this, the displacement of isolated structure is larger a lot than the non-isolation structure, and this will be detrimental to the safety of the building structure. Therefore, key measures should be considered to solve the large displacement problem of the isolated structures.

6.3.2 Isolated Frame Structure Equipped with Viscous Damper

For the purpose of solving the large displacement problem of the isolated structure excited by seismic vibration, a viscous damper is proposed and equipped on the floor 1 and the new model is shown in Fig. 6.29.

Fig. 6.27 The vibration response comparison of each story of the isolated and non-isolation structure

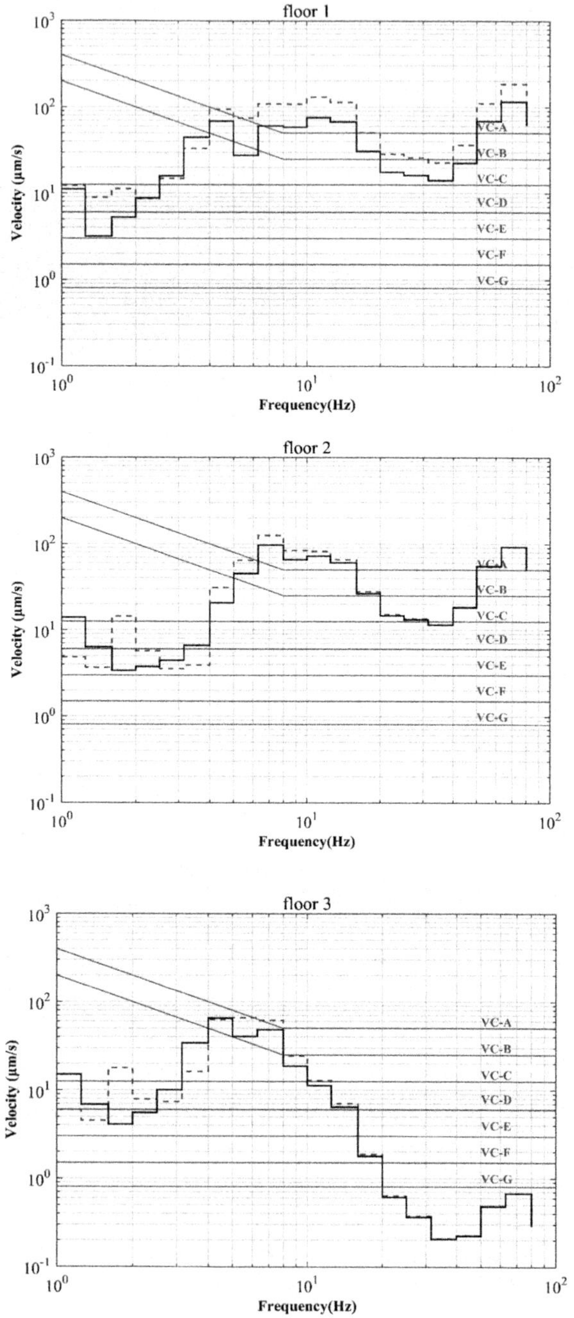

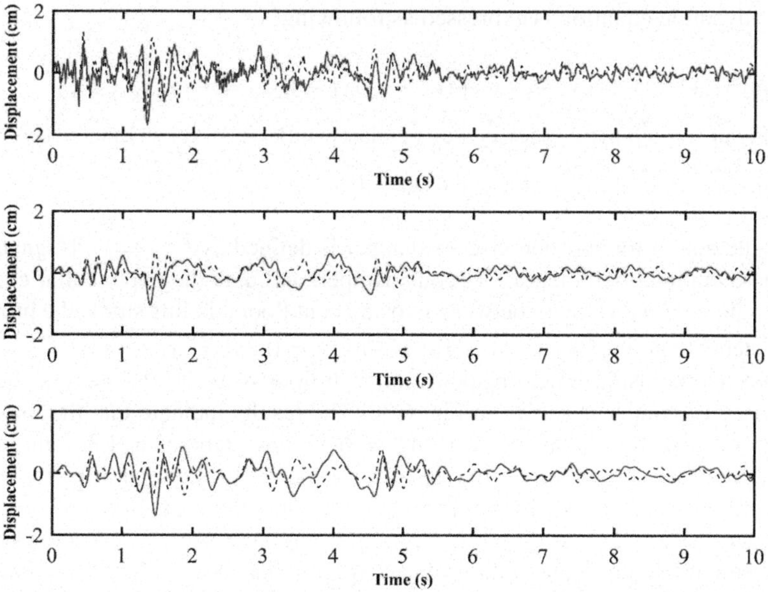

Fig. 6.28 The seismic displacement comparison of each story of the isolated and non-isolation structure

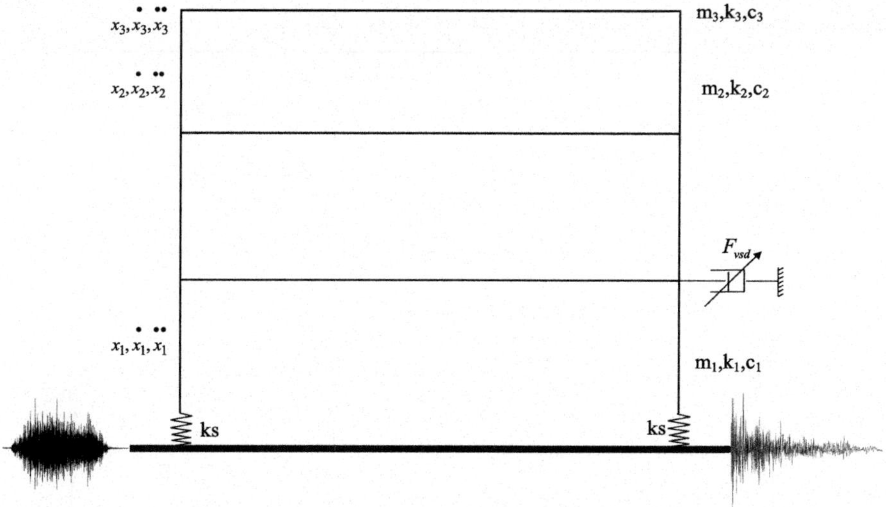

Fig. 6.29 Isolated structure equipped with viscous damper

The dynamic equation is expressed as following:

$$\begin{cases} \mathrm{m}_1 \ddot{x}_1 + (\mathrm{c}_1 + \mathrm{c}_2)\dot{x}_1 - \mathrm{c}_2 \dot{x}_2 + (\overline{\mathrm{k}_1} + \mathrm{k}_2)x_1 - \mathrm{k}_2 x_2 = -\mathrm{m}_1 \ddot{x}_g + F_{vsd} \\ \mathrm{m}_2 \ddot{x}_2 - \mathrm{c}_2 \dot{x}_1 + (\mathrm{c}_2 + \mathrm{c}_3)\dot{x}_2 - \mathrm{c}_3 \dot{x}_3 - \mathrm{k}_2 x_1 + (\mathrm{k}_2 + \mathrm{k}_3)x_2 - \mathrm{k}_3 x_3 = 0 \\ \mathrm{m}_3 \ddot{x}_3 - \mathrm{c}_3 \dot{x}_2 + \mathrm{c}_3 \dot{x}_3 - \mathrm{k}_3 x_2 + \mathrm{k}_3 x_3 = 0 \end{cases} \tag{6.35}$$

where the output force of the viscous damper is defined as $F_{vsd} = c|v|^{\alpha}\mathrm{sgn}(v)$, and c is the damping coefficient of viscous damper, and α is the exponential constant, and v is the relative velocity between two adjacent floors, in this study, the presented parameters are defined as $c = 1 \times 10^7$ N s/m, $\alpha = 0.2$, $v = \dot{x}_1 - \dot{x}_2$.

The seismic displacement response of each floor of isolated structure equipped with viscous damper is shown in Fig. 6.30. The results indicate that the maximum displacement of floor 1 of the structure with viscous damper is 1.33 cm, and the maximum of the structure without viscous damper is 1.69 cm, and the maximums of floor 2 are respectively 1.06 cm and 0.49 cm, and the maximums of floor 3 are respectively 1.38 cm and 0.66 cm. Seen from this, the seismic vibration induced displacement of each story of the isolated structure has been reduced and improved a lot, which indicates that the proposed strategy has a great significance for ensuring the safety of the isolated structures.

Figure 6.31 shows the displacement-force hysteretic curve of the viscous damper.

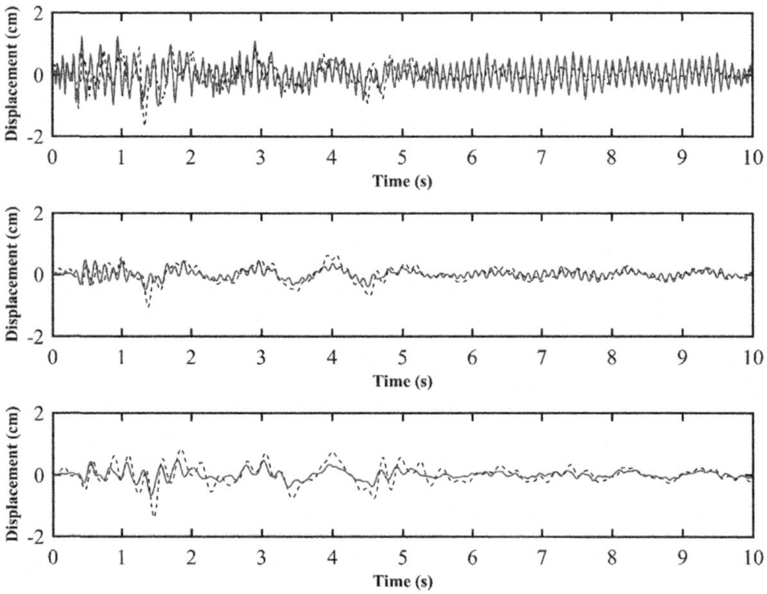

Fig. 6.30 The seismic displacement response of each floor of isolated structure equipped with viscous damper

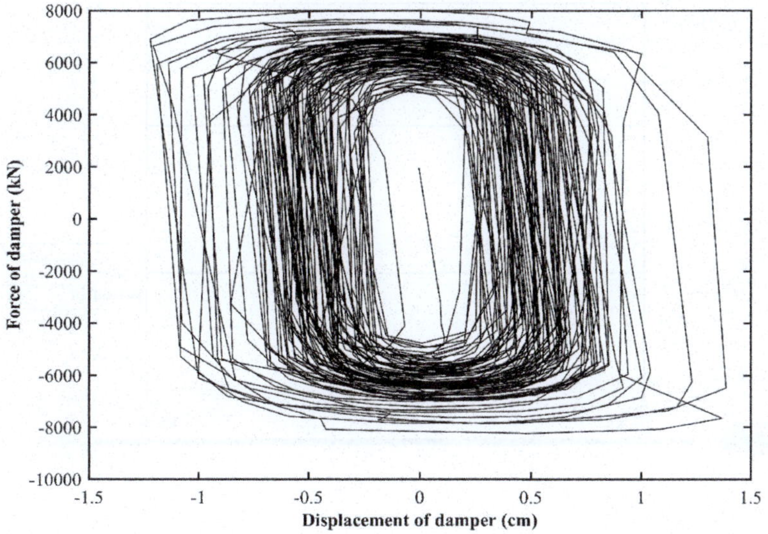

Fig. 6.31 Hysteretic curve of the viscous damper

6.3.3 Isolated Frame Structure Equipped with Active Control System

Due to the passive working mode of the viscous damper, the control effect of the seismic control is limited. For the purpose of improving the control effect, an active control strategy is carried out here, and the isolated structure equipped with active control system is shown in Fig. 6.32, and the $F_{\mathrm{a}}(\mathrm{t})$ is the output force of actuator.

The dynamic equation is expressed as following:

$$\begin{cases} \mathrm{m}_1 \ddot{x}_1 + (\mathrm{c}_1 + \mathrm{c}_2)\dot{x}_1 - \mathrm{c}_2 \dot{x}_2 + (\overline{\mathrm{k}_1} + \mathrm{k}_2)x_1 - \mathrm{k}_2 x_2 = -\mathrm{m}_1 \ddot{x}_g + F_a(t) \\ \mathrm{m}_2 \ddot{x}_2 - \mathrm{c}_2 \dot{x}_1 + (\mathrm{c}_2 + \mathrm{c}_3)\dot{x}_2 - \mathrm{c}_3 \dot{x}_3 - \mathrm{k}_2 x_1 + (\mathrm{k}_2 + \mathrm{k}_3)x_2 - \mathrm{k}_3 x_3 = 0 \\ \mathrm{m}_3 \ddot{x}_3 - \mathrm{c}_3 \dot{x}_2 + \mathrm{c}_3 \dot{x}_3 - \mathrm{k}_3 x_2 + \mathrm{k}_3 x_3 = 0 \end{cases} \quad (6.36)$$

In the active control system represented by Eq. (6.36), the state variables are $z_1 = x_1$, $z_2 = x_2$, $z_3 = x_3$, $z_4 = \dot{x}_1$, $z_5 = \dot{x}_2$ and $z_6 = \dot{x}_3$, and the state vector is $z = \{z_1, z_2, z_3, z_4, z_5, z_6\}^{\mathrm{T}}$. Additionally, the state space form is written as follows:

$$\dot{z} = \mathrm{A}z + \mathrm{B}_1 \ddot{x}_g + \mathrm{B}_2 F_a(t) \quad (6.37)$$

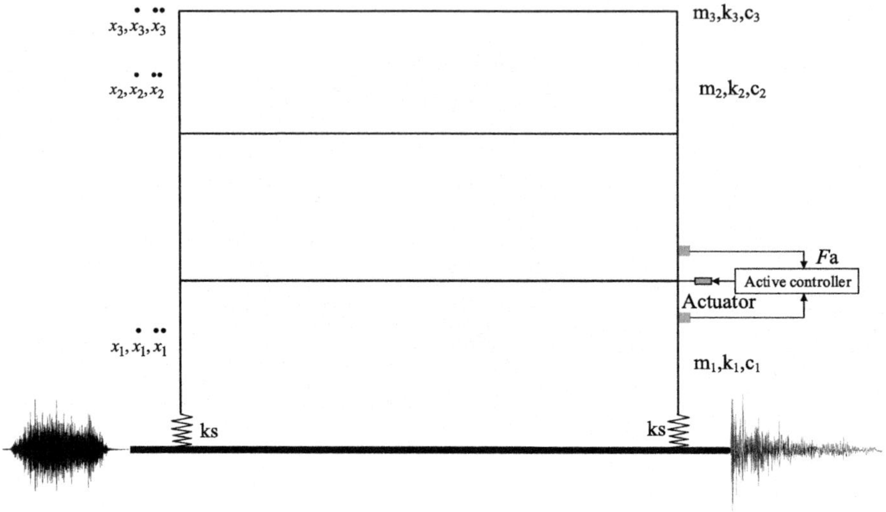

Fig. 6.32 Isolated structure equipped with active control system

where

$$
A = \begin{bmatrix} 0 & 0 & 0 & 1 & 0 & 0 \\ 0 & 0 & 0 & 0 & 1 & 0 \\ 0 & 0 & 0 & 0 & 0 & 1 \\ -\frac{k_1+k_2}{m_1} & \frac{k_2}{m_1} & 0 & -\frac{c_1+c_2}{m_1} & \frac{c_2}{m_1} & 0 \\ \frac{k_2}{m_2} & -\frac{k_2+k_3}{m_2} & \frac{k_3}{m_2} & \frac{c_2}{m_2} & -\frac{c_2+c_3}{m_2} & \frac{c_3}{m_2} \\ 0 & \frac{k_3}{m_3} & -\frac{k_3}{m_3} & 0 & \frac{c_3}{m_3} & -\frac{c_3}{m_3} \end{bmatrix},
$$

$$
B_1 = [-1, 0, 0, 0, 0, 0]^T, \text{ and } B_2 = [1/m_1, 0, 0, 0, 0, 0]^T
$$

In the active control, the vibration control output response is defined as:

$$
Y(t) = \left\{ \dot{x}_1, \dot{x}_2, \dot{x}_3 \right\} \tag{6.38}
$$

Equation (6.38) can be rewritten in the state space form as follows:

$$
Y(t) = C_1 z + D_{11} \ddot{x}_g + D_{12} F_a(t) \tag{6.39}
$$

where

$$\mathrm{C}_1 = \begin{bmatrix} 1\,0\,0\,0\,0\,0 \\ 0\,1\,0\,0\,0\,0 \\ 0\,0\,1\,0\,0\,0 \end{bmatrix},\ \mathrm{D}_{11} = [0, 0, 0]^{\mathrm{T}} \text{ and } \mathrm{D}_{12} = [0, 0, 0]^{\mathrm{T}}.$$

In this study, the following observation output is assumed:

$$y(t) = \left[\dot{x}_1, \dot{x}_2, \dot{x}_3\right]^{\mathrm{T}} \tag{6.40}$$

Equation (6.40) can be expressed in the form of a state space as follows:

$$y(t) = \mathrm{C}_2 z + \mathrm{D}_{21}\, \ddot{x}_g + \mathrm{D}_{22} F_a(t) \tag{6.41}$$

where

$$\mathrm{C}_2 = \begin{bmatrix} 0\,0\,0\,1\,0\,0 \\ 0\,0\,0\,0\,1\,0 \\ 0\,0\,0\,0\,0\,1 \end{bmatrix}, \mathrm{D}_{21} = [0, 0, 0]^{\mathrm{T}} \text{ and } \mathrm{D}_{22} = [0, 0, 0]^{\mathrm{T}}.$$

When the input noise $\varepsilon_1(t)$ for Eq. (6.37) and measurement noise $\varepsilon_2(t)$ for Eq. (6.41) must be considered, the state space equations and observation equations can be modified to:

$$\dot{z} = \mathrm{A}z + \mathrm{B}_1\, \ddot{x}_g + \mathrm{B}_2 F_a(t) + \varepsilon_1(t) \tag{6.42a}$$

$$y(t) = \mathrm{C}_2 z + \mathrm{D}_{21}\, \ddot{x}_g + \mathrm{D}_{22} F_a(t) + \varepsilon_2(t) \tag{6.42b}$$

where $\varepsilon_1(t)$ and $\varepsilon_2(t)$ are the Gaussian random white noise with zero mean.

The LQG optimal control theory is also used to eliminate the noise and obtain better active control.

$$F_a(t) = -K\hat{y}(t) \tag{6.43}$$

where K is the optimal feedback control gain generated by the LQR control algorithm, $\hat{y}(t)$ is the estimation of $y(t)$, which is the solution required to construct a Kalman filter as explained before.

The active control strategy using the PSO-LQG active control method is studied based on the PSO algorithm. The fitness function is defined as follows:

$$\max\left|\ddot{x}_1\right| \tag{6.44}$$

The parameters of the PSO algorithm are 100 particle populations and 50 iterations, with $c_1 = 2$, $c_2 = 1$, and $\omega = 0.99^t$ (t is the tth iteration). The search range is:

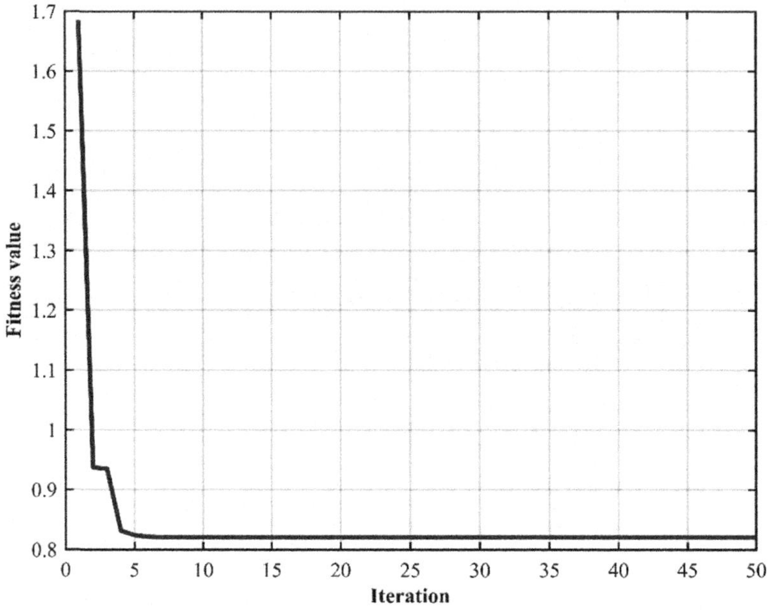

Fig. 6.33 Fitness convergence

$$
\begin{aligned}
&\left[1 \times 10^{-6}, 1 \times 10^{-6}, 1 \times 10^{-6}, 1 \times 10^{-6}, 1 \times 10^{-6}, 1 \times 10^{-6}, 1 \times 10^{-6}\right] \\
&\sim \left[1 \times 10^{6}, 1 \times 10^{6}, 1 \times 10^{6}, 1 \times 10^{6}, 1 \times 10^{6}, 1 \times 10^{6}\right]
\end{aligned}.
$$

The obtained fitness convergence curve is shown in Fig. 6.33, and the obtained gbest solution is:

$$
\begin{aligned}
&Q = diag\left[0.715\ 0.113\ 9.925\ 0.069\ 5.594\ 0.080\right] \times 10^{5}, \\
&R = 1 \times 10^{-6}.
\end{aligned}
$$

The seismic vibration induced displacement response of each floor of the isolated structure equipped with active control system is shown in Fig. 6.34. The black dashed line represents the isolated structure without active control, and the red solid line represents the isolated structure with active control. The results indicate that the maximum displacement of floor 1 of the structure without active control is 1.69 cm, and the maximum of the structure with active control is 0.95 cm, and the maximums of floor 2 are respectively 1.06 cm and 0.47 cm, and the maximums of floor 3 are respectively 1.38 cm and 0.53 cm.

The output force of actuator is shown in Fig. 6.35.

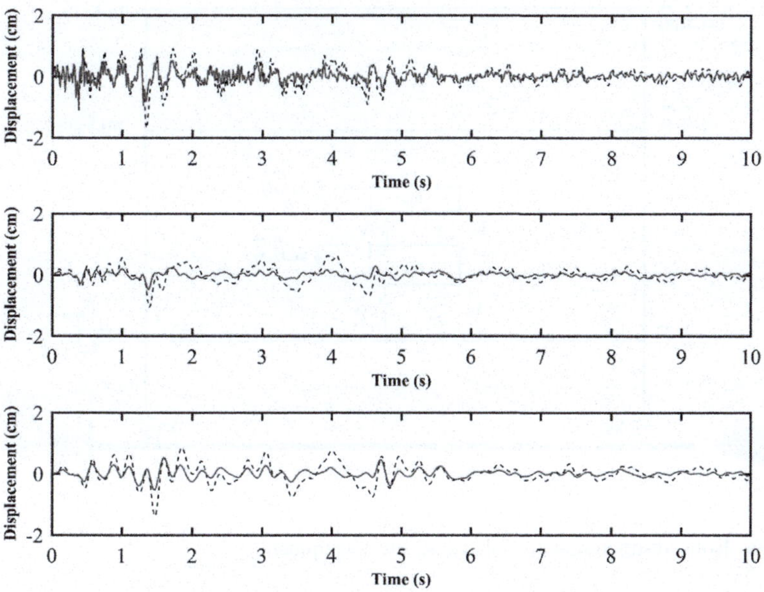

Fig. 6.34 The seismic vibration induced displacement response of each floor of the isolated structure equipped with active control

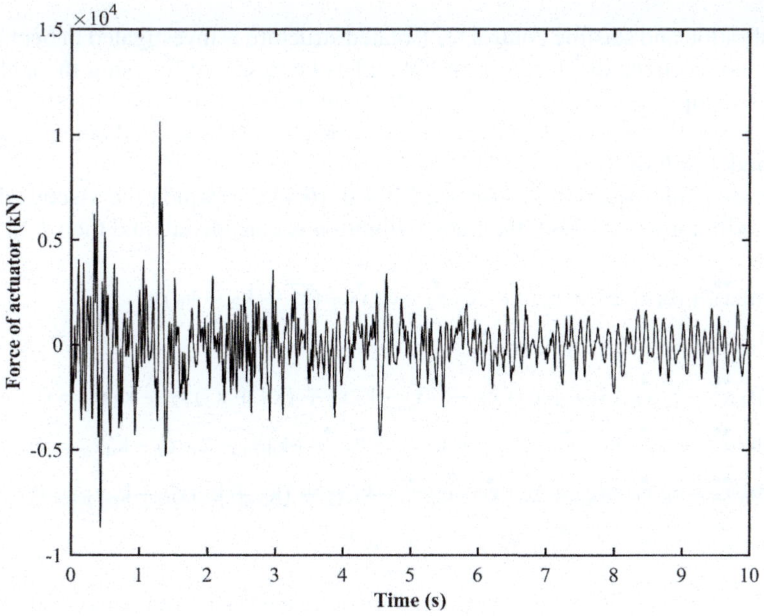

Fig. 6.35 The output force of actuator

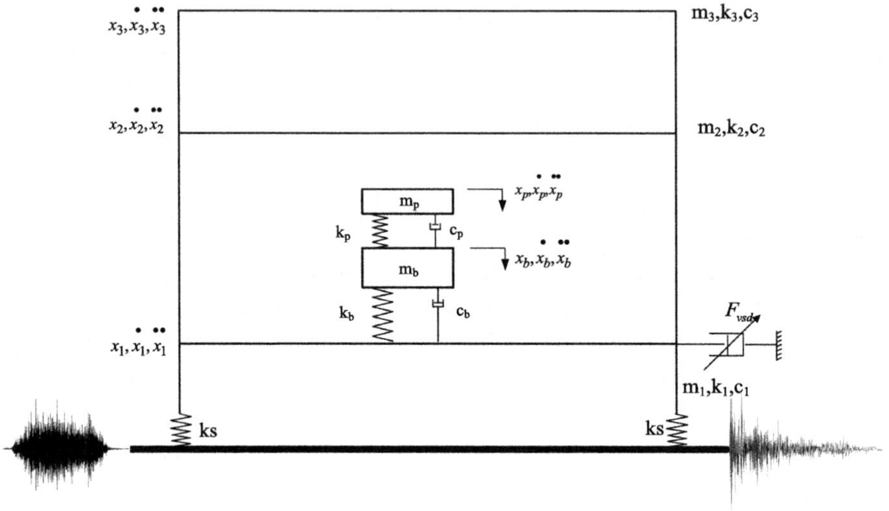

Fig. 6.36 Isolated structure with isolated sensitive equipment

6.3.4 *Isolated Frame Structure with Isolated Sensitive Equipment*

The vibration and seismic control for isolated structure is investigated in Sect. 6.3.1–6.3.4, and the large displacement problem has been solved by using the additional viscous damper and active control system. On basis of this, the investigation for isolated structure with isolated sensitive equipment will be carried out here, and the new model is described in Fig. 6.36.

The dynamic equation is expressed in Eq. (6.44), in which the viscous damper nearly will not work if only the transit vibration is considered, and then $F_{vsd} = 0$.

$$\begin{cases} \mathrm{m_1}\ddot{x}_1 + (\mathrm{c_1} + \mathrm{c_2} + \mathrm{c_b})\dot{x}_1 - \mathrm{c_2}\dot{x}_2 - \mathrm{c_b}\dot{x}_b + (\overline{\mathrm{k_1}} + \mathrm{k_2} + \mathrm{k_b})x_1 \\ \quad - \mathrm{k_2}x_2 - \mathrm{k_b}x_b = -\mathrm{m_1}\ddot{x}_t \\ \mathrm{m_2}\ddot{x}_2 - \mathrm{c_2}\dot{x}_1 + (\mathrm{c_2} + \mathrm{c_3})\dot{x}_2 - \mathrm{c_3}\dot{x}_3 - \mathrm{k_2}x_1 + (\mathrm{k_2} + \mathrm{k_3})x_2 - \mathrm{k_3}x_3 = 0 \\ \mathrm{m_3}\ddot{x}_3 - \mathrm{c_3}\dot{x}_2 + \mathrm{c_3}\dot{x}_3 - \mathrm{k_3}x_2 + \mathrm{k_3}x_3 + \mathrm{c_d}\dot{x}_3 + \mathrm{k_d}x_3 - \mathrm{c_d}\dot{x}_d - \mathrm{k_d}x_d = 0 \\ \mathrm{m_b}\ddot{x}_b - \mathrm{c_b}\dot{x}_1 + (\mathrm{c_b} + \mathrm{c_p})\dot{x}_b - \mathrm{c_p}\dot{x}_p - \mathrm{k_b}x_1 + (\mathrm{k_b} + \mathrm{k_p})x_b - \mathrm{k_p}x_p = 0 \\ \mathrm{m_p}\ddot{x}_p - \mathrm{c_p}\dot{x}_b + \mathrm{c_p}\dot{x}_p - \mathrm{k_p}x_b + \mathrm{k_p}x_p = 0 \end{cases} \tag{6.45}$$

Figure 6.37 shows the response comparison of the floor 1 of the isolated structure and the sensitive equipment induced by the transit vibration, and the blue dashed line represents the floor 1, and the black solid line represents the sensitive equipment. Seen from this, though the vibration of the isolated sensitive equipment has been

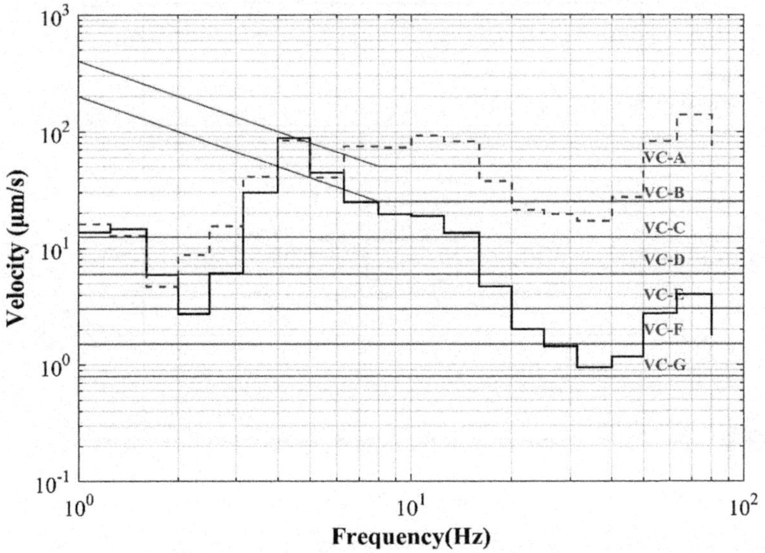

Fig. 6.37 The response comparison of the floor 1 of the isolated structure and the sensitive equipment induced by the transit vibration

improved a bit, the peak value is still existed and not reduced. For further suppressing the isolated vibration, an active control system will be performed for the sensitive equipment.

6.3.5 *Isolated Frame Structure with Actively Controlled Sensitive Equipment*

On the basis of the studies in Sect. 6.3.4, a proposed active control for the sensitive equipment is performed here, and the new model is described in Fig. 6.38.

$$\begin{cases} \mathrm{m_1}\ddot{x}_1 + (\mathrm{c_1}+\mathrm{c_2}+\mathrm{c_b})\dot{x}_1 - \mathrm{c_2}\dot{x}_2 - \mathrm{c_b}\dot{x}_b + (\overline{\mathrm{k_1}}+\mathrm{k_2}+\mathrm{k_b})x_1 \\ \quad - \mathrm{k_2}x_2 - \mathrm{k}_b x_b = -\mathrm{m_1}\ddot{x}_t \\ \mathrm{m_2}\ddot{x}_2 - \mathrm{c_2}\dot{x}_1 + (\mathrm{c_2}+\mathrm{c_3})\dot{x}_2 - \mathrm{c_3}\dot{x}_3 - \mathrm{k_2}x_1 + (\mathrm{k_2}+\mathrm{k_3})x_2 - \mathrm{k_3}x_3 = 0 \\ \mathrm{m_3}\ddot{x}_3 - \mathrm{c_3}\dot{x}_2 + \mathrm{c_3}\dot{x}_3 - \mathrm{k_3}x_2 + \mathrm{k_3}x_3 + \mathrm{c_d}\dot{x}_3 + \mathrm{k_d}x_3 - \mathrm{c_d}\dot{x}_d - \mathrm{k_d}x_d = 0 \\ \mathrm{m_b}\ddot{x}_b - \mathrm{c_b}\dot{x}_1 + (\mathrm{c_b}+\mathrm{c_p})\dot{x}_b - \mathrm{c_p}\dot{x}_p - \mathrm{k_b}x_1 + (\mathrm{k_b}+\mathrm{k_p})x_b - \mathrm{k_p}x_p = F_a(t) \\ \mathrm{m_p}\ddot{x}_p - \mathrm{c_p}\dot{x}_b + \mathrm{c_p}\dot{x}_p - \mathrm{k_p}x_b + \mathrm{k_p}x_p = -F_a(t) \end{cases} \tag{6.46}$$

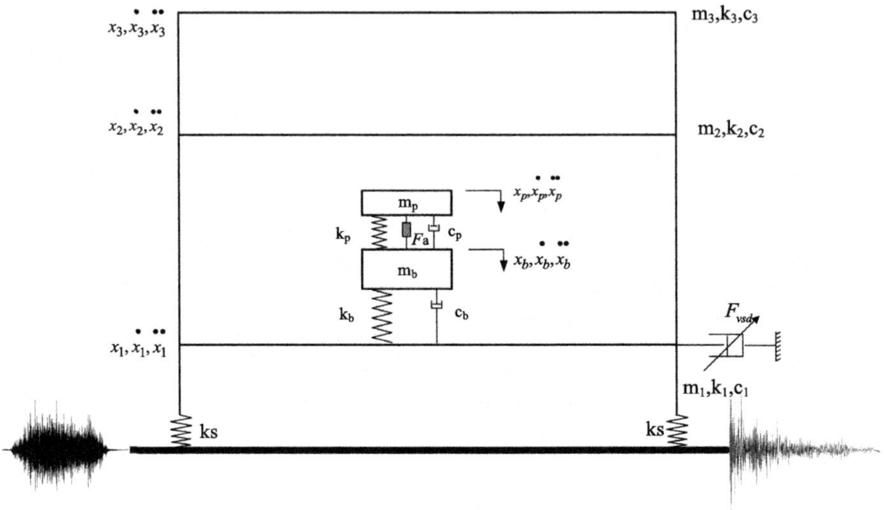

Fig. 6.38 Isolated structure with actively controlled sensitive equipment

In the active control system represented by Eq. (6.46), the state variables are $z_1 = x_1$, $z_2 = x_2$, $z_3 = x_3$, $z_4 = x_b$, $z_5 = x_p$, $z_6 = \dot{x}_1$, $z_7 = \dot{x}_2$, $z_8 = \dot{x}_3$, $z_9 = \dot{x}_b$ and $z_{10} = \dot{x}_p$, and the state vector is $z = \{z_1, z_2, z_3, z_4, z_5, z_6, z_7, z_8, z_9, z_{10}\}^{\mathrm{T}}$. The state space form is rewritten as follows:

$$\dot{z} = \mathrm{A}z + \mathrm{B}_1 \ddot{x}_t + \mathrm{B}_2 F_a(t) \tag{6.47}$$

where

$$A = \begin{bmatrix} 0 & 0 & 0 & 1 & 0 & 0 \\ 0 & 0 & 0 & 0 & 1 & 0 \\ 0 & 0 & 0 & 0 & 0 & 1 \\ -\frac{\mathrm{k}_1+\mathrm{k}_2}{\mathrm{m}_1} & \frac{\mathrm{k}_2}{\mathrm{m}_1} & 0 & -\frac{\mathrm{c}_1+\mathrm{c}_2}{\mathrm{m}_1} & \frac{\mathrm{c}_2}{\mathrm{m}_1} & 0 \\ \frac{\mathrm{k}_2}{\mathrm{m}_2} & -\frac{\mathrm{k}_2+\mathrm{k}_3}{\mathrm{m}_2} & \frac{\mathrm{k}_3}{\mathrm{m}_2} & \frac{\mathrm{c}_2}{\mathrm{m}_2} & -\frac{\mathrm{c}_2+\mathrm{c}_3}{\mathrm{m}_2} & \frac{\mathrm{c}_3}{\mathrm{m}_2} \\ 0 & \frac{\mathrm{k}_3}{E_3} & -\frac{\mathrm{k}_3}{\mathrm{m}_3} & 0 & \frac{\mathrm{c}_3}{\mathrm{m}_3} & -\frac{\mathrm{c}_3}{\mathrm{m}_3} \end{bmatrix}$$

$\mathrm{B}_1 = [-1, 0, 0, 0, 0, 0, 0, 0, 0, 0]^{\mathrm{T}}$, and $\mathrm{B}_2 = [0, 0, 0, 0, 0, 0, 0, 0, 1/\mathrm{m}_\mathrm{b}, -1/\mathrm{m}_\mathrm{p}]^{\mathrm{T}}$.

In the active control, the vibration control output is defined as:

$$Y(t) = \left\{\dot{x}_\mathrm{p}\right\} \tag{6.48}$$

Equation (6.48) can be rewritten in the state space form as follows:

$$Y(t) = \mathrm{C}_1 z + \mathrm{D}_{11}\ddot{x}_t + \mathrm{D}_{12} F_a(t) \tag{6.49}$$

where $\mathrm{C}_1 = \begin{bmatrix} 0\,0\,0\,0\,0\,0\,0\,0\,0\,0\,1 \end{bmatrix}$, $\mathrm{D}_{11} = [0]$, and $\mathrm{D}_{12} = [0]$.

The following observation output is assumed here:

$$y(t) = \left[\dot{x}_1, \dot{x}_2, \dot{x}_3, \dot{x}_\mathrm{b}, \dot{x}_\mathrm{p}\right]^\mathrm{T} \tag{6.50}$$

Equation (6.50) can be expressed in the form of a state space as follows:

$$y(t) = \mathrm{C}_2 z + \mathrm{D}_{21}\ddot{x}_t + \mathrm{D}_{22} F_a(t) \tag{6.51}$$

where

$$\mathrm{C}_2 = \begin{bmatrix} 0\,0\,0\,0\,0\,1\,0\,0\,0\,0 \\ 0\,0\,0\,0\,0\,0\,1\,0\,0\,0 \\ 0\,0\,0\,0\,0\,0\,0\,1\,0\,0 \\ 0\,0\,0\,0\,0\,0\,0\,0\,1\,0 \\ 0\,0\,0\,0\,0\,0\,0\,0\,0\,1 \end{bmatrix},$$

$$\mathrm{D}_{21} = [0, 0, 0, 0, 0]^\mathrm{T} \text{ and } \mathrm{D}_{22} = [0, 0, 0, 0, 0]^\mathrm{T}.$$

When the input noise $\varepsilon_1(t)$ for Eq. (6.46) and measurement noise $\varepsilon_2(t)$ for Eq. (6.51) must be considered, the state space equations and observation equations can be modified to:

$$\dot{z} = \mathrm{A}z + \mathrm{B}_1\ddot{x}_t + \mathrm{B}_2 F_a(t) + \varepsilon_1(t) \tag{6.52a}$$

$$y(t) = \mathrm{C}_2 z + \mathrm{D}_{21}\ddot{x}_t + \mathrm{D}_{22} F_a(t) + \varepsilon_2(t) \tag{6.52b}$$

The LQG optimal control theory is also used to eliminate the noise.

$$F_a(t) = -K\hat{y}(t) \tag{6.53}$$

The PSO-LQG active control method is also studied here based on the PSO algorithm. The fitness function is considered as follows:

$$\max\left|\dot{x}_\mathrm{p}\right| \tag{6.54}$$

The parameters of the PSO algorithm are 100 particle populations and 50 iterations, with $c_1 = 2$, $c_2 = 1$, and $\omega = 0.99^t$ (t is the tth iteration). The search range is:

$$\begin{array}{l}\left[1\times10^{-6},1\times10^{-6},1\times10^{-6},1\times10^{-6},1\times10^{-6},1\times10^{-6},1\times10^{-6}\right]\\ \quad\sim\left[1\times10^{6},1\times10^{6},1\times10^{6},1\times10^{6},1\times10^{6},1\times10^{6}\right]\end{array}.$$

The obtained fitness convergence curve is shown in Fig. 6.39, and the obtained gbest solution is:

$$Q = diag\left[6.062\ 3.121\ 1.980\ 2.987\ 2.847\ 2.433\ 9.438\ 5.732\ 0.047\ 7.046\right]\times 10^5,$$
$$R = 6.938\times 10^{-5}$$

Figure 6.40 shows the response comparison of the floor 1 and the actively controlled sensitive equipment induced by the transit vibration, and the blue dashed line represents the sensitive equipment with passive isolation system, and the black solid line represents the actively controlled sensitive equipment. Seen from this, the active control system can improve the isolation effect a lot, and the vibration in the whole frequency band can be reduced and suppressed. Figure 6.41 gives the output force of the actuator.

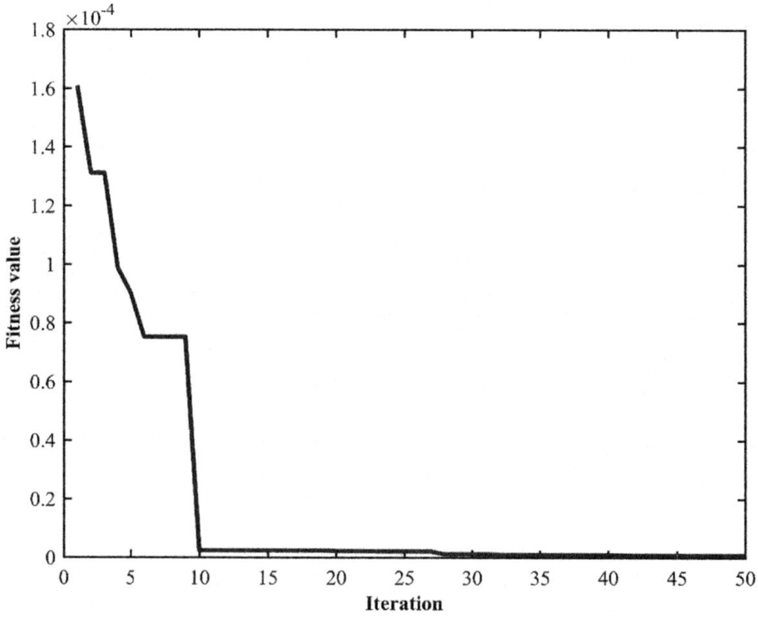

Fig. 6.39 Fitness convergence

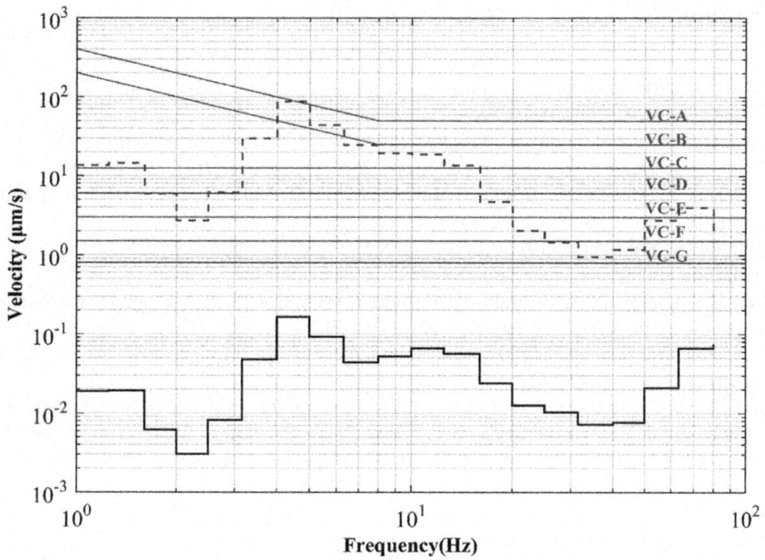

Fig. 6.40 Response comparison of the floor 1 and the actively controlled sensitive equipment induced by the transit vibration

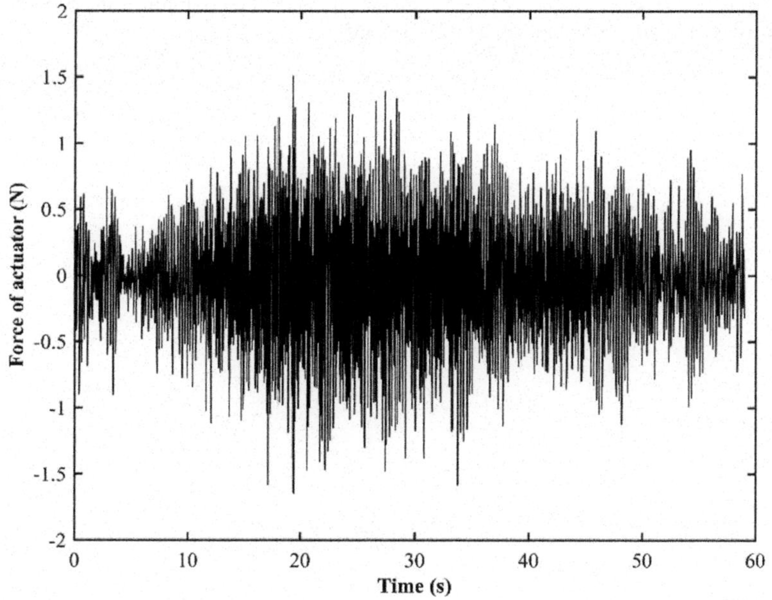

Fig. 6.41 The output force of actuator

References

1. Ming TYG (2006) Analysis on control of wind induced vibration of a super-tall building with TMD. J Vib Shock 25(2):16–19
2. Nagarajaiah S, Varadarajan N (2005) Short time Fourier transform algorithm for wind response control of buildings with variable stiffness TMD. Eng Struct 27(3):431–441
3. Hu X, Hu S (2000) Researches on application of TMD to seismic control. Earthq Eng Eng Vib 20(2):112–116
4. Ankireddi S, Yang HTY (1996) Simple ATMD control methodology for tall buildings subject to wind loads. J Struct Eng 122(1):83–91
5. Guclu R, Yazici H (2008) Vibration control of a structure with ATMD against earthquake using fuzzy logic controllers. J Sound Vib 318(1):36–49
6. Adhikari R, Yamaguchi H (1997) Sliding mode control of buildings with ATMD. Earthquake Eng Struct Dynam 26(4):409–422
7. Cao L, Li A, Chen X et al (2010) Vibration serviceability control of a long-span floor in large station room under crowd-induced excitation. China Civil Eng J 2010(S1):334–340
8. Murray TM, Allen D, Ungar E (1997) Steel design guide series 11. Floor vibrations due to human activity
9. Hao Y, Qi H, Liu S et al (2022) Study of noise and vibration impacts to buildings due to urban rail transit and mitigation measures. Sustainability 14(5):3119
10. Shi L X, Zhang N (2011) The simulation analysis on vibration of a museum building nearby induced by urban subway transit. Adv Mater Res 243:3427–3431
11. Zhang X, Le Cao D (2006) Investigation on the vibration effect induced by rail transit system on urban environment. Appl Adv Technol Transp, pp 117–122
12. Crispino M, D'apuzzo M (2001) Measurement and prediction of traffic-induced vibrations in a heritage building. J Sound Vib 246(2):319–335

Chapter 7
Passive and Active Control Using Refined FEM Analysis

Abstract In this chapter, passive and active control using refined FEM analysis is performed out. Firstly, passive control using refined FEM analysis with single-stage and two-stage vibration isolation system is studied. Based on this, active control using refined FEM analysis is investigated, and a finite element calculation strategy for active control is proposed in FEM environment by using ANSYS. Followed by, additional base isolation, additional viscous damper, and additional active vibration control is adopted in the performance improvement for vibration isolation base.

7.1 Passive Control Using Refined FEM Analysis

7.1.1 Single-Stage Vibration Isolation System

Figure 7.1 shows the single-stage vibration isolation system. The bottom of the vibration isolation system is fixed to the ground, or the foundation mass is regarded as infinite with respect to the sensitive equipment.

Consider the single-stage vibration isolation system of the sensitive equipment shown in Fig. 7.1. The interference load is taken as the ground vibration input, and the vibration circle frequency is defined as $\omega = 2\pi f$. The vibration displacement, velocity, and acceleration of the sensitive equipment are x, $\dot{x}$ and $\ddot{x}$, respectively, and the system's equation of motion is:

$$\mathrm{m}\dot{x} + \mathrm{c}\dot{x} + \mathrm{k}x = \mathrm{k}x_g + \mathrm{c}\dot{x}_g \tag{7.1}$$

The analytical calculation of the single-stage vibration isolation system's passive control based on MATLAB/SIMULINK is shown in Fig. 7.2. The ground vibration disturbances are simple harmonic displacement excitation with an amplitude x_g of 1×10^{-3} m, a frequency of 4 Hz, a velocity $\dot{x}_g$ defined as the derivative of the displacement excitation x_g, a sensitive equipment mass of m = 1000 kg, a vibration control system vertical stiffness of k = 3.94×10^4 N/m, a damping ratio of ζ = 0.02, and a damping coefficient c calculated from the damping ratio.

W. Huang and J. Xu, *Optimized Engineering Vibration Isolation, Absorption and Control*, Springer Tracts in Civil Engineering, https://doi.org/10.1007/978-981-99-2213-0_7

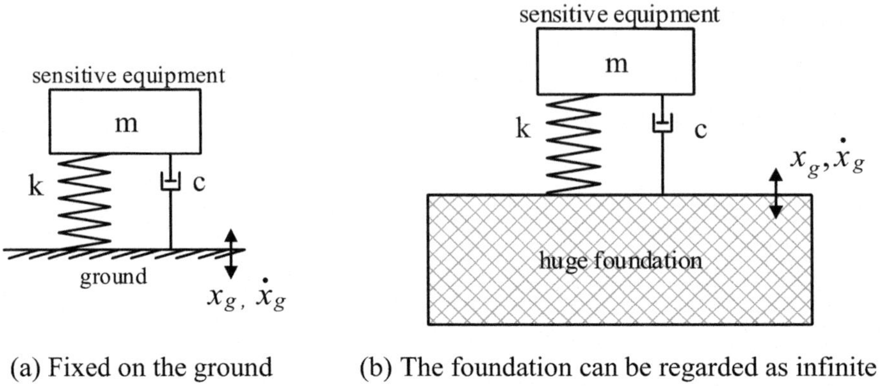

Fig. 7.1 Single-stage vibration isolation system

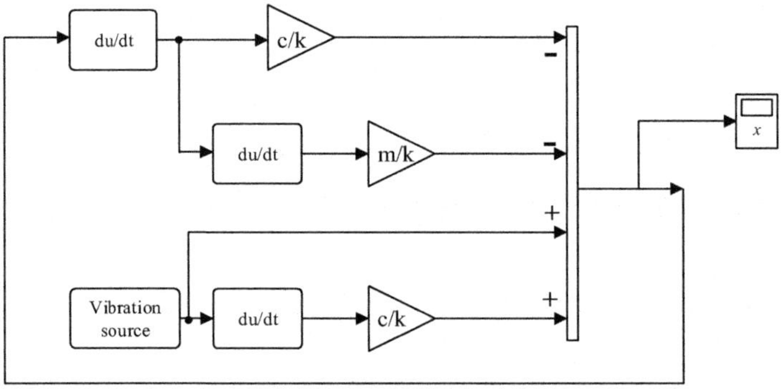

Fig. 7.2 MATLAB/SIMULINK analytical calculation for passive control of single-stage vibration isolation system

The finite element calculation was carried out using the ANSYS/APDL for the passive control system. Figure 7.3 depicts the finite element model of the single-stage passive control. The sensitive equipment was considered as a mass21 single point mass element with a solid65 solid element, a combin14 spring damper element, and the same interference excitation and vibration control system parameters.

Figure 7.4 compares analytical and finite element calculation results, indicating that the error between the two objects is small. In addition, the calculation results of the solid element and the point mass element are very close, which lays the foundation for the refined analysis of the next two-stage vibration control system.

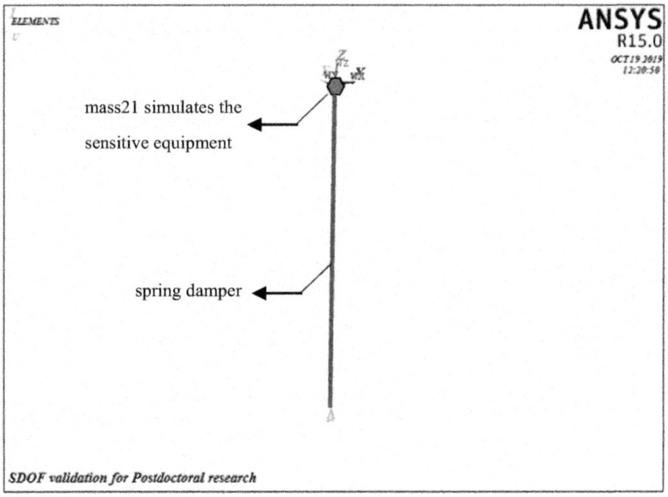

(a) Simulation of the sensitive equipment using the point mass element

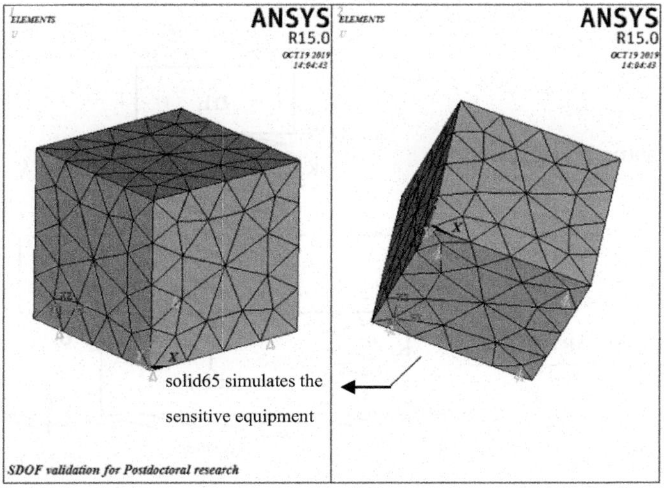

(b) Simulation of the sensitive equipment using the solid element

Fig. 7.3 ANSYS/APDL finite element modeling of the single-stage passive control

7.1.2 Two-Stage Vibration Isolation System

Figure 7.5 shows the two-stage vibration isolation system. The supporting foundation of this system with respect to the control object cannot be ignored. Otherwise, there will be a secondary support system. The vibration isolation system of the sensitive equipment is called the primary system.

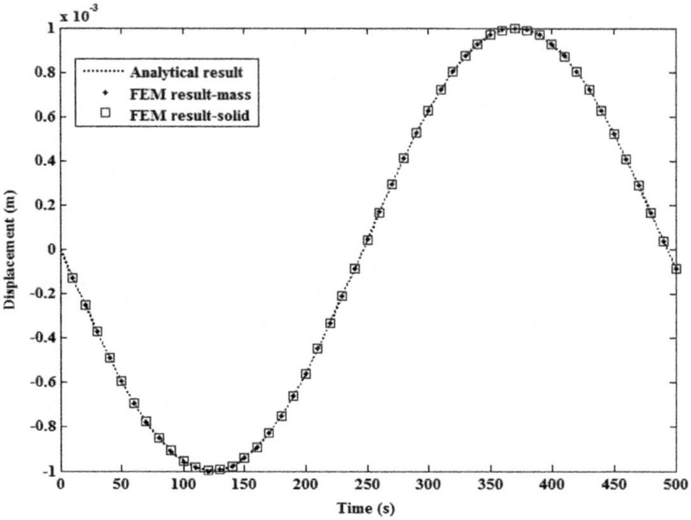

Fig. 7.4 Comparison between the analytical and the FEM results of the single-stage passive control

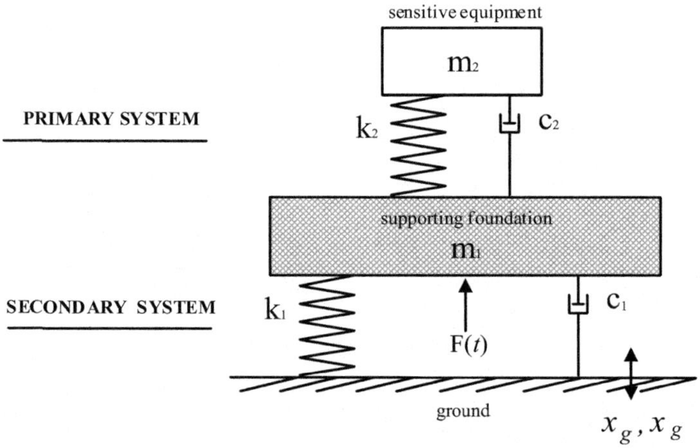

Fig. 7.5 Two-stage vibration isolation system

Consider the two-stage vibration isolation system of the sensitive equipment shown in Fig. 7.5, where m_2 is the sensitive equipment mass, k_2 and c_2 are the stiffness and damping of the primary vibration isolation system, respectively, m_1, k_1, and c_1 are the mass, stiffness, and damping of the secondary vibration isolation system, respectively. The vibration load is consistent with the single-stage system, and the displacement, velocity, and acceleration of sensitive equipment are x_2, $\dot{x}_2$ and

$\ddot{x}_2$, respectively, the displacement, velocity, and acceleration of the supporting foundation are x_1, $\dot{x}_1$ and $\ddot{x}_1$, respectively, and the vibration load excited on the supporting foundation is $F(t)$. The equation of motion of this system is defined as follows:

$$\begin{cases} m_1\ddot{x}_1 - c_2\dot{x}_2 + (c_2 + c_1)\dot{x}_1 - k_2x_2 + (k_2 + k_1)x_1 = k_1x_g + c_1\dot{x}_g + F(t) \\ m_2\ddot{x}_2 + c_2\dot{x}_2 - c_2\dot{x}_1 + k_2x_2 - k_2x_1 = 0 \end{cases} \tag{7.2}$$

The state variables are set to $z_1 = x_1$, $z_2 = x_2$, $z_3 = \dot{x}_1$, and $z_4 = \dot{x}_2$, and the state vector is defined as $z = [z_1, z_2, z_3, z_4]^T$ (7.3). Equation (7.2) can be transformed into the following state space form:

$$\begin{cases} \dot{z}(t) = \mathbf{A}z(t) + \mathbf{B}U(t) \\ y(t) = \mathbf{C}z(t) + \mathbf{D}U(t) \end{cases} \tag{7.3}$$

where

$$\mathbf{A} = \begin{bmatrix} 0 & 0 & 1 & 0 \\ 0 & 0 & 0 & 1 \\ -\frac{k_1+k_2}{m_1} & \frac{k_2}{m_1} & -\frac{c_1+c_2}{m_1} & \frac{c_2}{m_1} \\ \frac{k_2}{m_2} & -\frac{k_2}{m_2} & \frac{c_2}{m_2} & -\frac{c_2}{m_2} \end{bmatrix}, \mathbf{B} = \begin{bmatrix} 0 & 0 & 0 \\ 0 & 0 & 0 \\ \frac{1}{m_1} & \frac{k_1}{m_1} & \frac{c_1}{m_1} \\ 0 & 0 & 0 \end{bmatrix}, \mathbf{C} = \begin{bmatrix} 0\,1\,0\,0 \\ 0\,0\,0\,1 \end{bmatrix},$$

$$\mathbf{D} = \begin{bmatrix} 0\,0\,0 \\ 0\,0\,0 \end{bmatrix}, \text{ and } U(t) = \begin{bmatrix} F(t) & x_g(t) & \dot{x}_g(t) \end{bmatrix}^T$$

Figure 7.6 shows the analytical modeling of the two-stage vibration isolation system's passive control based on MATLAB/SIMULINK. If only the vibration load $F(t)$ acting on the support foundation is considered, and the ground vibration excitation x_g, $\dot{x}_g$ are not considered, then the amplitude of $F(t)$ is 100 N, the frequency is $f = 0.6$ Hz, the sensitive equipment mass is $m_2 = 100$ kg, the supporting foundation mass is $m_1 = 1200$ kg, the secondary system's stiffness and damping are $k_1 = 1 \times 10^6$ N/m, $c_1 = 1.6 \times 10^4$ N s/m, respectively, and the primary system's stiffness and damping are $k_2 = 1.5 \times 10^4$ N/m and $c_2 = 1 \times 10^3$ N s/m, respectively.

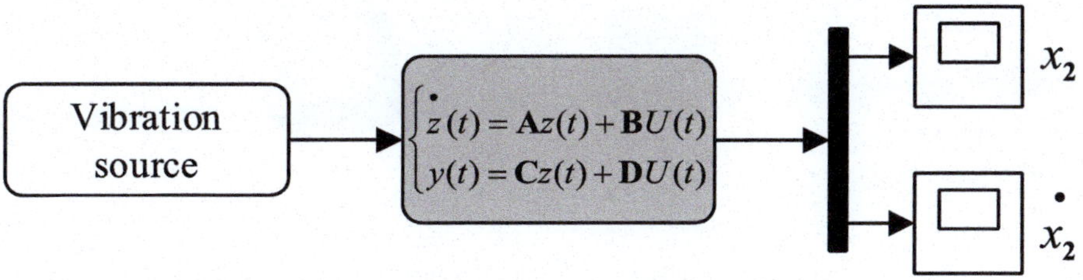

Fig. 7.6 MATLAB/SIMULINK analytical calculation for passive control of the two-stage vibration isolation system

The finite element calculation for this two-stage system was performed using ANSYS/APDL. The finite element model of the two-stage passive control system is shown in Fig. 7.7. The sensitive equipment and foundation were considered as point mass element (mass21) and solid element (solid65), respectively. Additionally, combin14 was used to simulate the spring damper element, and the interference excitation and vibration control system parameters are the same as above.

Figure 7.8 compares the analytical and finite element calculation results, which show that the errors between the two are small and the fitting result is very good. In addition, the calculation results of the solid and mass elements are very close.

The refined analysis of the single-stage and two-stage passive control lay the foundation for the subsequent active control research based on the finite element method.

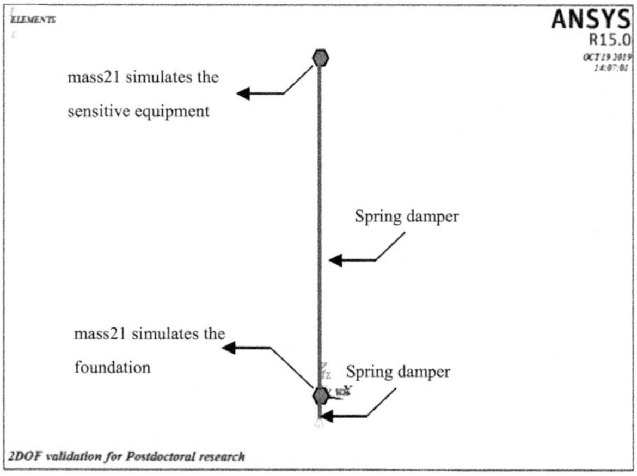

(a) Equipment and foundation simulation using the point mass element

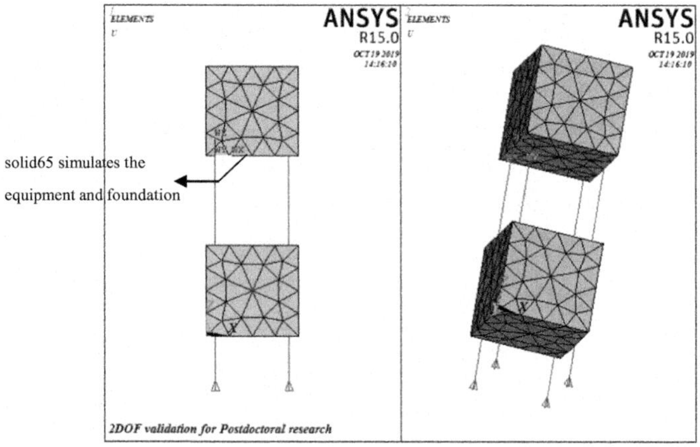

(b) Equipment and foundation simulation using the solid element

Fig. 7.7 ANSYS/APDL finite element calculation of the two-stage passive control

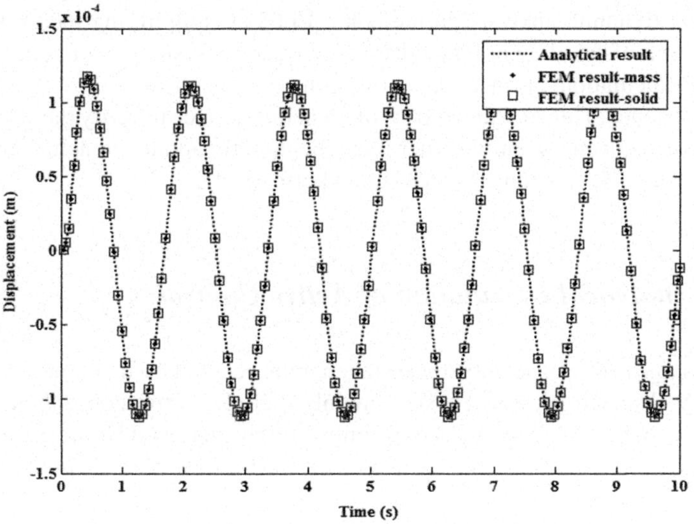

Fig. 7.8 Comparison between the analytical and finite element results of the two-stage passive control

This chapter used analytical and finite element calculations to model the single-stage and two-stage passive control systems. The results indicate that the errors between both methods are small, which verifies their reliability and highlights the way for active control using refined FEM analysis.

7.2 Active Control Using Refined FEM Analysis

In Sect. 7.1, analytical and finite element calculations were carried out for single-stage and two-stage passive control systems. Practical engineering cases often include a secondary supporting structure for sensitive equipment or the foundation mass with respect to the sensitive equipment mass cannot be ignored. In other words, the vibration control system is mostly a two-stage system.

Passive control cannot be autonomously adjusted based on the changes in the external vibration sources due to the fixed parameters [1, 2]. In recent years, scholars have extensively studied active control in vehicle suspension [3], building structure [4], sensitive precision equipment [5], and other fields.

Simplified systems or analytical calculations based on mass points often fail to reflect the actual state. Thereafter, when the computed active control is added to the actual engineering structure, there will be significant differences, especially for the abnormal structure, which will cause repeated debugging or even diverged systems upon completing the active control installation.

Currently, the research on active control is mainly focused on control theories and optimization algorithms, including LQR/LQG [6], PID [7], and H_2/H_∞ [8]. When performing dynamic analysis on the control object, simplifying the system into a single-degree-of-freedom or multi-degree-of-freedom system is common practice in analytical calculations. Hence, the active control research was carried out based on these approaches. This research type is of great importance to verify the active control algorithm effectiveness, but whether it can be directly applied to actual engineering and the control effect cannot be determined effectively.

7.2.1 Analytical Calculation of Active Control

The two-stage active control system is taken into account based on Fig. 7.5, as shown in Fig. 7.9. The active control system mainly includes active controllers, actuators, and sensors, where $F_a(t)$ is the active control force generated by the actuator.

The dynamic equation can be expressed as follows regardless of the ground vibration excitation:

$$\begin{cases} m_1 \ddot{x}_1 + k_1 x_1 + c_1 \dot{x}_1 - c_2(\dot{x}_2 - \dot{x}_1) - k_2(x_2 - x_1) = F(t) - F_a(t) \\ m_2 \ddot{x}_2 + c_2(\dot{x}_2 - \dot{x}_1) + k_2(x_2 - x_1) = F_a(t) \end{cases} \tag{7.4}$$

The state variables are $z_1 = x_1, z_2 = x_2, z_3 = \dot{x}_1$, and $z_4 = \dot{x}_2$, and the state vector is $z(t) = [z_1, z_2, z_3, z_4]^{\mathrm{T}}$. Equation (7.4) can be transformed into the following state space function:

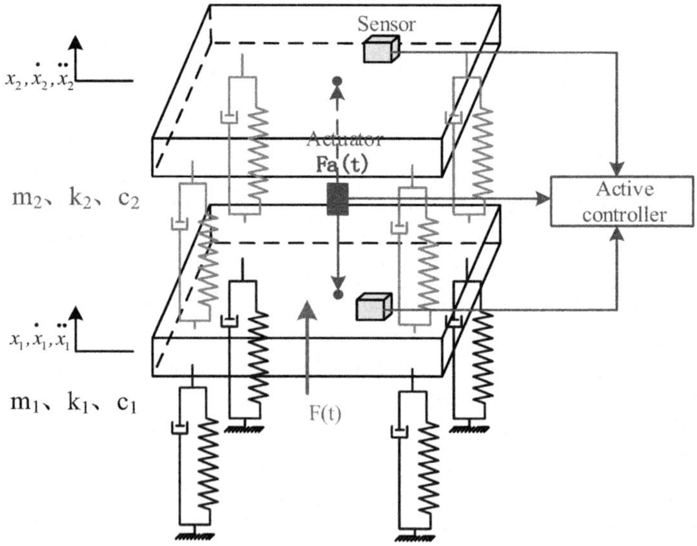

Fig. 7.9 Two-stage active control system

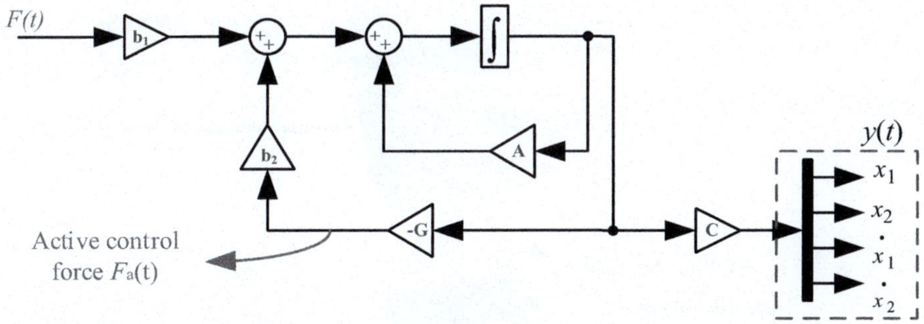

Fig. 7.10 LQR active control

$$\begin{cases} \dot{z}(t) = \mathbf{A}z(t) + \mathbf{b}_1 F(t) + \mathbf{b}_2 F_a(t) \\ y(t) = \mathbf{C}z(t) + \mathbf{d}_1 F(t) + \mathbf{d}_2 F_a(t) \end{cases} \tag{7.5}$$

where

$$\mathrm{A} = \begin{bmatrix} 0 & 0 & 1 & 0 \\ 0 & 0 & 0 & 1 \\ -\frac{k_1+k_2}{m_1} & \frac{k_2}{m_1} & -\frac{c_1+c_2}{m_1} & \frac{c_2}{m_1} \\ \frac{k_2}{m_2} & -\frac{k_2}{m_2} & \frac{c_2}{m_2} & -\frac{c_2}{m_2} \end{bmatrix},\ \mathbf{b_1} = \begin{bmatrix} 0 \\ 0 \\ \frac{1}{m_1} \\ 0 \end{bmatrix},\ \mathbf{b_2} = \begin{bmatrix} 0 \\ 0 \\ -\frac{1}{m_1} \\ \frac{1}{m_2} \end{bmatrix},$$

$$\mathbf{C} = \begin{bmatrix} 1\,0\,0\,0 \\ 0\,1\,0\,0 \\ 0\,0\,1\,0 \\ 0\,0\,0\,1 \end{bmatrix},\ \mathrm{d}_1 = \begin{bmatrix} 0 \\ 0 \\ 0 \\ 0 \end{bmatrix},\ \text{and } \mathbf{d_2} = \begin{bmatrix} 0 \\ 0 \\ 0 \\ 0 \end{bmatrix}$$

The active controller is computing using the LQR control theory. Thus, the feedback gain $\boldsymbol{G}$ is determined first, and then the active control force $F_a(t) = -\boldsymbol{G}z(t)$ is calculated. The computation process is shown in Fig. 7.10.

The vibration isolation system parameters are $m_1 = 1200$ kg, $k_1 = 1 \times 10^6$ N/m, $c_1 = 1.6 \times 10^4$ N s/m, $m_2 = 600$ kg, $k_2 = 1.5 \times 10^4$ N/m, $c_2 = 1.0 \times 10^3$ N s/m, and $F(t) = 100 \times \sin(2 \times 3.14 \times 0.6)$ N. The weight matrix in the LQR control was defined as $\boldsymbol{Q} = \text{diag}[6.121, 7.281, 4.476, 1.308] \times 10^6$ and $R = 0.05$, and the active control was programmed and performed using MATLAB/SIMULINK. The active controller is $\boldsymbol{G} = [-2.4138, 0.4251, -0.2132, 0.4735] \times 10^4$.

7.2.2 Finite Element Calculation of Active Control

The finite element model of Fig. 7.9 was developed using ANSYS/APDL, with the dimensions of m_1 and m_2 being $1.0 \times 1.0 \times 1.0$ m. The solid element solid65 was

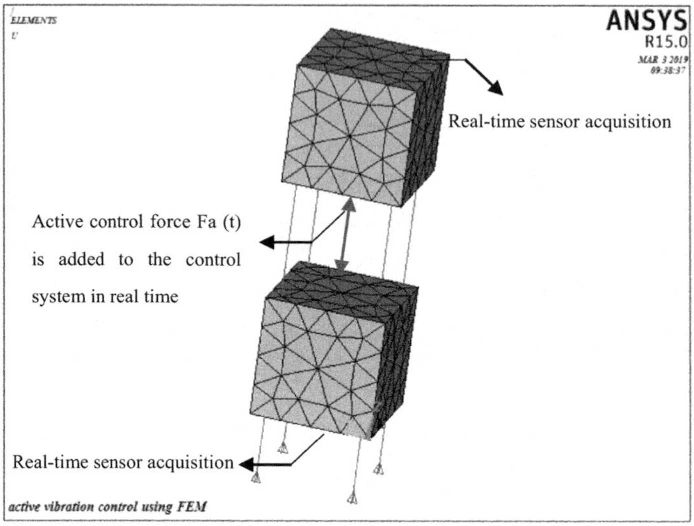

Fig. 7.11 Finite element model of the active control

adopted for modeling the sensitive equipment and foundation (Sect. 7.1 has verified that the calculation error between the point mass element and the solid element is small; however, the solid element can simulate the system's real geometry, so the solid element is used), and the combin14 element was utilized for the spring damper. There are 4 combin14 elements in the primary system to model the primary vibration isolators' arrangement and 4 combin14 elements in the secondary system to simulate the secondary vibration isolators' arrangement or the stiffness of the soil foundation. The finite element model is shown in Fig. 7.11.

Figure 7.12 shows the modal calculation results, the 1st- to 6th-order represents the overall vibration isolation characteristics, and the 7th and higher orders represent the local characteristics of the sensitive equipment and foundation. Figure 7.13 depicts a comparison between the analytical and finite element vibration response results of a point on m_2 without active control. In general, the results indicate that the error is small, the fitting result is good, and the proposed strategy is reliable. Hence, the active control calculation can be continued along this technical route.

Since the active control is based on the real-time response and forces feedback, it must be calculated using ANSYS' time-history transient analysis module. According to the state variables in Sect. 7.1.2, the vibration objects of this study are x_1, $\dot{x}_1$, x_2, and $\dot{x}_2$, and the vibration points are A and B in Fig. 7.14. In the designed analysis program, ANSYS can obtain these variables in real-time and input them to the active controller $\boldsymbol{G}$, from which the active control force F_a(t) can be calculated in real-time. Additionally, the force is directly applied and excited on the system. The action points are C and D, as shown in Fig. 7.14. The detailed calculation process is shown in Fig. 7.15.

```
*** MASS SUMMARY BY ELEMENT TYPE ***

TYPE      MASS
  1    1200.00
  2    600.000

Range of element maximum matrix coefficients in global coordinates
Maximum = 2.912895241E+10 at element 369.
Minimum = 3750 at element 1088.

 *** ELEMENT MATRIX FORMULATION TIMES
TYPE    NUMBER   ENAME        TOTAL CP   AVE CP

  1       529   SOLID65        0.062    0.000118
  2       552   SOLID65        0.062    0.000113
  3         8   COMBIN14       0.000    0.000000
```

(a) Element characteristics

***** PARTICIPATION FACTOR CALCULATION ***** Z DIRECTION

MODE	FREQUENCY	PERIOD	PARTIC. FACTOR	RATIO	EFFECTIVE MASS
1	0.775206	1.2900	0.71380E-05	0.000000	0.509517E-10
2	0.775234	1.2899	0.44856E-04	0.000001	0.201205E-08
3	0.789695	1.2663	25.235	0.739896	636.793
4	4.54541	0.22000	0.68504E-03	0.000020	0.469277E-06
5	4.54572	0.21999	0.21893E-02	0.000064	0.479300E-05
6	4.62971	0.21600	34.106	1.000000	1163.21
7	2317.71	0.43146E-03	0.14767E-05	0.000000	0.218072E-11
8	2346.82	0.42611E-03	0.72884E-06	0.000000	0.531213E-12
9	2943.45	0.33974E-03	0.73598E-06	0.000000	0.541667E-12
10	2946.31	0.33941E-03	-0.25768E-06	0.000000	0.663971E-13
11	2957.05	0.33817E-03	0.12949E-05	0.000000	0.167669E-11
12	2967.80	0.33695E-03	0.28357E-05	0.000000	0.804140E-11
13	2979.09	0.33567E-03	0.97560E-06	0.000000	0.951791E-12
14	2986.91	0.33479E-03	0.20291E-05	0.000000	0.411739E-11
15	3206.58	0.31186E-03	-0.21180E-04	0.000001	0.448583E-09
16	3210.12	0.31152E-03	0.67789E-04	0.000002	0.459530E-08
17	3246.54	0.30802E-03	-0.22187E-09	0.000000	0.492260E-19
18	3293.19	0.30366E-03	0.10155E-09	0.000000	0.103117E-19
19	3351.56	0.29837E-03	0.43047E-04	0.000001	0.185300E-08
20	3356.78	0.29790E-03	-0.22880E-04	0.000001	0.523506E-09
sum					1800.00

The designed vertical frequency of the primary vibration isolation system is 0.79Hz!

(b) Modal frequencies

Fig. 7.12 Modal calculation results

Figures 7.16 and 7.17 compare the system's response using the analytical and finite element methods without and with active control, respectively. The results show that active control can significantly improve the system's vibration response, reflecting active control's role. Moreover, it shows that it is feasible to implement active control based on the finite element approach.

Figure 7.18 shows the system response's analytical and finite element solutions under active control. By combining Figs. 7.16 and 7.17, it can see that both of them can induce the active control effect. The active control using the finite element method is based on the active controller obtained by the analytical solution, and the calculated result is better than the uncontrolled state. The response curve law and trend of both objects are consistent, which confirms the validity and reliability of the finite element solution. It is worth mentioning that the approximation degree of the system response's analytical solution and finite element solution in the active control state is somewhat different from that in the uncontrolled state (Fig. 7.13). This is because the active control is based on real-time pickup and feedback of vibration

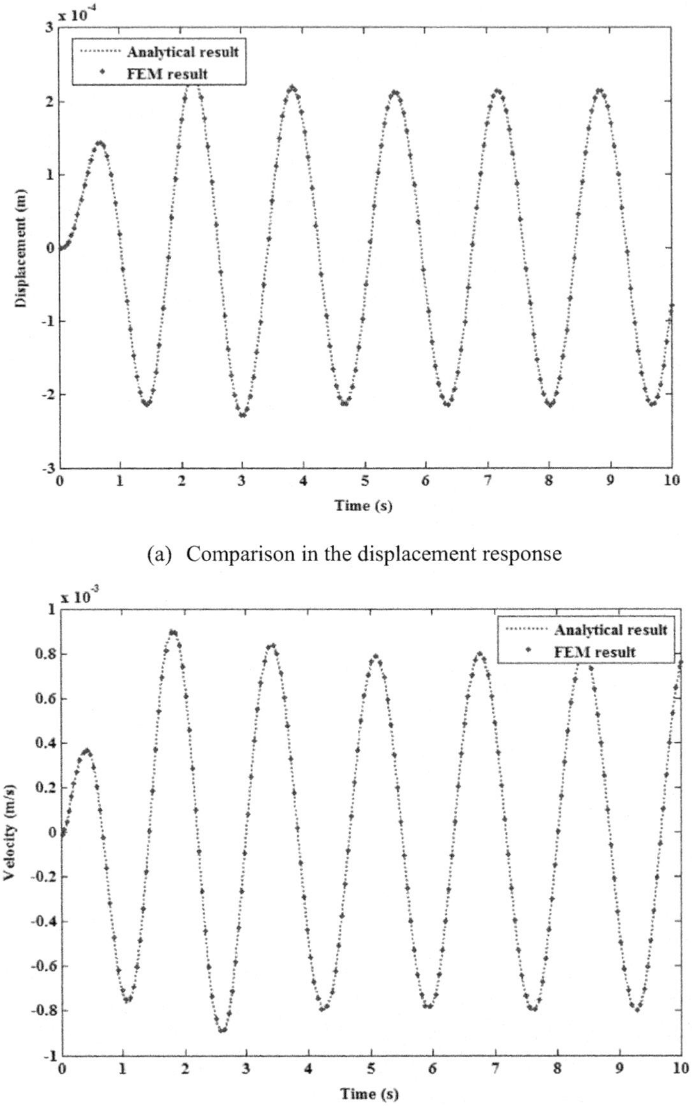

(a) Comparison in the displacement response

(b) Comparison in the velocity response

Fig. 7.13 Comparison between the analytical and finite element solutions without active control

response, the object of the analytical solution is a single point mass, but the finite element solution is any node picked when the entity is divided into elements. Each node has a vibration mode contribution, leading to a definite response difference.

This section proposes a novel analysis strategy of active control calculation based on the finite element method, and analytical and finite element calculations without

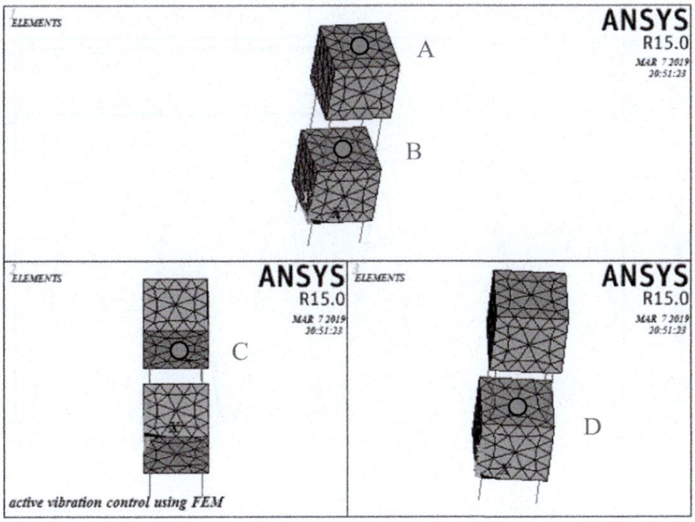

Fig. 7.14 Pickup points and active control force action points in the finite element model

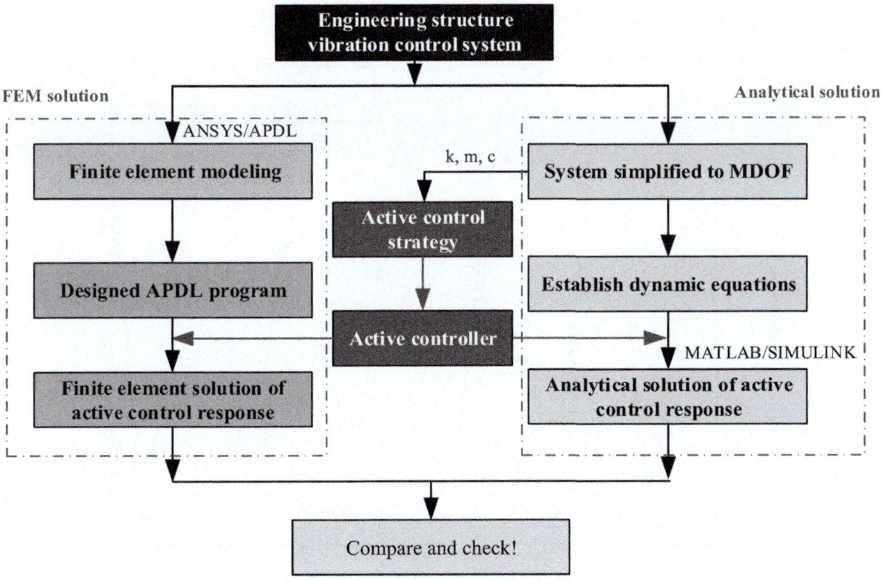

Fig. 7.15 Finite element calculation diagram for active control

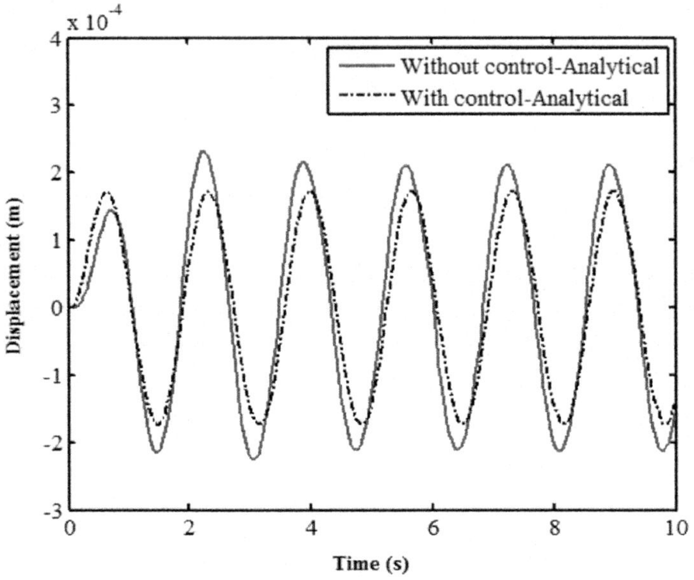

(a) Displacement response comparison

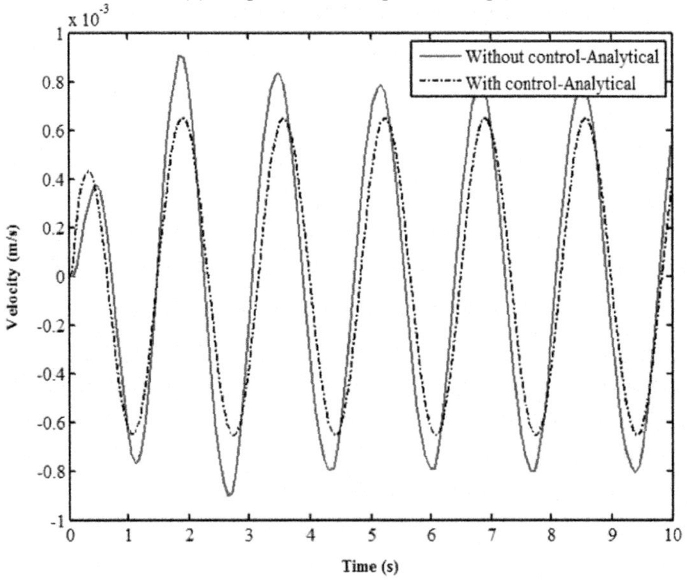

(b) Velocity response comparison

Fig. 7.16 Comparison between the analytical solutions for estimating the system response with and without active control

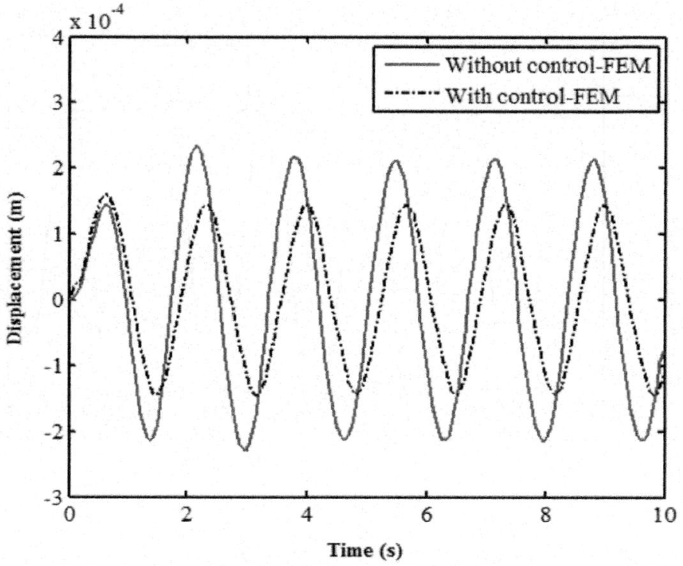

(a) Displacement response comparison

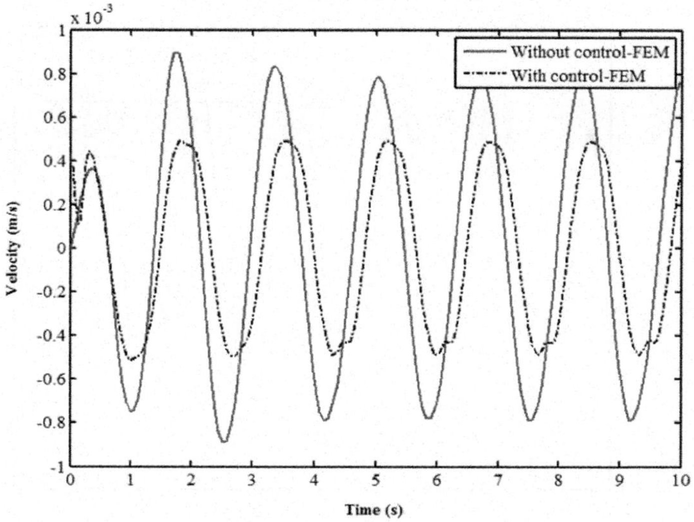

(b) Velocity response comparison

Fig. 7.17 Comparison between the FEM solutions for estimating the system response with and without active control

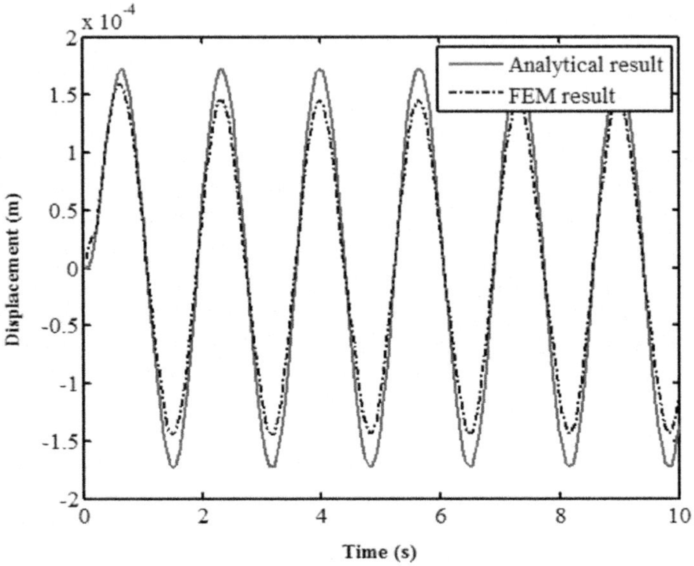

(a) Displacement response comparison

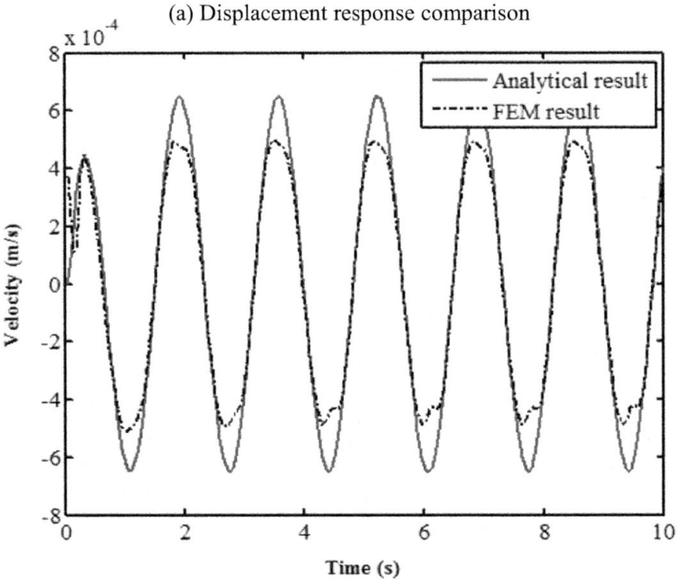

(b) Displacement response comparison

Fig. 7.18 Comparison between the analytical and finite element solutions for modeling the system response with active control active control and with active control are carried out, respectively. The results verify the rationality of the finite element solution. Furthermore, this study provides an effective calculation method for the active control of actual engineering structures. Compared to the simplified single-degree-of-freedom or multi-degree-of-freedom systems, this proposed strategy can better reflect the true state of the structure. In actual engineering, the analysis and finite element calculation of active control should be carried out simultaneously. The former can guide the latter, but the latter is more convincing. The active control theories are diverse, and the research field around the finite element and analytical methods is still vast.

7.3 Performance Improvement of Vibration Isolation Base Using FEM

The vibration isolation base station is widely used in the vibration control of power equipment and precision equipment. The system mainly includes the platen (with underhanging), vibration isolators, and necessary limit devices. The vibration isolators can be linear steel spring dampers or nonlinear air spring vibration isolators. With modern vibration isolation technology and new materials development, smart devices such as the aforementioned magnetorheological fluid, eddy current damper, and active piezoelectric ceramics are gradually being used [9, 10].

In actual projects, the vibration isolation abutment causes horizontal, vertical, and torsional vibrations under environmental vibration and other sudden disturbances (such as explosion, earthquake, and construction dynamic compaction). Ensuring the safety of the base station and the vibration isolation equipment under large disturbances, as well as effectively exerting the performance of the vibration isolation abutment, is a critical issue.

This section carries out additional base vibration isolation, viscous dampers, and active control research for the designed vibration isolation base station to improve its vibration isolation performance.

7.3.1 Design of Vibration Isolation Abutment

The reinforced concrete vibration isolation abutment shown in Fig. 7.19 considers the decoupling of the degrees of freedom based on the superposition of the mass and stiffness. The size of the platen is $1 \times 1 \times 0.1$ m, and the size of the underhanging is $0.5 \times 0.5 \times 0.2$ m. Steel springs were used as vibration isolators to avoid the interference of damping in the subsequent research, where the steel spring vibration isolators do not consider damping. The system's vertical design frequency is 3.5 Hz, and the horizontal stiffness is 0.7 times that of the vertical stiffness. In the ANSYS environment, the abutment is simulated using a solid65 element, and the steel spring is

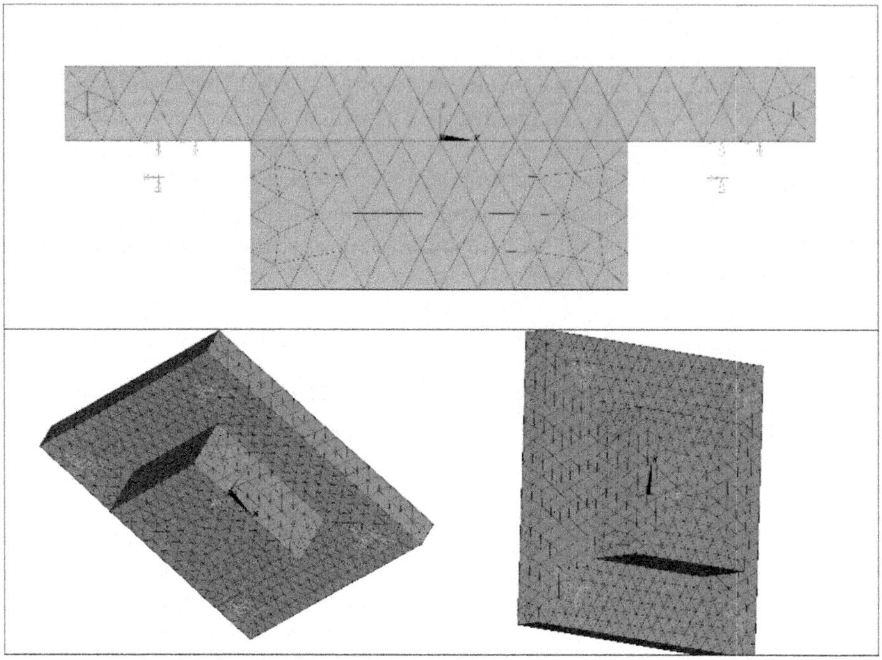

Fig. 7.19 Numerical calculation model of vibration isolation abutment

simulated using a combin14 element [11]. Figure 7.20 shows the system's horizontal and vertical modes of vibration.

Consider the random white noise acceleration input shown in Fig. 7.21, the amplitude range is [−1, 1] m/s^2, the sampling frequency is 128 Hz, and it acts on the vibration isolation base station. Figure 7.22 shows the vertical frequency response of a point in the middle of the table.

7.3.2 *Additional Base Isolation*

When using the vibration isolation base station, in case of sudden explosion, earthquake, dynamic compaction and other serious vibration, we can consider installing base vibration isolators (such as rubber vibration isolation bearings) in series under the steel spring vibration isolator and use larger displacement to reduce and dissipate the abutment energy. The mechanical model of the vibration isolation support can be simplified to two horizontal nonlinear springs, dampers, and vertical linear springs. In the ANSYS program, the combin14 element can be used to simulate the vertical stiffness, and the combin40 element can be utilized to model both horizontal directions [11]. Figure 7.23 shows the mechanics of combin40, and Fig. 7.24 depicts a schematic diagram of the additional base vibration isolation.

Fig. 7.20 Horizontal and vertical modes of vibration isolation abutment

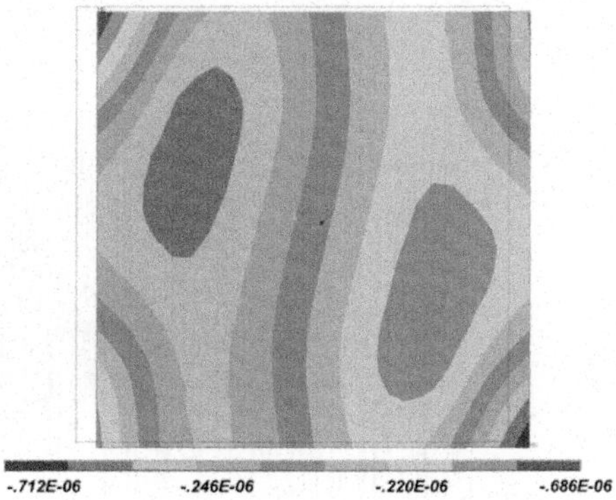

(a) Translation along the x direction (2.92 Hz)

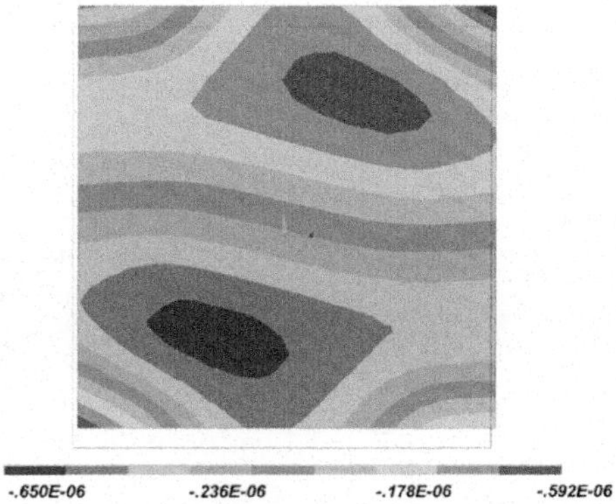

(b) Translation along the y direction (2.92 Hz)

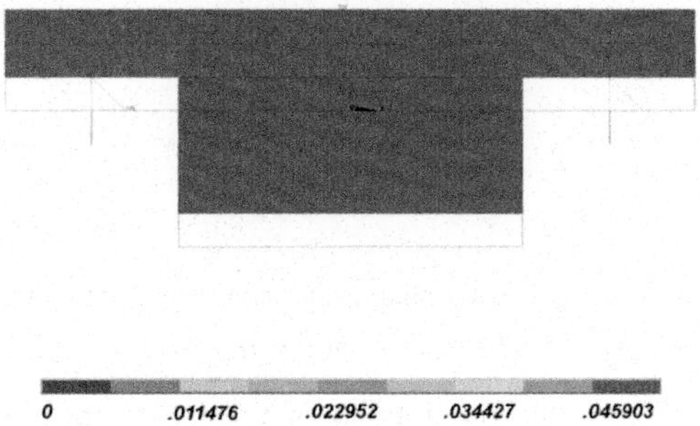

(c) Translation along the z direction (3.5 Hz)

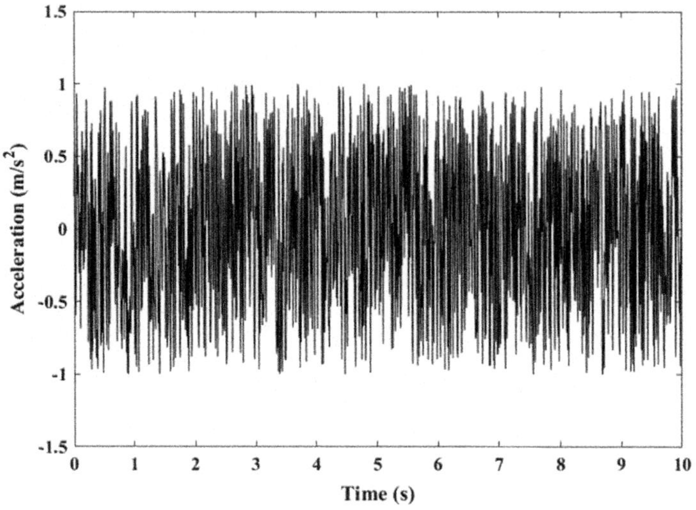

Fig. 7.21 Random white noise acceleration input

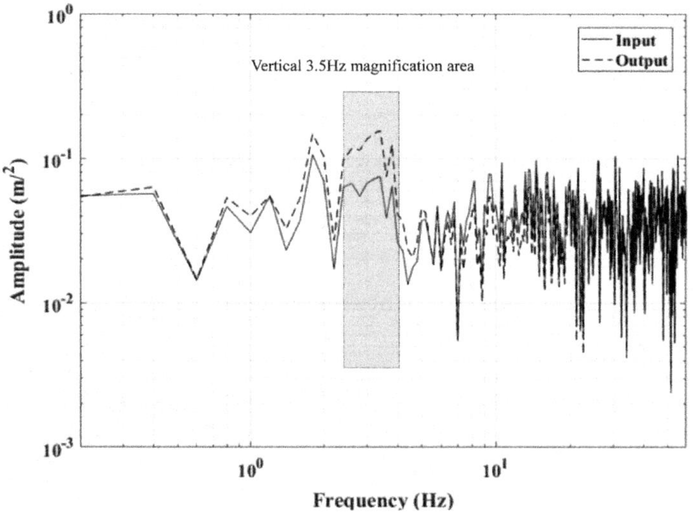

Fig. 7.22 Vertical frequency domain response of a point in the middle of the table

The basic parameters of the base vibration isolation support in the horizontal direction are stiffness before yielding (K_u), stiffness after yielding (K_d), yield force (Q_d), and damping ratio. The primary real constants in the combin40 element are K_1, C, m, GAP, FSLIDE, and K_2. Figure 7.22 shows the calculation and selection method of the real constants, where $K_2 = K_d$, $K_1 = K_u - K_d$, FSLIDE = Q_d, GAP = 0 (the contact surface is not open), and C is the damping coefficient. The support

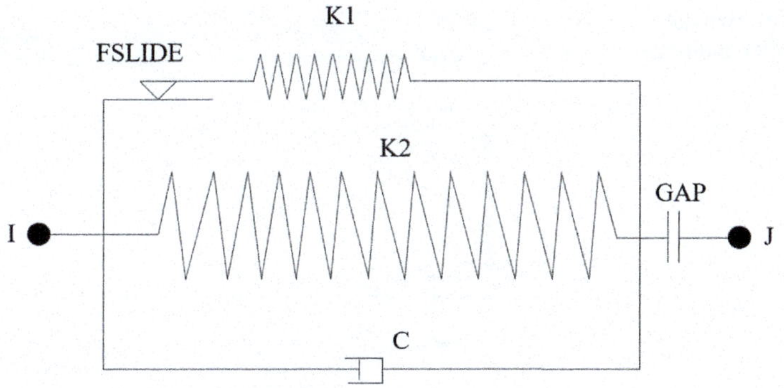

Fig. 7.23 Combin40 mechanics principle

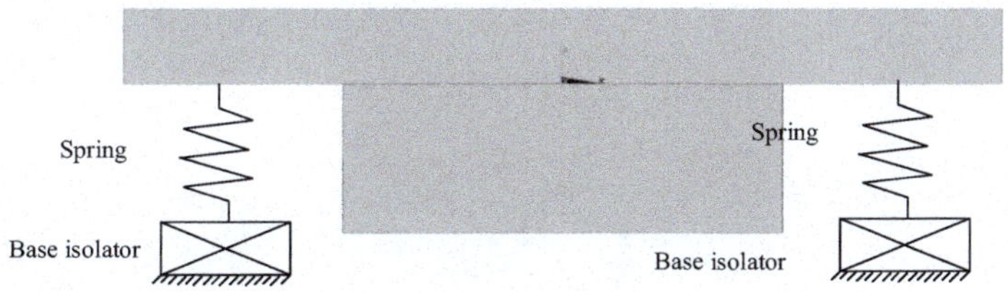

Fig. 7.24 Schematic diagram of the additional base vibration isolation

parameters selected in this study are a vertical stiffness of 2400 kN/mm, pre-yield stiffness of 15.29 kN/mm, post-yield stiffness of 1.18 kN/mm, yield force of 63 kN, and equivalent damping ratio of 23.

Figure 7.25 shows the horizontal response with additional isolation base. It can be seen that the abutment dissipates energy by generating a large displacement, and its acceleration response is significantly improved.

7.3.3 Additional Viscous Damper

Base vibration isolation requires large displacement to dissipate energy, where it is difficult to effectively control the displacement within the allowable range. In this case, viscous dampers can be considered. The typical mechanical model of the velocity-type viscous damper is as follows:

$$F = C|v|^n \mathrm{sgn}(v) \tag{7.6}$$

where F is the damping force, C is the damping coefficient, v is the relative speed, n is the speed index that ranges between 0.1 and 2.0.

Fig. 7.25 Horizontal response after additional base isolation

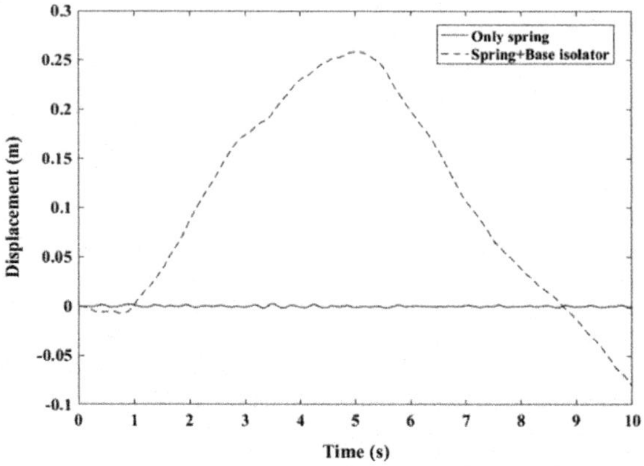

(a) Displacement response

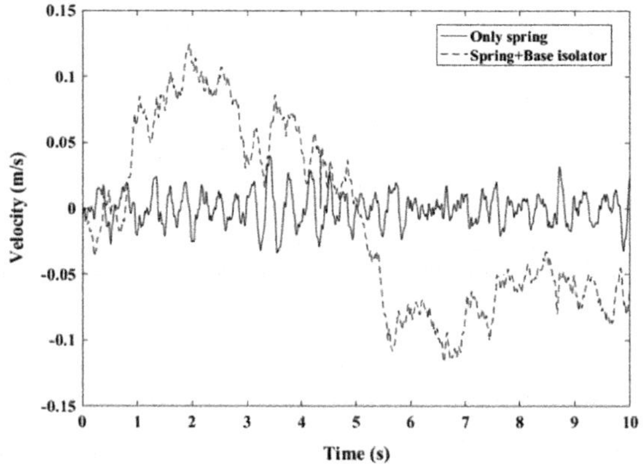

(b) Velocity response

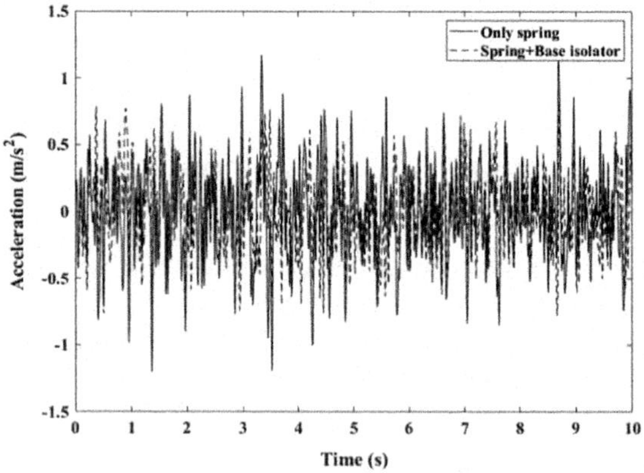

(c) Acceleration response

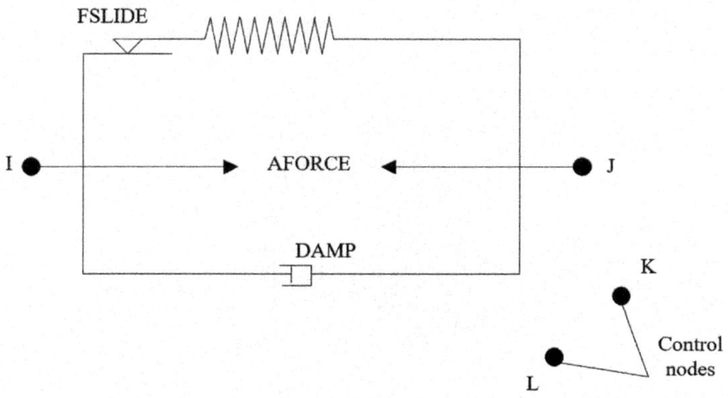

Fig. 7.26 Mechanical principle diagram of combin37

The combin37 element in ANSYS simulates the velocity-type viscous damper of the mechanical model in Eq. (7.6) through complicated parameter settings [11, 12]. The combin37 element is a nonlinear one-dimensional element, which consists of 2 element active nodes and optional 2 control nodes. The complex nonlinear behavior of the entire element is mainly generated using the control nodes. If the control nodes are not set, the element is the same as the general spring damper (Fig. 7.26).

In order to achieve the characteristics of the viscous damper, the element options are defined as follows:

KEYOPT(1) = 2,
KEYOPT(2) = 1,
KEYOPT(3) = 1 or 2 (1 represents the application of the x-direction damper, and 2 represents the application of the y-direction damper);
KEYOPT(4) = 1,
KEYOPT(5) = 1,
KEYOPT(6) = 2,
KEYOPT(9) = 0.

In this study, a viscous damper was applied in the x and y directions of the vibration isolation base, with a damping coefficient of 1000 N/(s m) and an exponential constant of 0.4. Figure 7.27 shows the displacement-damping force hysteresis curve of the viscous damper. In general, the x and y directions are the same due to the symmetry of the structure. It can be seen from Fig. 7.28 that the addition of a viscous damper can effectively reduce the horizontal displacement and deformation of the vibration isolation base and significantly decrease its velocity and acceleration response. It can be seen from Fig. 7.29 that once the horizontal viscous dampers are applied in both x and y directions, the vertical response of the vibration isolation abutment is not affected.

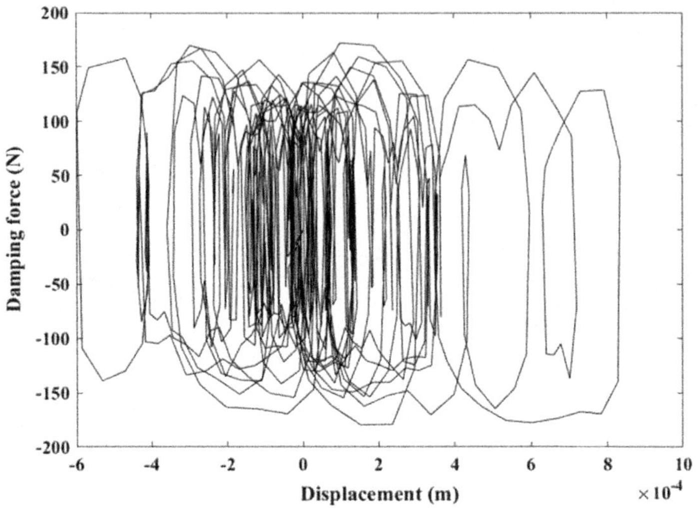

Fig. 7.27 Displacement-damping force hysteresis curve of the viscous damper

7.3.4 Additional Active Vibration Control

The vibration isolation platform mentioned above has been improved by utilizing the base vibration isolation and viscous dampers. These two methods are passive; hence, they rely on the inherent characteristics of the design base vibration isolator or viscous damper. When the system is designed, its structural parameters are fixed and cannot fully adapt to changes in environmental vibrations, such as amplitude changes, frequency changes, or excitation forms. Additionally, active control can be considered, which mainly consists of sensors, active controllers, and actuators.

Generally, when performing active control analysis on a controlled object, the system is often simplified into a single-degree-of-freedom or multi-degree-of-freedom system before carrying out active vibration control research. The simplified system cannot reflect the system's actual state. Moreover, there will be significant differences when the active control system is added to the actual engineering structure, especially the special-shaped structure, which causes repeated debugging upon the installation of the active control device and even diverges the system's actual response. Therefore, this research carries out active vibration control based on finite element analysis, as shown in Fig. 7.30. Previous research has pointed out that active control using FEM is effective, whereas compared to the simplified single-degree-of-freedom or multi-degree-of-freedom systems, it can better reflect the actual state of the structure [13].

Fig. 7.28 Comparison between the horizontal responses after introducing the viscous damper

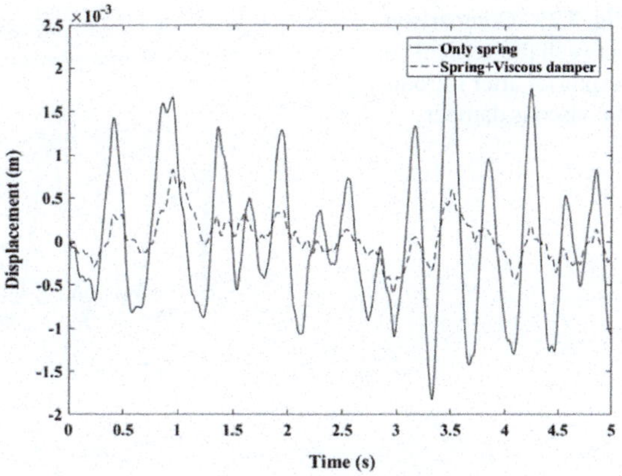

(a) Displacement response

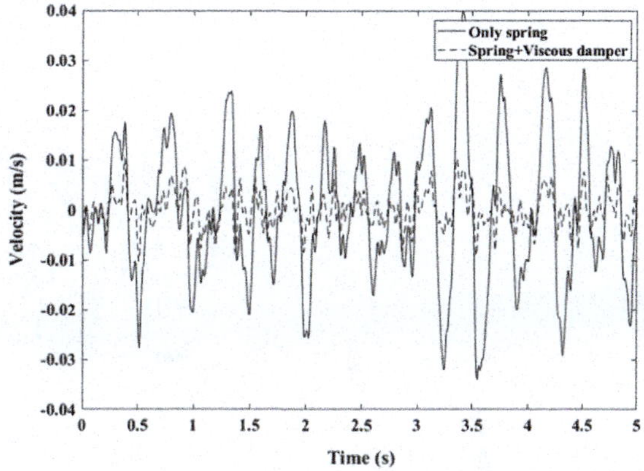

(b) Velocity response

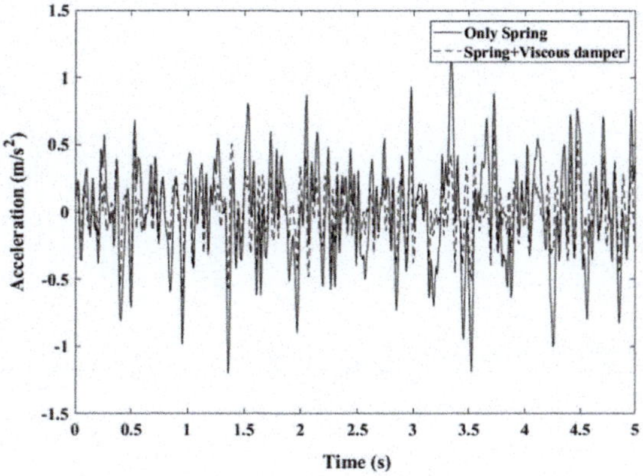

(c) Acceleration response

Fig. 7.29 Comparison between the vertical responses after introducing the viscous damper

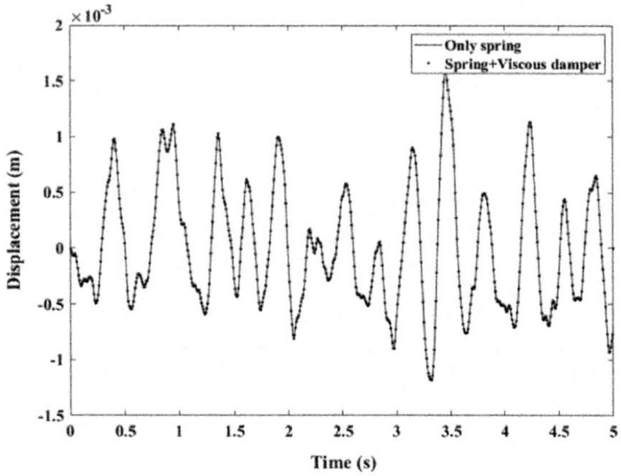

(a) Displacement response

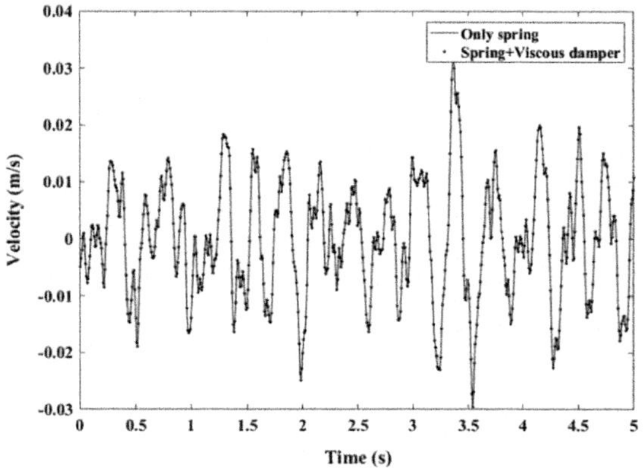

(b) Velocity response

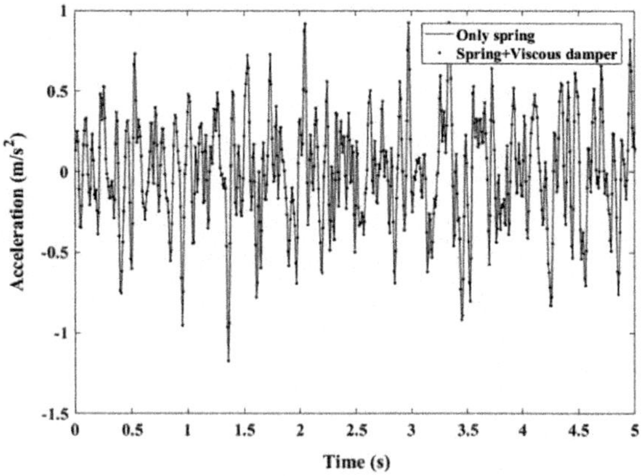

(c) Acceleration response

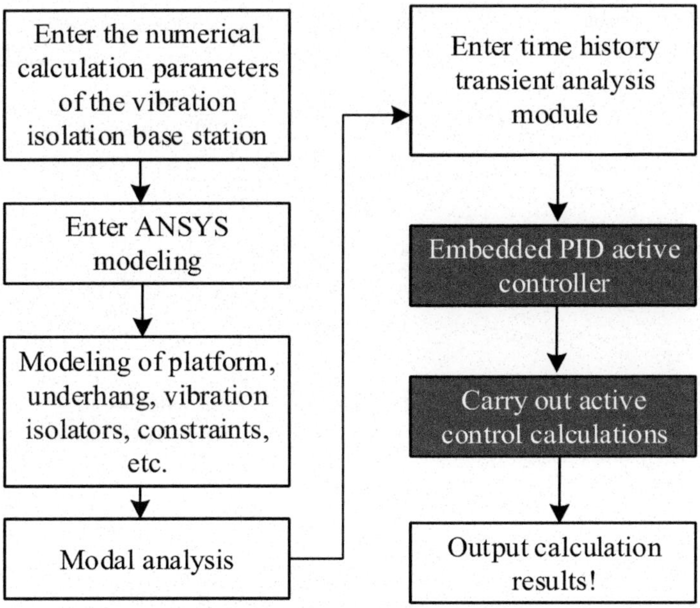

Fig. 7.30 Active control flowchart based on ANSYS/APDL

In this study, an active control system was applied in the x and y directions of the vibration isolation base station. As shown in Fig. 7.31, the active controller adopts PID control. It can be seen from Fig. 7.32 that using an active control system can effectively reduce the horizontal displacement, velocity, and acceleration responses in the base. Moreover, its vertical response is not affected. Figure 7.33 shows the active control's actuator output.

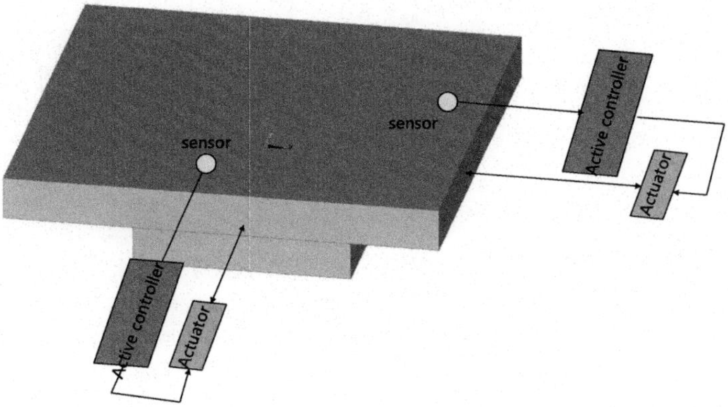

Fig. 7.31 Vibration isolation platform with additional active control

Fig. 7.32 Comparison between the horizontal responses after introducing active control

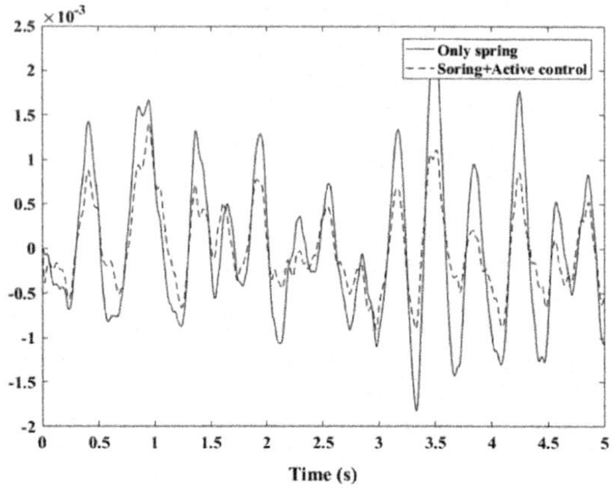

(a) Displacement response

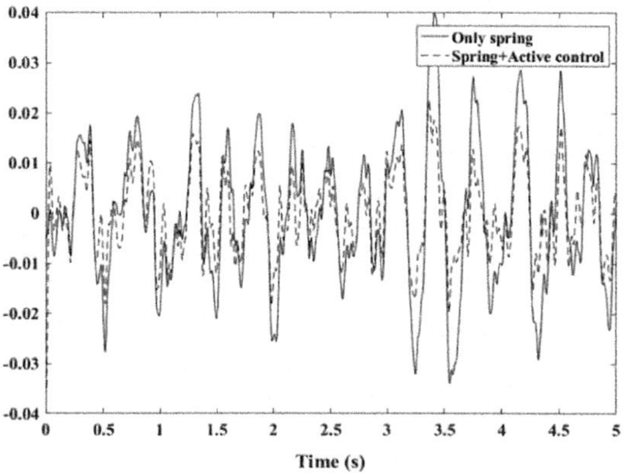

(b) Velocity response

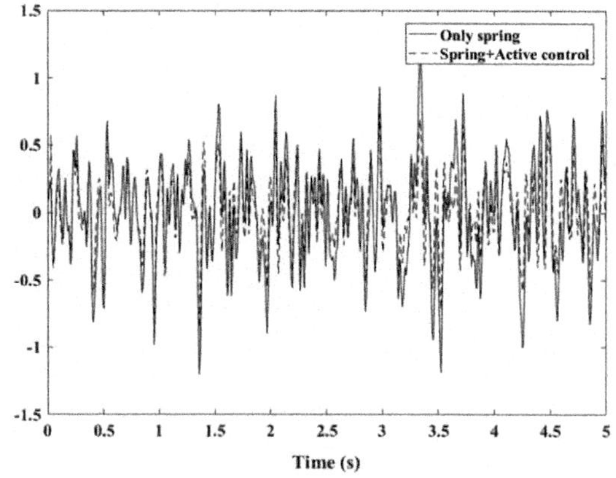

(c) Acceleration response

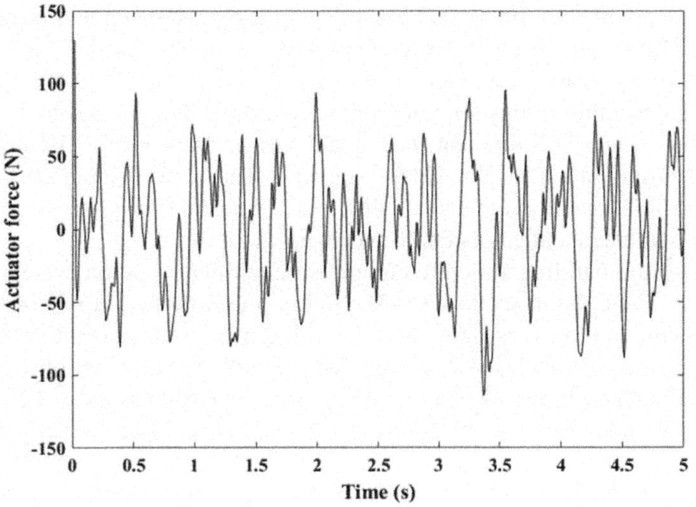

Fig. 7.33 Output of actively controlled actuator

Traditional vibration isolation abutments mainly consider the impacts of ground pulsation, pedestrians, and power equipment disturbances. The vibration isolation abutments and upper equipment safety after sudden and accidental significant disturbances is essential. In addition, high-value equipment and equipment that may cause secondary pollution and disasters should be considered.

This research has conducted a systematic study on the performance improvement of the vibration isolation base station. With the development of vibration isolation technology, some new smart materials and new vibration control devices will continue to appear, and hopefully, new multifunctional composite vibration isolation devices will be developed as soon as possible and then applied to modern engineering vibration control.

In the follow-up work, the author will focus on intelligent control (such as piezoelectric materials) and conduct more in-depth research.

7.4 Proposed Refined Numerical Calculation Model for Building Structure-Equipment

In Chap. 6, the coupling vibration control of building structure equipment is investigated, the theoretical vibration control equations are given, and the research on building structure vibration caused by power equipment and the impact of building structure vibration on the normal use of precision equipment is carried out. In this study, only the vertical vibration control response of the vibrating system is considered. In actual cases, the translational motion in the x and y directions and rotation of the system are considered; thus, the dynamic equation of the system is very complex. In practical projects, it is possible to use sophisticated numerical methods such as FEM (using various software packages, including ANSYS, SAP2000, and ABAQUS). This chapter carries out a refined analysis of equipment vibration control. Therefore, the author of this section proposes a refined numerical calculation model for building structural equipment. The model was developed using the ANSYS platform and programmed in the APDL language. The formed modular language is convenient for system parameter modification, A standardized model that can be used by researchers and engineers (Benchmark model).

Consider the building structure equipment coupling vibration system as shown in Fig. 7.1. The column section is 0.3 × 0.3 m, the floor height is 3 m, the floor's length, width, and thickness are 4.3 × 4.3 × 0.2 m, and the concrete grade is C35. The mass, three-way stiffness, and three-way damping coefficient of the two-stage power equipment vibration isolation system are $m_p = 100$ kg, $m_b = 1200$ kg, $k_p = 1.5 \times 10^4$ N/m, $k_b = 1.0 \times 10^6$ N/m, $c_p = 1.0 \times 10^3$ N s/m, $c_b = 1.6 \times 10^4$ N s/m, respectively.

When modeling in ANSYS, columns, floors, and equipment were simulated using the Solid65 element, and the combin14 element was used for the vibration isolator (including stiffness and damping). When modeling the two-stage equipment vibration isolation system, the actual size was also considered, and the three-dimensional size of the first level system is 1.5 × 1.5 × 0.213 m, the three-dimensional dimension of the second level system is 0.5 × 0.5 × 0.16 m. Indeed, when the quality of the two-level system changes, the quality can be changed by setting different densities in the APDL language. The finite element model is shown in Fig. 7.34, and the modal calculation results are shown in Fig. 7.35.

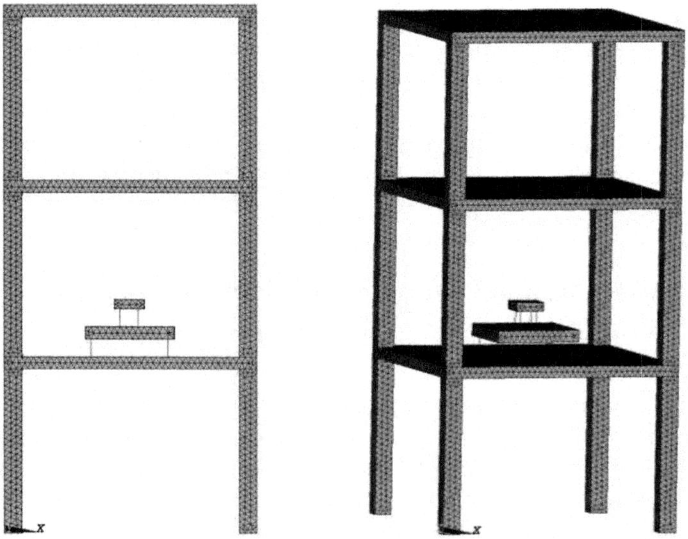

Fig. 7.34 Numerical calculation model of building structure-equipment coupling vibration

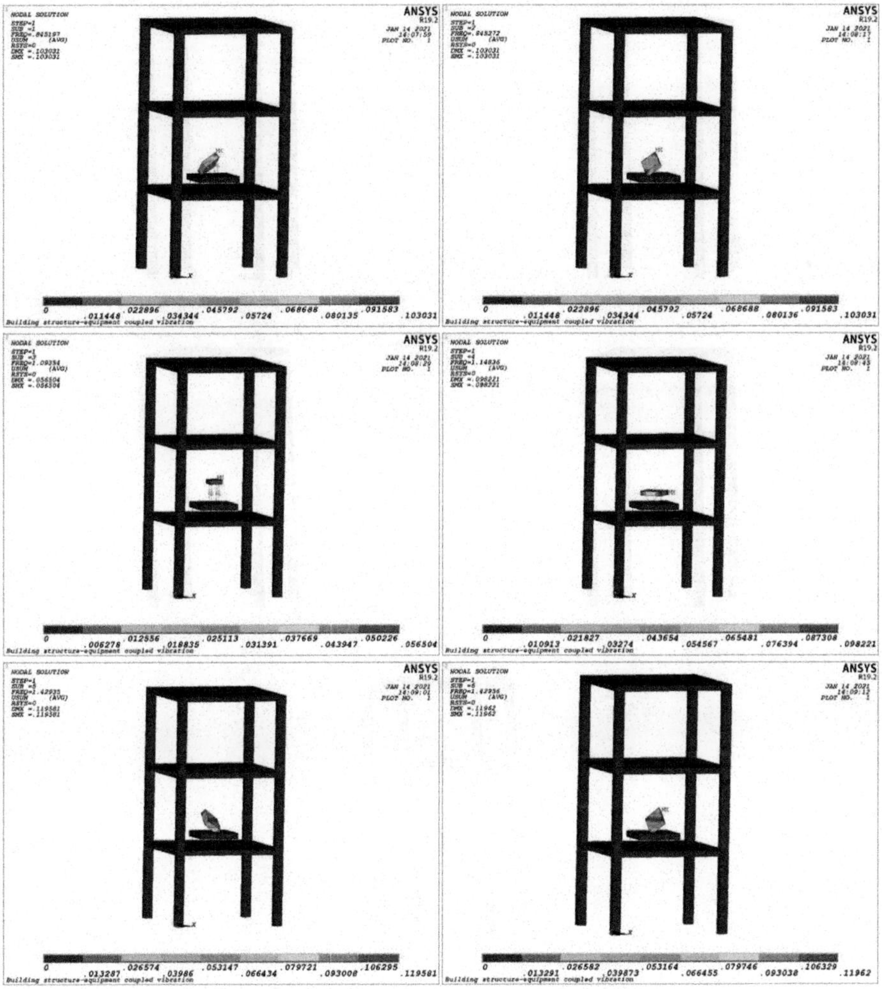

Fig. 7.35 Modes calculation results of building structure-equipment coupling vibration

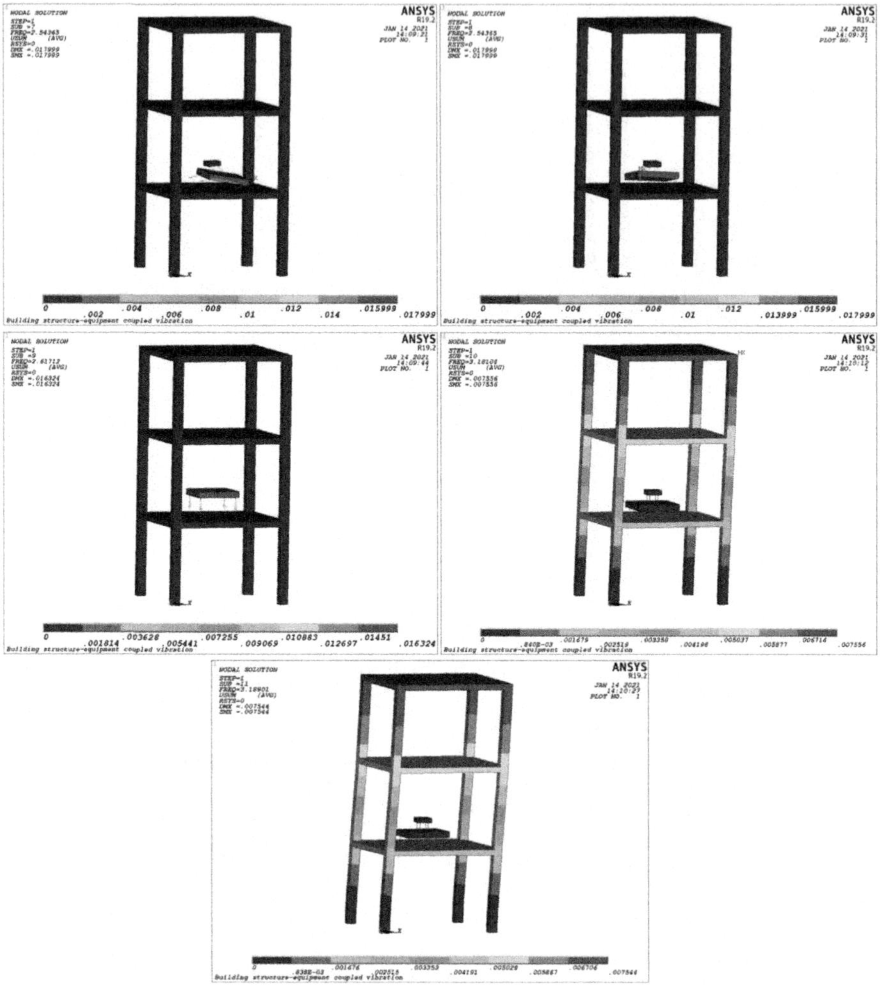

Fig. 7.35 (continued)

References

1. Daley S, Hätönen J, Owens DH (2006) Active vibration isolation in a "smart spring" mount using a repetitive control approach. Control Eng Pract 14(9):991–997
2. Harris CM, Piersol AG (2002) Harris' shock and vibration handbook. McGraw-Hill, New York
3. Yoshimura T, Kume A, Kurimoto M et al (2001) Construction of an active suspension system of a quarter car model using the concept of sliding mode control. J Sound Vib 239(2):187–199
4. Pourzeynali S, Lavasani HH, Modarayi AH (2007) Active control of high rise building structures using fuzzy logic and genetic algorithms. Eng Struct 29(3):346–357
5. Brownjohn J, Pavic A (2006) Vibration control of ultra-sensitive facilities. Proc Inst Civil Eng Struct Build 159(5):295–306

6. Gokcek C, Kabamba PT, Meerkov SM (2001) An LQR/LQG theory for systems with saturating actuators. IEEE Trans Autom Control 46(10):1529–1542
7. Chen CK, Kuo HH, Yan JJ et al (2009) GA-based PID active queue management control design for a class of TCP communication networks. Expert Syst Appl 36(2):1903–1913
8. Orukpe PE, Zheng X, Jaimoukha IM et al (2008) Model predictive control based on mixed $\mathcal{H}2/\mathcal{H}\infty$ control approach for active vibration control of railway vehicles. Veh Syst Dyn 46(S1):151–160
9. Xu J et al (2016) Key technology of vibration control in industrial engineering. China Construction Industry Press, Beijing
10. Xu J (2016) Building vibration engineering handbook, 2nd edn. China Construction Industry Press, Beijing
11. Wang XM (2007) ANSYS numerical analysis of engineering structure. People's Communications Press, Beijing
12. Liu HL, Lan HY (2011) Application of Ansys in type selection of dampers for large-span bridges. Technol Highway Transp 6:44–48
13. Huang W, Xu J et al (2019) Active vibration control using finite element method strategy. Sci Technol Eng 19(24):102–106

Chapter 8
Decoupled Passive and Active Control

Abstract In this chapter, decoupled passive and active control strategies are performed out. Firstly, four different arrangements of vibration isolators for vibration isolation base are investigated, Followed by, decoupling using counter coincidence of mass and stiffness and natural frequency decoupling with loaded mass by using double-chamber air spring are carried out. Based on this, a sensing-control-actuation system was introduced based on the air-floating passive control system decoupled from the load to achieve active control. Finally, a typical engineering example is introduced here, and a systematic research for the presented techniques in this chapter is carried out.

8.1 Vibration Isolation Performance Influence of Vibration Isolator Arrangement

Frequently, the vibration control system consists of numerous vibration isolation units with multi-point support. The vibration control of the controlled object is transformed into the attitude or position control of the tabletop and then manifests itself as the control of multiple supporting points. The vibration isolator arrangement determines the system's characteristics. The platform's position is unique when three vibration isolators are utilized since the three points can determine a plane. When there are more than three vibration isolators, the tabletop's position is identified using the three vibration isolators that perform the main bearing load. In addition, when the posture varies, so does the load distribution, causing bearing load redistribution and a change in posture.

8.1.1 Four Different Arrangements of Vibration Isolators

The research object includes a high-rigid base plate, a vibration isolator, and others. The vibration isolator consists of stiffness and damping in parallel (nonlinearity is

W. Huang and J. Xu, *Optimized Engineering Vibration Isolation, Absorption and Control*, Springer Tracts in Civil Engineering, https://doi.org/10.1007/978-981-99-2213-0_8

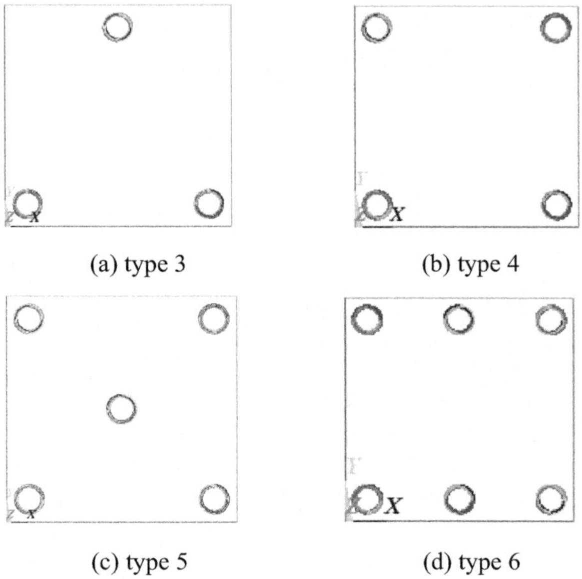

Fig. 8.1 Vibration isolation platforms with four different vibration isolators

neglected for stiffness and damping). As depicted in Fig. 8.1, the vibration isolators are arranged in triangular 3-point support (type 3), rectangular 4-point support (type 4), 5-point support (type 5), and rectangular 6-point support (type 6). The plate dimensions are 2 × 2 × 0.2 m, the designed vertical frequency of the vibration isolation system is 3.5 Hz, and the horizontal stiffness is 0.7 times the vertical stiffness. Figure 8.2 illustrates the finite element calculation model.

8.1.2 Modal Characteristics

The modal results indicated that the 1st to 6th-order of the four different posture vibration isolation platforms is the overall vibration mode of the vibration isolation platform, containing translation along the x, y, and z-axis and rotation around the x, y, and z-axis. It belongs to the overall vibration mode of the vibration isolation system. The 7th and higher orders are local vibration modes of the platform (Table 8.1 displays the 1st to 6th local vibration modes of the type 4 vibration isolation platform, which are the overall 7th to 12th modal frequencies), so only the first 6 modes of the overall characteristics of the vibration isolation system are investigated here.

Figure 8.3 and the modal calculation results indicate that the first and second orders are mainly translational motions along the x and y-directions, and the third order is vertical vibration along the z-direction, among which the third order of type 5 is rotation mode around the z-direction. Moreover, the vertical vibration occurs in the 4th-order. The 4th to 6th-order modal frequencies of type 4 are higher than

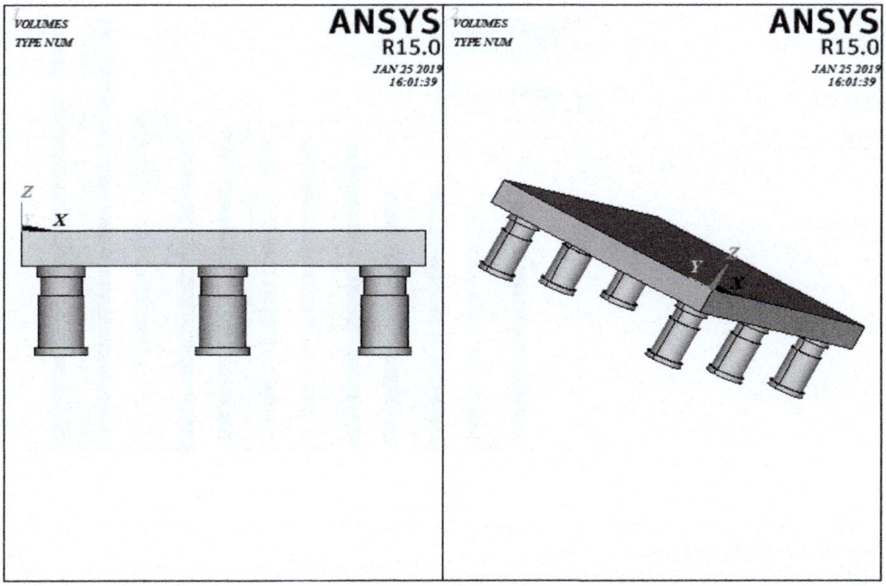

Fig. 8.2 Finite element calculation model of the vibration isolation system

Table 8.1 The 1st to 6th-order (overall 7th to 12th-order) modal frequencies of Type 4 vibration isolation platform

Local modes	Overall modes	Frequency (Hz)
1	7	196.61
2	8	286.30
3	9	315.58
4	10	478.01
5	11	482.71
6	12	780.29

the other arrangements, while the overall modal frequency of type 5 is smaller. This shows that the type 4 or type 6 vibration isolation methods commonly used in actual engineering may not be the most practical vibration isolation design. Type 3 belongs to the three-point support type. Since the three points determine a plane, although the position and posture of the vibration isolation platform can be unique, the system design is not optimal, as further elaborated below.

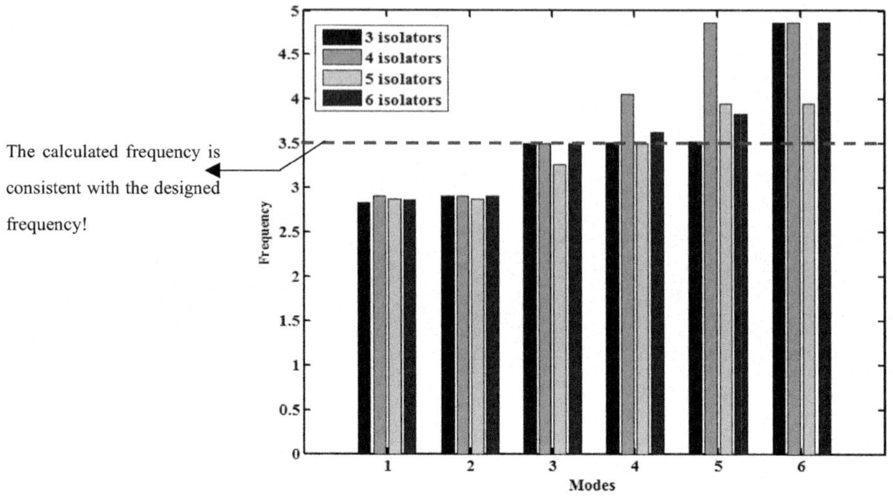

Fig. 8.3 Comparison of the first 6th-order modal frequencies of vibration isolation platform with four different vibration isolators

8.1.3 Harmonic Response Characteristics

Harmonic response analysis was conducted on four vibration isolation systems with a sweep frequency range of 1–10 Hz. Besides, accelerations along the x, y, and z-directions were applied at one m/s^2, and the midpoint response of the plate's surface was extracted. Consider that the vibration isolator damping will suppress the resonance peak, resulting in an inaccurate evaluation of the inherent vibration isolation system characteristics. Hence, the vibration isolator damping is neglected while the specific structural damping is considered.

Figure 8.4 indicates that the response of type 5 is maximum in the x-direction, and types 6 and 4 are minimum. Moreover, types 3 and 4 have other peaks besides the main peak point. The response of type 5 is maximum in the y-direction, while that of types 4 and 6 have other peaks. In the z-direction response, the prominent peaks of the four types are not much different, but type 4 has significant other peaks. Therefore, overall, the smoothness of the frequency response curve in the type 5 vibration isolation system is better than the other three methods.

8.1.4 White Noise Response Characteristics

The response characteristics of the four vibration isolation systems under random white noise were studied herein. The given white noise acceleration signal is located in the interval $[-1, 1]$ m/s^2, as shown in Fig. 8.5.

Fig. 8.4 Comparison between the harmonic response of four different vibration isolator arrangements

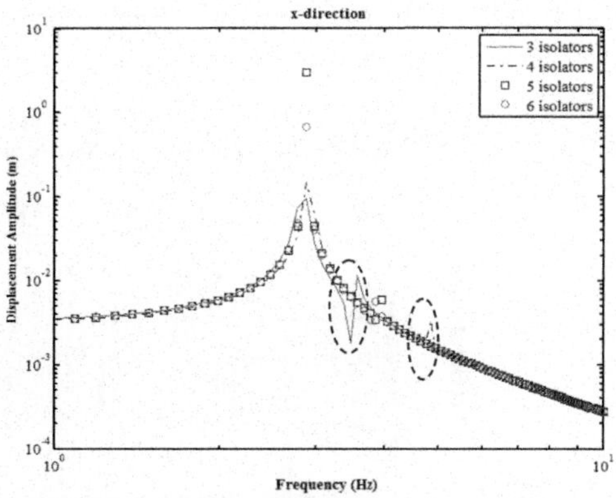

(a) x direction

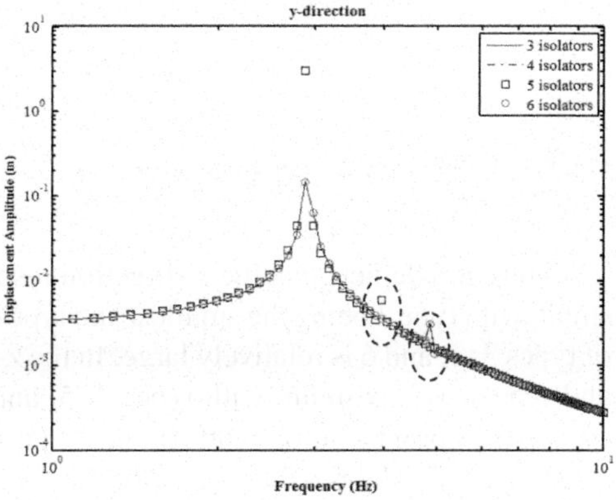

(b) y direction

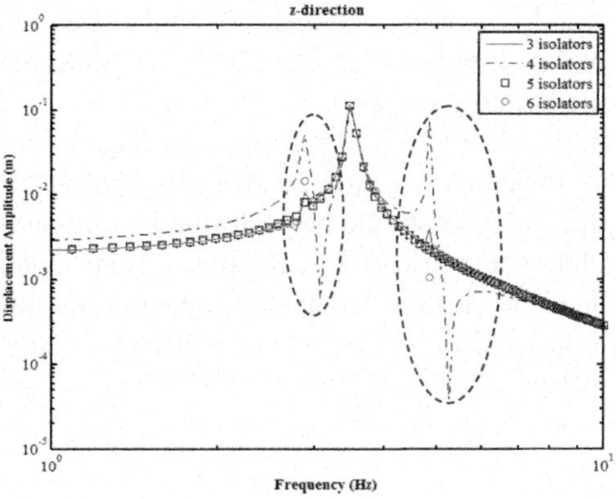

(c) z direction

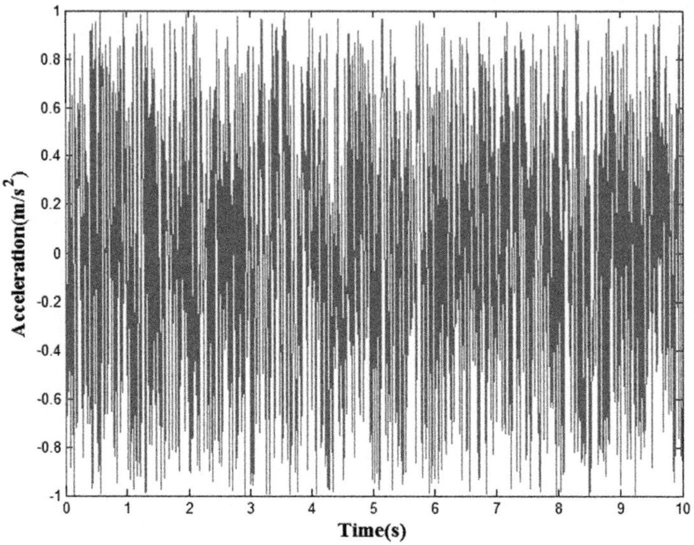

Fig. 8.5 White noise acceleration signal

Figure 8.6 depicts that the x-direction responses of types 3, 5, and 6 are relatively small, with type 3 being the smallest and type 4 being larger. The y-direction response of types 3, 4, and 6 is relatively large. In the z-direction response, type 5 is minimum, and type 4 is maximum, with types 3, 5, and 6 being between them. The common type 4 vibration isolation platform is not an ideal approach.

Figure 8.7 indicates the FFT response analysis for the vibration isolation platform with four types of arranged different vibration isolators under white noise acceleration excitation. It can be seen that the vibration isolation system resonates around 3.5 Hz, which can effectively filter the higher frequency excitation and reflect the importance of the vibration isolation design. In addition, the amplitude and the spectrum curve of the x- and y-direction are the same, and the difference is minimal, but the amplitude in the z-direction of the type 4 arrangement is significantly higher than the others.

In actual vibration control engineering, detailed optimization design and calculation analysis should be performed based on the vibration source characteristics, base plate geometric characteristics, vibration isolator arrangement, system decoupling, and others. Given that the four-point support application is much more common, a decoupling system study will be presented in Sect. 8.2 to eliminate the adverse effects of the coupling.

Fig. 8.6 Comparison between the RMS values of the white noise for the vibration isolation platform with four different vibration isolator arrangements

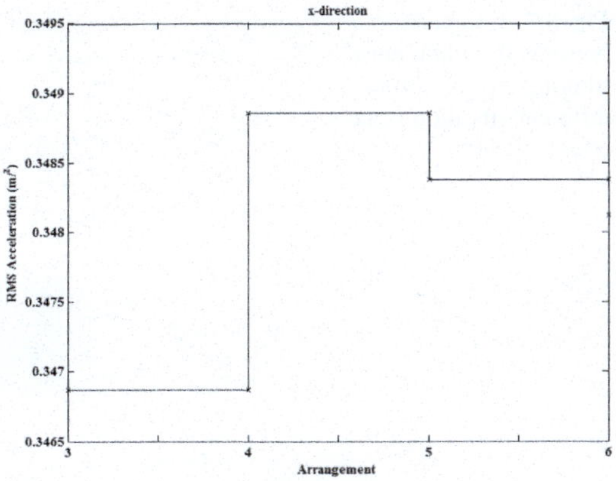

(a) x-direction

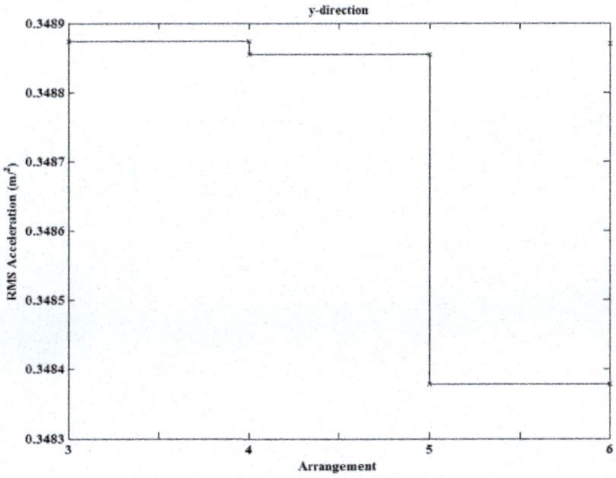

(b) y-direction

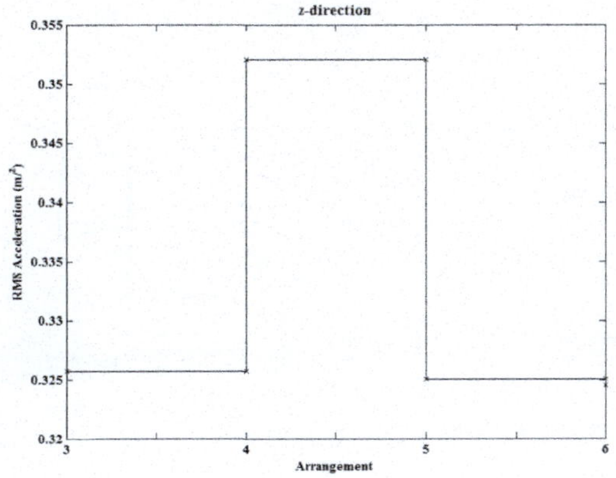

(c) z-direction

Fig. 8.7 Comparison between the white noise filtering effects of four different vibration isolator arrangements

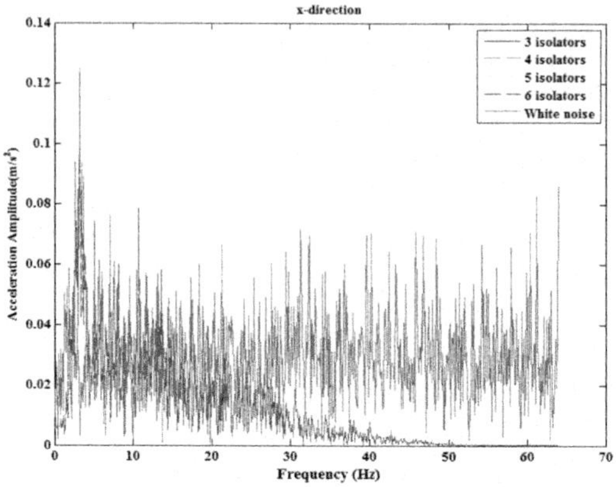

(a) x-direction

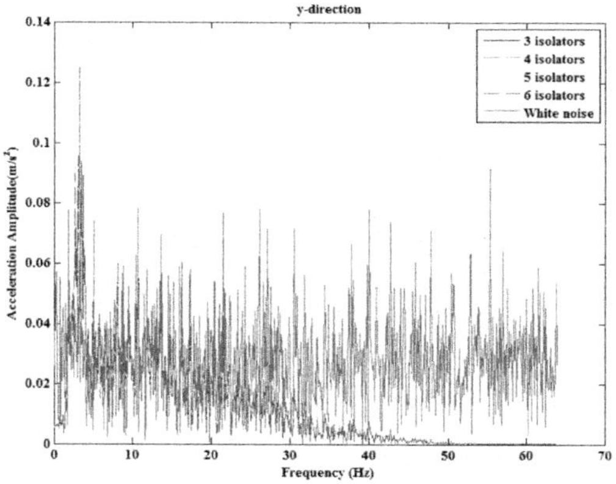

(b) y-direction

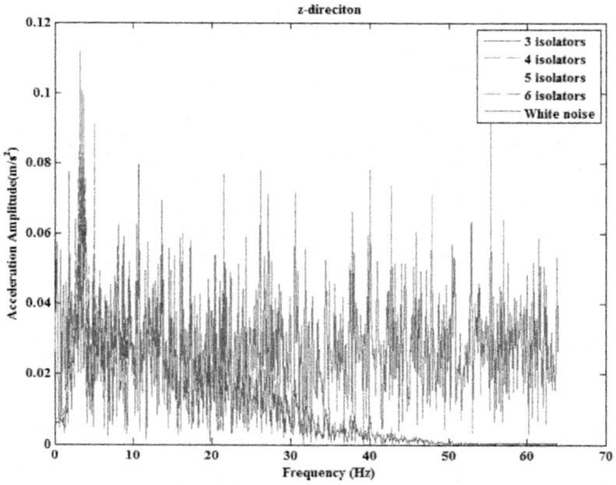

(c) z-direction

8.2 Decoupled Passive and Active Control

Section 8.1 investigates the characteristics of the vibration isolation system with different vibration isolator arrangements. The results show that the vibration isolation performance of the commonly used four-point support is not optimal. However, considering that the four-point support is simple, convenient, and familiar in an application, the shortcomings must be overcome as much as possible during the process. Based on the previous calculation results, it is believed that the adverse effects caused by the coupling of degrees of freedom are more pronounced. Therefore, the vibration isolation effect from the perspective of system decoupling can be improved.

For the traditional passive control system of steel spring damper, its natural frequency changes with the loaded mass, i.e., once the system is designed, the vibration isolation performance for a specific vibration isolation object is fixed, which is not conducive to actual engineering. Therefore, a vibration control system decoupled from the load is urgently needed.

In engineering vibration control, the adverse effects due to the coupling of degrees of freedom should be reduced as much as possible, such as reducing the center of mass to get the vibration isolation system close to the uncoupled state [1–3]. The linear elastic vibration isolation system's natural frequency changes with the loaded mass. Once the system is designed, its vibration isolation characteristics are fixed. In actual engineering projects, producing low frequencies below 2 Hz using a linear steel spring is challenging. The dual-chamber air spring effectively solves the coupling problem between natural frequency and load and can achieve low-frequency control [4].

8.2.1 Decoupling Using Counter Coincidence of Mass and Stiffness

Figure 8.8 exhibits two vibration isolation platforms: (a) the vibration isolators are located under the platform; (b) the vibration isolators are embedded into the middle of the vibration isolation platform, i.e., the center of mass is on the same plane as the force of the vibration isolator that provides external stiffness (mass-stiffness coincidence), and the former is a non-decoupled system, the latter is a decoupled system.

The vibration isolation platform is a $2 \times 2 \times 0.2$ m marble base plate with dimensions. Additionally, the vibration isolator is a steel spring damper whose nonlinear effect on the system is ignored. The designed vertical frequency of the vibration isolation system is 3.5 Hz. Figure 8.9 displays the analysis result of the harmonic response. The frequency range is 1–10 Hz and the acceleration is applied in the x, y, and z directions. The results indicate that the peaks of the x- and y-direction harmonic responses in the decoupled state are below those in the non-decoupled state. The z-direction harmonic response is significantly smaller than the uncoupled

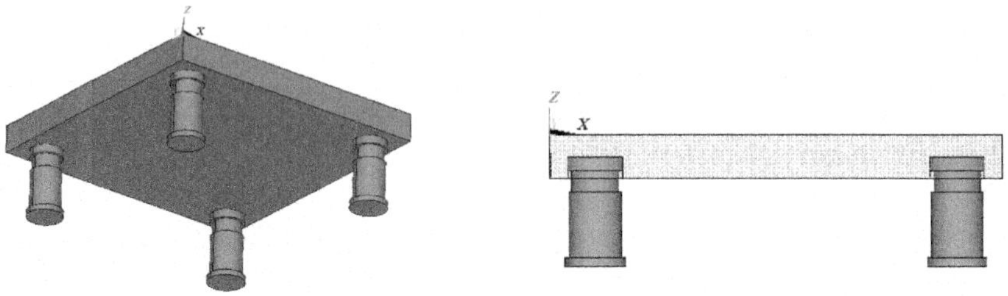

(a) mass-spring without coincidence (b) mass-spring with coincidence

Fig. 8.8 Vibration isolation platforms with two types of the relative position of mass and stiffness

state before the resonance peak, and the decoupled system can avoid the other easy resonance points. Besides, the main resonance peak occurs due to the coupling of degrees of freedom. Therefore, the system should be as close as possible to achieve a non-coupling system in actual engineering.

8.2.2 Natural Frequency Decoupling with Loaded Mass

For the vibration isolation system of the steel spring damper described in Sect. 8.2.1, the natural frequency decreases with the increase in the load mass. On the other hand, a low natural frequency must be designed for low-frequency-based micro-vibration control. In this case, the system will be very soft and has much deformation, which makes applying it practically difficult. The air spring floating vibration control system can solve the problem of low-frequency control. The design of the dual-chamber air spring (Fig. 8.10) [4] can effectively solve the problem of the natural frequency in the vibration isolator being decoupled from the load mass.

The dynamic model of the double-chamber air spring can be established according to the following equations [4]:

$$\delta \dot{P}_t = -\frac{nRT\rho C_r}{V_{t0}}\delta P_t + \frac{nRT\rho C_r}{V_{t0}}\delta P_b - \frac{nA_p P_0}{V_{t0}}\delta v_p + \frac{nA_p P_0}{V_{t0}}\delta v_{\text{base}} + \frac{nRT\rho}{V_{t0}}Q_t \tag{8.1}$$

$$\delta \dot{P}_b = -\frac{nRT\rho C_r}{V_b}\delta P_b + \frac{nRT\rho C_r}{V_b}\delta P_t + \frac{nRT\rho}{V_b}Q_b \tag{8.2}$$

$$\delta \dot{v}_p = \frac{A_p}{m_p}\delta P_t + \frac{1}{m_p}F_d \tag{8.3}$$

where P_t is the air pressure of the upper chamber, P_b is the air pressure of the lower chamber, v_p is the vibration velocity of load, v_{base} is the velocity excitation acting on the bottom of the air spring, Q_t is the airflow of the upper chamber, Q_b is the airflow

Fig. 8.9 Comparison between the harmonic response under the decoupled and non-decoupled states

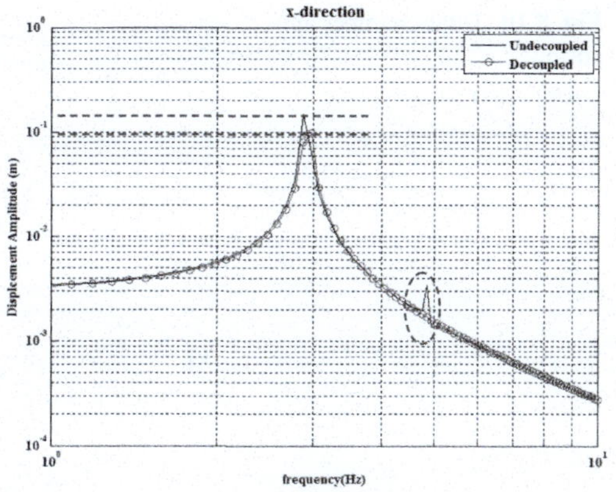

(a) x direction

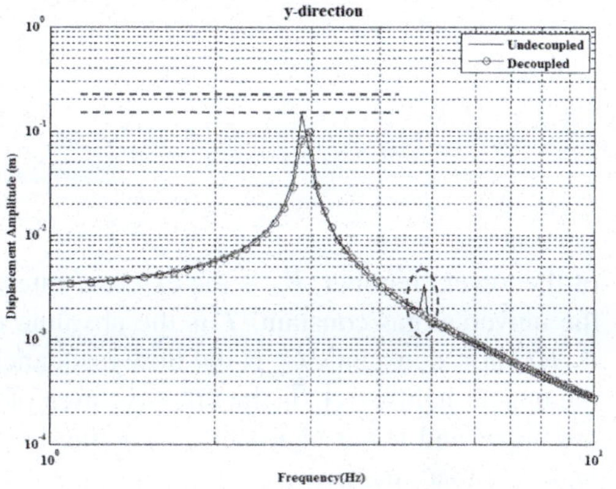

(b) y direction

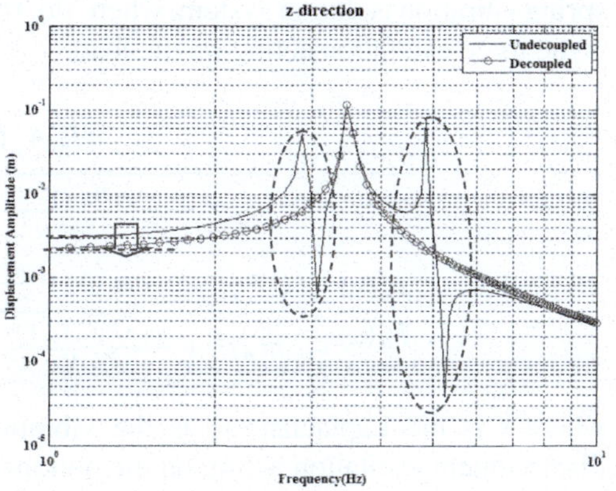

(c) z direction

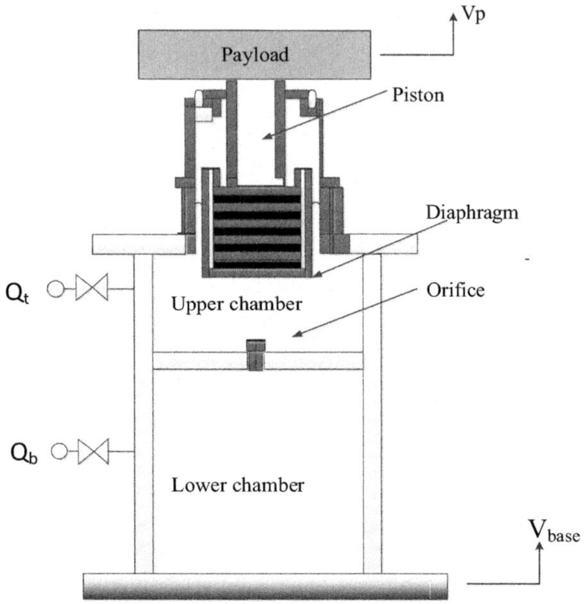

Fig. 8.10 Dual-chamber air spring

of the lower chamber, F_d is the external interference; n is the variability index; R is the universal gas constant; T is the absolute temperature, ρ is the air density ($\rho = P_t/RT$, non-constant), C_r is the flow resistance constant, V_{t0} is the initial volume of the upper chamber, A_p is the effective area of the piston, P_0 is the initial pressure of the upper and lower chambers, V_b is the initial volume of the lower chamber, and m_p is the load mass.

Equations (8.4) and (8.5) highlight the transfer function of the dual-chamber air spring vibration isolation system, where Eq. (8.4) is the base excitation, and Eq. (8.5) is the force excitation acting on the load.

$$\frac{x_p(s)}{x_{\text{base}}(s)} = \frac{nP_0A_p^2\left(s + \frac{nP_0C_r}{V_b}\right)}{m_ps^3 + \frac{m_pnP_0C_rV_{t0}+V_b}{(V_{t0}V_b)}s^2 + \frac{nP_0A_p^2}{V_{t0}}s + \frac{nP_0^2C_rA_p^2}{(V_{t0}V_b)}} \tag{8.4}$$

$$\frac{x_p(s)}{F_d(s)} = \frac{s + \frac{nP_0C_rV_{t0}+V_b}{(V_{t0}V_b)}}{m_ps^3 + \frac{m_pnP_0C_rV_{t0}+V_b}{(V_{t0}V_b)}s^2 + \frac{nP_0A_p^2}{V_{t0}}s + \frac{nP_0^2C_rA_p^2}{(V_{t0}V_b)}} \tag{8.5}$$

where s is the Laplacian; x_p is the vibration displacement of load; x_{base} is the displacement excitation acting on the bottom of the air spring.

Figure 8.11 shows the transmission rate curve, with its natural frequency calculated as 0.9 Hz. The simple calculation method of the natural frequency in the double-chamber air spring [5] can be calculated using Eq. (8.6). When the effective height of the chamber remains constant, its natural frequency is unchanged, i.e., it is not dependent on the load mass.

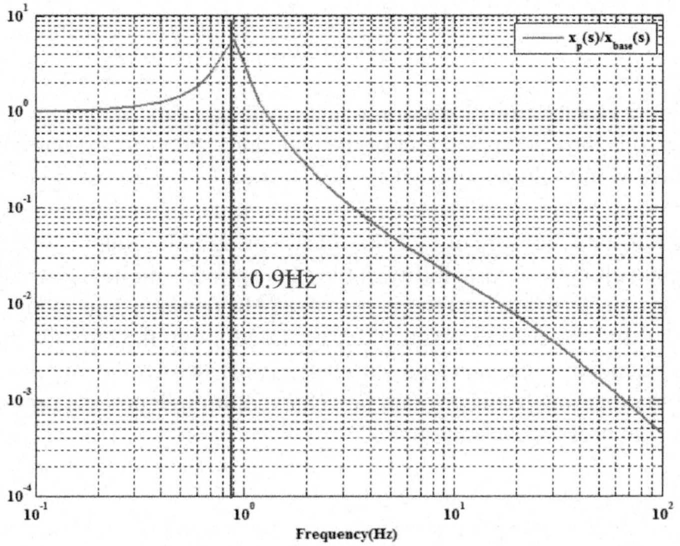

(a) $x_p(s)/x_{base}(s)$

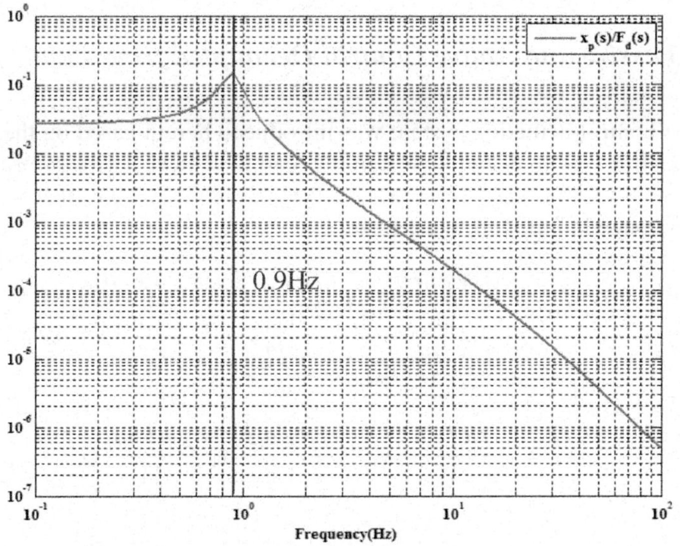

(b) $x_p(s)/F_d(s)$

Fig. 8.11 Transmission rate of double-chamber air spring

$$\omega_n = \sqrt{\frac{g\gamma}{h}} \tag{8.6}$$

where g is the acceleration of gravity, γ is the heat capacity ratio of air, and the value is 1.4 here, h is the effective height of the chamber, and ω_n is the natural frequency.

Equation (8.6) highlights that if $\gamma = 1.4$, then $h = \frac{g\gamma}{\omega_n^2}$. This indicates that the effective height of the dual-chamber air spring at the natural frequency ω_n is only 1.4 times the static deformation of the linear spring at the same natural frequency. However, the ultimate deformability of the steel spring is limited. For example, if the steel spring reaches the natural frequency of 0.9 Hz, its static deformation is 306 mm, which is hardly achieved in practical cases. In addition, the effective height of the air spring equals the volume of the air chamber divided by the effective area of the piston so that the low frequency can be achieved through process optimization and increasing the piston area, and others.

Equation (8.6) provides a basis for the design of the double-chamber air spring, with its natural frequency calculation being simple. The C. Erin method can be mutually verified with the actual test curve of the air spring, and the combination of both can realize the characteristics of the air spring for continuous optimization.

8.2.3 Passive and Active Control Decoupled from Load

A sensing-control-actuation system was introduced herein based on the air-floating passive control system decoupled from the load to achieve active control.

If the state variables are $\delta P_t = x_1$, $\delta P_b = x_2$, and $\delta V_p = x_3$, then Eqs. (8.1) and (8.3) are changed to the following form:

$$\begin{aligned} \dot{x}_1 = & -\frac{nRT\rho C_r}{V_{t0}}x_1 + \frac{nRT\rho C_r}{V_{t0}}x_2 - \frac{nA_pP_0}{V_{t0}}x_3 \\ & + \frac{nA_pP_0}{V_{t0}}\delta V_{\text{base}} + \frac{nRT\rho}{V_{t0}}Q_t \end{aligned} \tag{8.7}$$

$$\dot{x}_2 = -\frac{nRT\rho C_r}{V_b}x_2 + \frac{nRT\rho C_r}{V_b}x_1 + \frac{nRT\rho}{V_b}Q_b \tag{8.8}$$

$$\dot{x}_3 = \frac{A_p}{m_p}x_1 + \frac{1}{m_p}F_d \tag{8.9}$$

Additionally, Eqs. (8.7), (8.8), (8.9) changes into the following state space equation form:

$$\begin{bmatrix} \dot{x}_1 \\ \dot{x}_2 \\ \dot{x}_3 \end{bmatrix} = \begin{bmatrix} A_{11}\rho & A_{12}\rho & A_{13} \\ A_{21} & A_{22}\rho & 0 \\ A_{31} & 0 & 0 \end{bmatrix} \begin{pmatrix} x_1 \\ x_2 \\ x_3 \end{pmatrix} + \begin{bmatrix} B_{11} & 0 \\ 0 & 0 \\ 0 & B_{23} \end{bmatrix} \begin{pmatrix} V_{\text{base}} \\ F_d \end{pmatrix} \tag{8.10}$$

where $A_{11} = -\frac{nRTC_r}{V_{t0}}$, $A_{12} = \frac{nRTC_r}{V_{t0}}$, $A_{13} = -\frac{nA_pP_0}{V_{t0}}$, $A_{21} = -\frac{nRTC_r}{V_b}$, $A_{22} = \frac{nRTC_r}{V_b}$, $A_{31} = \frac{A_p}{m_p}$, $B_{11} = \frac{nA_pP_0}{V_{t0}}$, $B_{23} = \frac{1}{m_p}$, $\rho = \frac{P_t}{RT}$ (time-related variables), and $P_t = P_0 + \delta P_t$ (time-related variables).

The following equation can be obtained by assuming full-state observation output:

$$y = \begin{pmatrix} y_1 \\ y_2 \\ y_3 \end{pmatrix} = \begin{bmatrix} 1 & 0 & 0 \\ 0 & 1 & 0 \\ 0 & 0 & 1 \end{bmatrix} \begin{pmatrix} x_1 \\ x_2 \\ x_3 \end{pmatrix} + \begin{bmatrix} 0 & 0 \\ 0 & 0 \\ 0 & 0 \end{bmatrix} \begin{pmatrix} V_{\text{base}} \\ F_d \end{pmatrix} \tag{8.11}$$

Combining Eqs. (8.10) and (8.11) results in the following equation:

$$\begin{cases} \dot{x} = \mathbf{A}x + \mathbf{B}U \\ y = \mathbf{C}x + \mathbf{D}U \end{cases} \tag{8.12}$$

where $\mathbf{A} = \begin{bmatrix} A_{11}\rho & A_{12}\rho & A_{13} \\ A_{21} & A_{22}\rho & 0 \\ A_{31} & 0 & 0 \end{bmatrix}$, $\mathbf{B} = \begin{bmatrix} B_{11} & 0 \\ 0 & 0 \\ 0 & B_{23} \end{bmatrix}$, $\mathbf{C} = \begin{bmatrix} 1 & 0 & 0 \\ 0 & 1 & 0 \\ 0 & 0 & 1 \end{bmatrix}$, $\mathbf{D} = \begin{bmatrix} 0 & 0 \\ 0 & 0 \\ 0 & 0 \end{bmatrix}$, $x = \begin{pmatrix} x_1 \\ x_2 \\ x_3 \end{pmatrix}$, $\dot{x} = \begin{bmatrix} \dot{x}_1 \\ \dot{x}_2 \\ \dot{x}_3 \end{bmatrix}$, $y = \begin{pmatrix} y_1 \\ y_2 \\ y_3 \end{pmatrix}$, and $U = \begin{pmatrix} V_{\text{base}} \\ F_d \end{pmatrix}$.

The parameter values of the air spring vibration isolation system are listed in Table 8.2, and the dynamic model of the double-chamber air spring considering the dynamic diaphragm effect is as follows (Fig. 8.12).

When active control is introduced into the system, active control force F_a acts on the load, which can offset the external vibration load. Hence, Eq. (8.9) is changed to the following form:

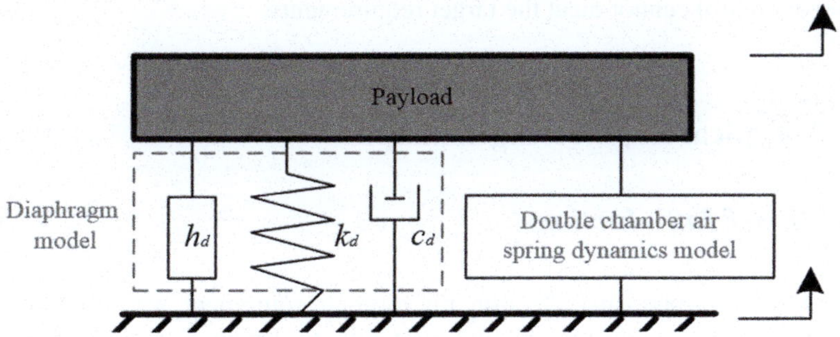

Fig. 8.12 Air spring model considering diaphragm effect

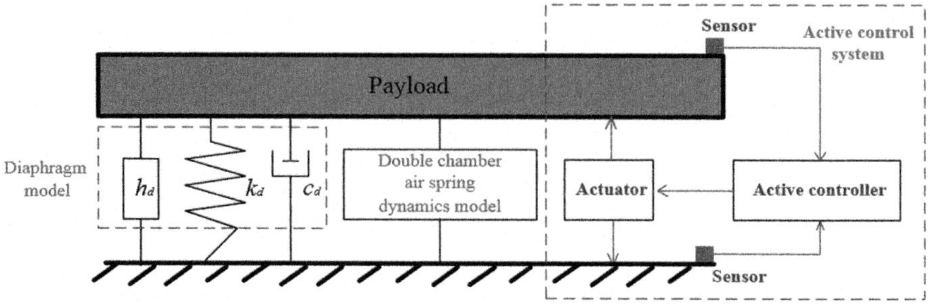

Fig. 8.13 Air spring dynamic model with active control

$$\dot{x}_3 = \frac{A_p}{m_p}x_1 + \frac{1}{m_p}F_d - \frac{1}{m_p}F_a \tag{8.13}$$

Active control considers the PID approach, where it's proportional, integral, and derivative control factors can be tuned using the classical theory or optimized using algorithms such as PSO and GA [6, 7]. This study defines them as − 500, − 500, and 500. The external vibration loads are considered random white noise load F_d with an amplitude of 1 N acting on the load mass, and displacement excitation X_{base} and velocity excitation V_{base} that are random white noise loads acting on the bottom of the air spring with amplitudes of 1×10^{-6} m and 1×10^{-6} m/s, respectively. Figure 8.13 describes the model after introducing active control. Figure 8.14 displays the response comparison between the active and non-active control systems. The results demonstrated that active control can significantly improve the passive control effect. Figure 8.15 represents the output force of the actuator. The system's response can be further improved by enhancing the active control effect and increasing the actuator output range. However, in the actual engineering project, the economy should be considered. After all, it is not necessary to blindly pursue control effects.

In actual industrial engineering, if the vibration control is based on linear steel spring dampers (the system's natural frequency is above 2 Hz), the system decoupling must be achieved as much as possible. If the air spring-based floating system is used for vibration control (the natural system frequency is below 2 Hz), it can be decoupled from the load based on system decoupling. Active control can be used when the effect of passive control cannot meet the target requirements.

8.3 A Typical Engineering Example

8.3.1 Brief Introductions

A group of ultra-precision electron microscope equipment was used in micro-vibration control in a research institute in China to design an anti-micro-vibration

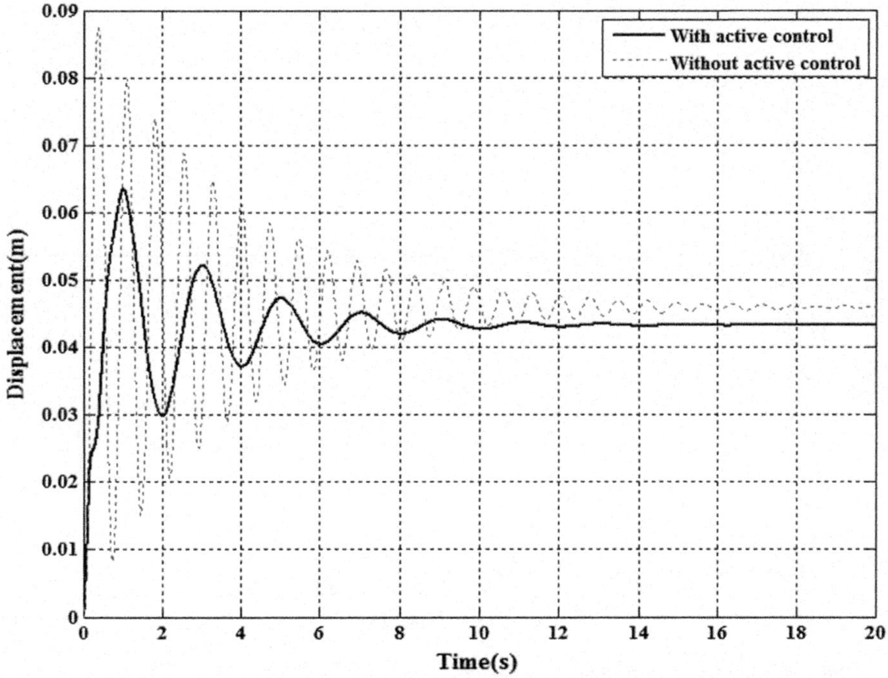

Fig. 8.14 Comparison between response with and without active control system

large volume of concrete foundation and passive and active control system for sensitive equipment. The layout of the concrete foundation is shown in Fig. 8.16, and the foundation type is a pool type, as shown in Fig. 8.17.

8.3.2 Key Technologies Application

Select a pool-type large-volume reinforced concrete foundation with the foundations shown in Figs. 8.8, and 8.16, 6, 4, and 2 m outer dimensions and 5, 3, and 0.6 m inner dimensions. The foundation soil is considered elastic, and the simulation was calculated according to a three-direction soil spring. The parameters were selected according to the code for dynamic machine foundation design (GB50040-96) [8]. According to the foundation survey report, the embankment under the foundation is sand, silt, and clay from top to bottom. The bearing capacities of the foundation are 150, 100, and 100 kPa, and the thicknesses are 1.5, 1.0, and 3.0 m.

Provision 8.3.3.2 of the code for the design of dynamic machine foundation (GB50040):

The compressive stiffness coefficient of a natural foundation can be determined according to the following provisions:

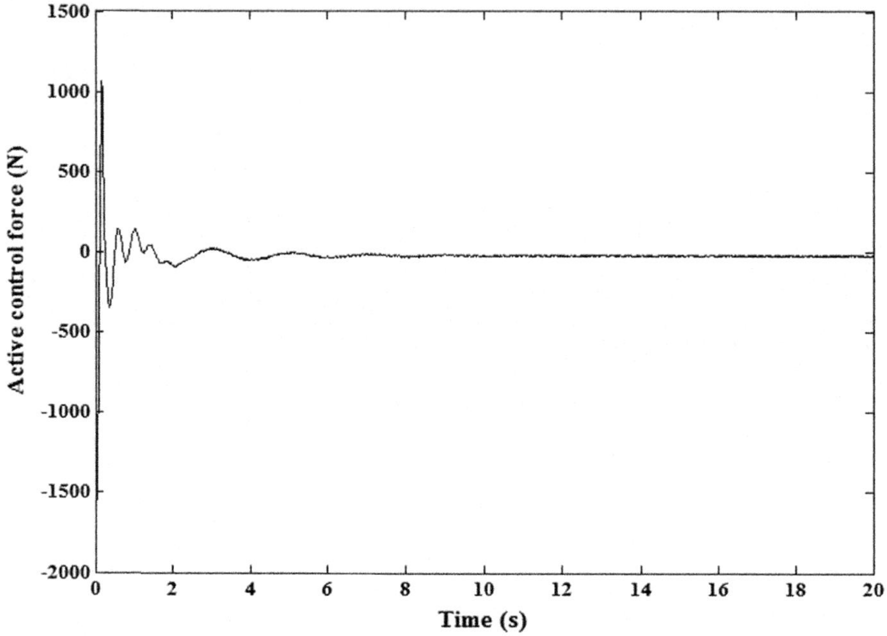

Fig. 8.15 Output force of the actuator in active control

Table 8.2 Parameter values of the air spring vibration isolation system

Parameter item	Value
d	$6.1 \times 10^{-}$ m
l	7.27×10^{-3} m
m_p	110 kg
N	1.4
A_p	1.85×10^{-3} m^2
C_r	2.85×10^{-8} $m^3/(s{\cdot}Pa)$
P_{atm}	$101{,}325$ Pa
R	286.9 $Pa{\cdot}m^3/(kg{\cdot}K)$
T	298 K
V_b	4.18×10^{-4} m^3
V_{t0}	7.32×10^{-5} m^3
μ	1.824×10^{-5} $Pa{\cdot}s$
k_d	8170 N/m
c_d	60 $N{\cdot}s/m$
h_d	0

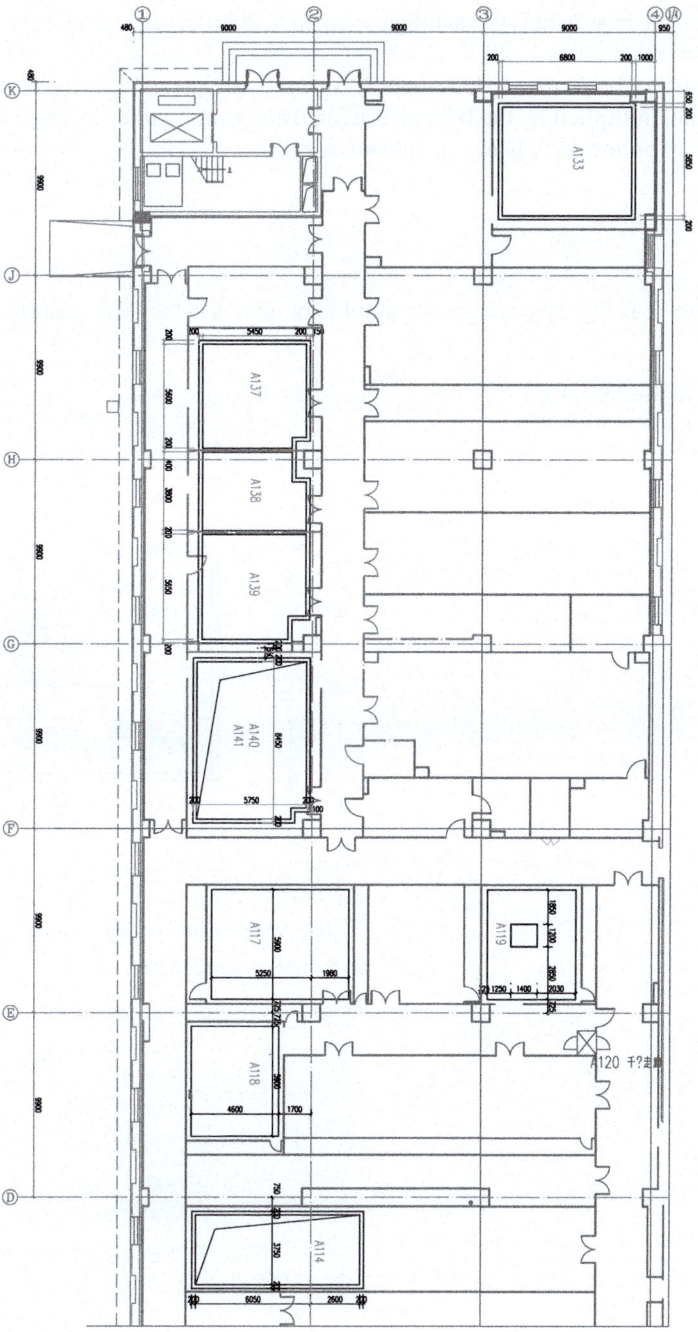

Fig. 8.16 The layout of the large volume of the concrete foundation in an actual engineering

(1) When the base area is greater than or equal to 20 m^2, it can be adopted according to Table 3.
(2) When the base area is less than 20 m^2, the compressive stiffness coefficient can be multiplied by the base area-related correction factor in Table 3, with the correction factor β_r being calculated as follows:

$$\beta_r = \sqrt[3]{\frac{20}{A}} \tag{8.3.3.2}$$

where β_r is the base area correction factor, and A is the base area (m^2).

Fig. 8.17 Active control and the large volume of concrete foundation design drawing in an actual engineering

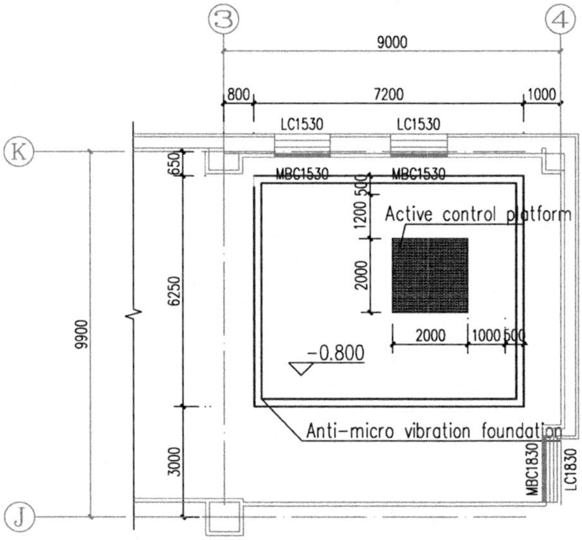

(a) Active control design drawing

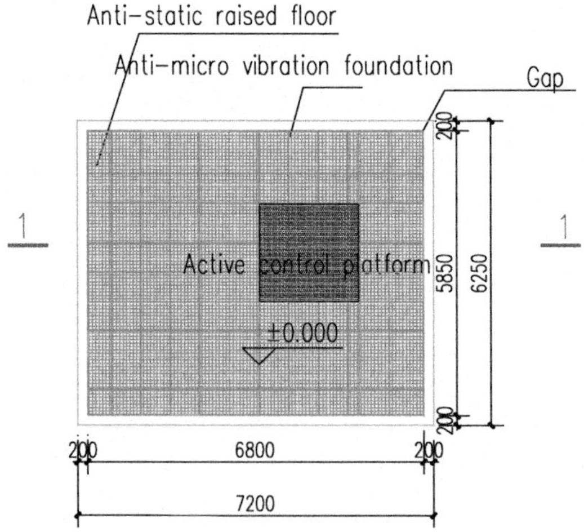

(b) Anti-static electricity floor drawing

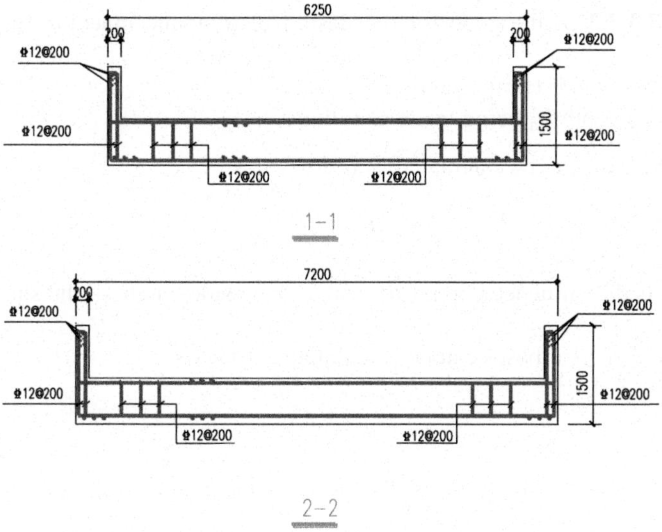

(c) Large volume of concrete foundation design drawing

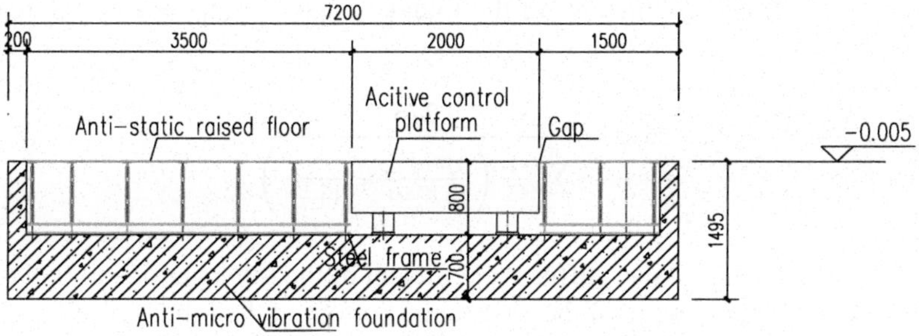

(d) Overall design drawing of active control device, anti-static electricity floor and large volume of concrete foundation

Fig. 8.17 (continued)

Table 3 Compression stiffness coefficient Cz of natural foundation (kN/m^3)

Standard value of foundation-bearing capacity f_k (kPa)	Types of soil		
	Clay	Silt	Sand
300	66,000	59,000	52,000
250	55,000	49,000	44,000
200	45,000	40,000	36,000
150	35,000	31,000	28,000
100	25,000	22,000	18,000
80	18,000	16,000	

Provision 8.3.3.3 of the code for the design of dynamic machine foundation (GB50040-96):

The foundation soil is composed of different soil layers at the bottom of the base, with its influence depth h_{d} being taken as follows:

(1) The square base can be calculated as follows:

$$h_d=2d \tag{8.3.3.3.1}$$

where h_d is the influence depth (m), and d is the side length of the square base (m).

(2) The other shapes of bases can be calculated as follows:

$$h_d=2\sqrt{A} \tag{8.3.3.3.2}$$

Provision 8.3.3.4 of the code for the design of dynamic machine foundation (GB50040-96):

Within the depth range of the foundation soil affected by the base (Fig. 8.18), the compressive stiffness coefficients of the foundation soil composed of different soil layers can be calculated as follows:

$$C_{\mathrm{z}} = \frac{2/3}{\sum_{i=1}^{\mathrm{n}} \frac{1}{C_{zi}} \left(\frac{1}{1+\frac{2h_{i-1}}{h_{\mathrm{d}}}} - \frac{1}{1+\frac{2h_i}{h_{\mathrm{d}}}} \right)} \tag{8.3.4.4}$$

where C_{zi} is the compressive stiffness coefficient of the i-th layer of soil (kN/m^3), h_i is the depth from the bottom of the base to the bottom of the i-th layer of foundation soil (m), and h_{i-1} is the depth from the bottom of the base to the bottom of the $i-1$-th layer of soil (m).

Provision 8.3.3.5 of the code for the design of dynamic machine foundation (GB50040-96): The flexural, shear, and torsional stiffness coefficients of natural foundations can be calculated according to the following formulas:

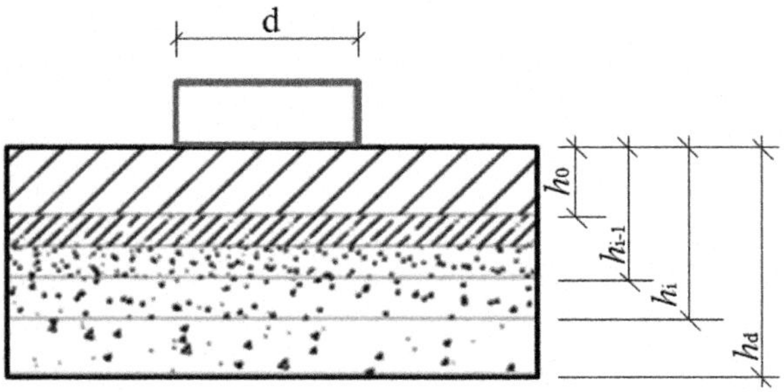

Fig. 8.18 Layered soil foundation

$$C_{\varphi}=2.15C_z \tag{8.3.3.5.1}$$

$$C_x=0.70C_z \tag{8.3.3.5.2}$$

$$C_{\phi}=1.05C_z \tag{8.3.3.5.3}$$

where C_{φ} is the flexural stiffness coefficient of natural foundation (kN/m^3), C_x is the shear stiffness coefficient of natural foundation (kN/m^3), and C_{ϕ} is the torsional stiffness coefficient of natural foundation (kN/m^3).

Provision 8.3.3.9 of the code for design of dynamic machine foundation (GB50040-96):

The calculation of the damping ratio for the natural foundation shall meet the following requirements.

Equation (8.3.3.9.1)–The vertical damping ratio can be calculated according to the following equation:

(1) Clay

$$\zeta_z = \frac{0.16}{\sqrt{\overline{m}}} \tag{8.3.3.9.1}$$

$$\overline{m} = \frac{m}{\rho A\sqrt{A}} \tag{8.3.3.9.2}$$

(2) Sand and silt

$$\zeta_z = \frac{0.11}{\sqrt{\overline{m}}} \tag{8.3.3.9.3}$$

where ζ_z is the vertical damping ratio of natural foundation, $\overline{m}$ is the mass ratio of the base set, m is the mass of the base set, and ρ is the density of foundation soil.

The mass of the large concrete foundation volume m_1 is 97500 kg. The stiffness coefficient k_1 and the damping coefficient c_1 were calculated according to the above equations. The sensitive equipment's length, width, and height are 2, 2, and 0.2 m, and the mass m_2 is 1958 kg. Air floating vibration control was adopted in the vibration control system. The natural vertical frequency of the primary stage vibration isolation system is 1.0 Hz under steady state, the vertical stiffness k_2 is 7.722×10^4 N/m, the damping coefficient c_2 is 3.688×10^3 N·s/m (damping ratio $\zeta = 15\%$), and the horizontal stiffness and damping coefficient are 0.7 times of the vertical ones. *When designing the vibration control system, the system decoupling design should be carried out first* (*see* Fig. 8.19).

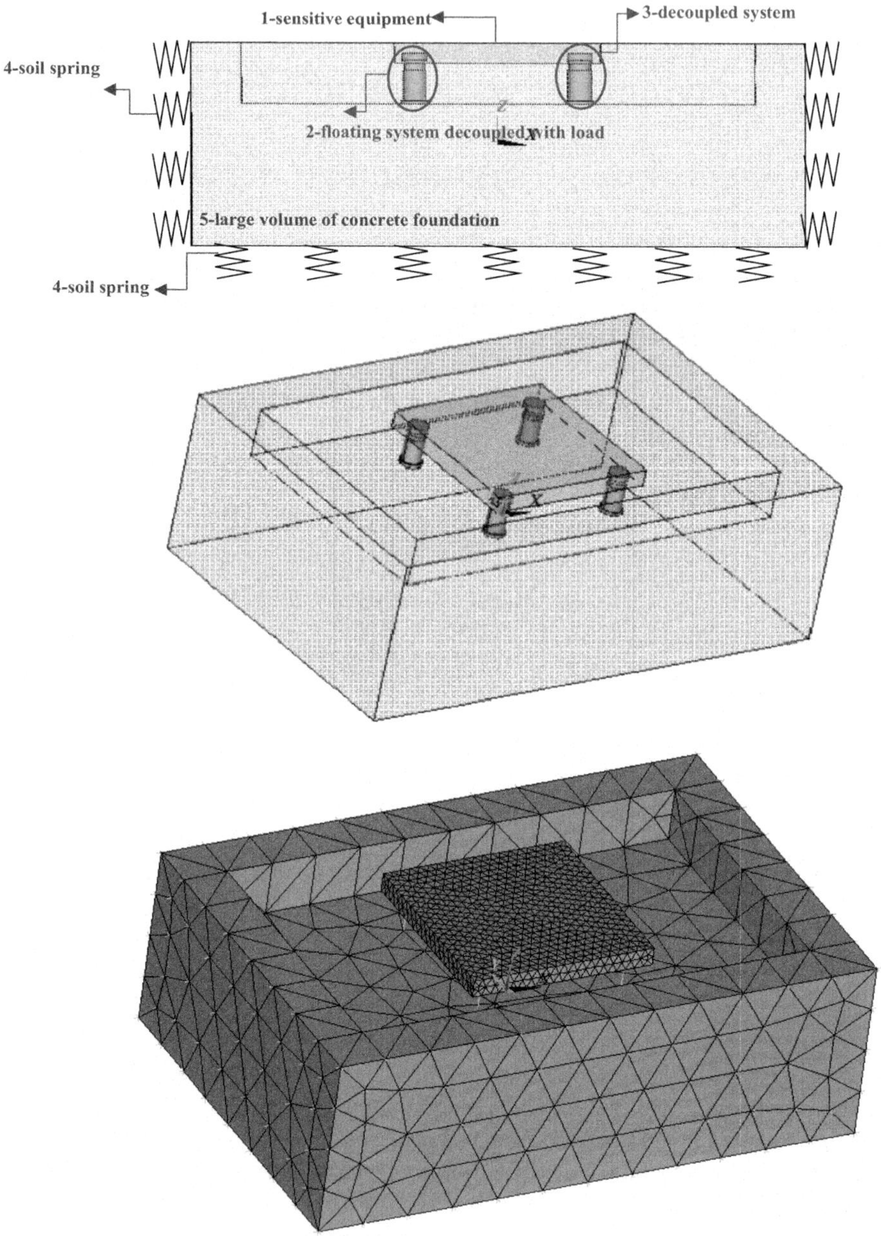

Fig. 8.19 Sensitive equipment-decoupling air floating control system-large volume of the concrete foundation-soil spring integrated refined design model

The reinforced concrete in the finite element analysis utilizes a solid65 element, with its parameters being selected according to C30, and the spring damper adopts a combin14 element. The finite element analysis using ANSYS/APDL and the analytical calculation using MATLAB/SIMULINK were adopted based on the discussion in Chap. 7. The analytical calculation simplifies the large volume of reinforced concrete foundation and sensitive equipment into a two-stage single mass point system.

Figure 8.19 shows an integrated refined design model for the sensitive equipment-decoupling air floating control system-large volume for the concrete foundation-soil spring.

It can be seen from the calculation results in Figs. 8.20 to 8.21 that the calculated vertical natural frequency according to the code for the design of dynamic machine foundation (GB50040-96) is 17.87 Hz, and the designed vertical design frequency of the isolation system of sensitive equipment is 1.0 Hz. The two results are almost identical to the analysis ones, which satisfies the accuracy of the refined calculation.

Figure 8.22 shows the measured vertical ground acceleration as the model calculation input, with a sampling frequency of 256 Hz.

The finite element calculation method for active control is the same as the strategy in Chap. 7. The active control uses the LQR controller. Figure 8.23 shows the process of sensitive equipment-air floating passive and active control system-large volume for the concrete foundation-soil spring integrated vibration control refined design. Since the vibration input of this system is acceleration, Eq. (7.2) can be rewritten as follows:

$$\begin{cases} m_1 \ddot{x}_1 + k_1 x_1 + c_1 \dot{x}_1 - c_2\left(\dot{x}_2 - \dot{x}_1\right) - k_2(x_2 - x_1) = m_1 \ddot{x}_g(t) - F_a(t) \\ m_2 \ddot{x}_2 + c_2\left(\dot{x}_2 - \dot{x}_1\right) + k_2(x_2 - x_1) = m_2 \ddot{x}_g(t) + F_a(t) \end{cases} \tag{8.24}$$

where $\ddot{x}_g(t)$ is the input acceleration, $z_1 = x_1$, $z_2 = x_2$, $z_3 = \dot{x}_1$, and $z_4 = \dot{x}_2$ are the state parameters, and $z(t) = [z_1, z_2, z_3, z_4]^{\mathrm{T}}$ is the state vector.

Equation (8.24) can be transformed into the following state space equation:

$$\begin{cases} \dot{z}(t) = \mathrm{A}z(t) + \mathrm{b}_1 \ddot{x}_g(t) + \mathrm{b}_2 F_a(t) \\ y(t) = \mathrm{C}z(t) + \mathrm{d}_1 \ddot{x}_g(t) + \mathrm{d}_2 F_a(t) \end{cases} \tag{8.25}$$

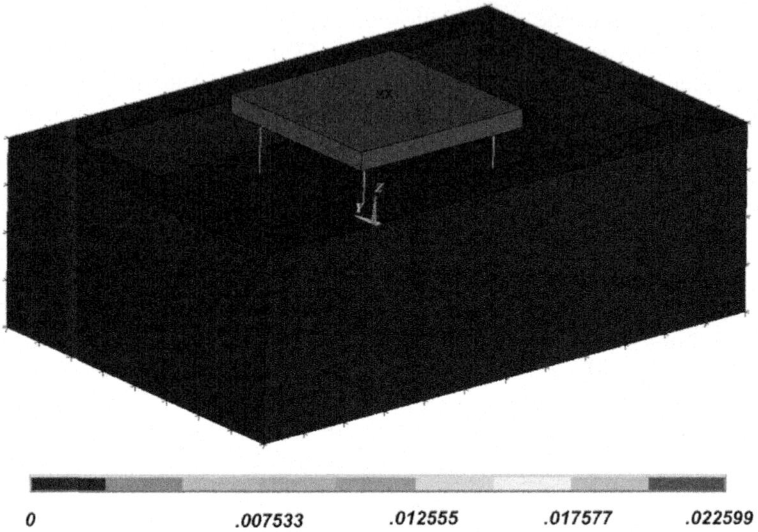

(a) Pure vertical vibration mode of sensitive equipment (1.0 Hz)

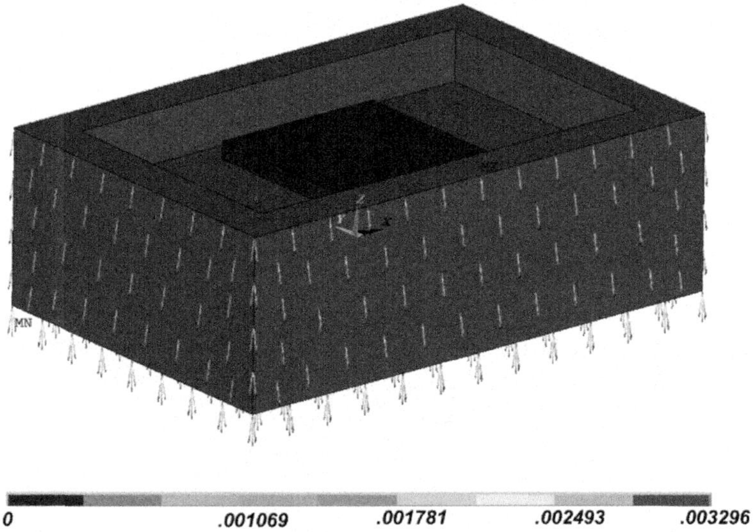

(b) Pure vertical vibration mode of concrete foundation (17.85 Hz)

Fig. 8.20 Pure vertical vibration mode of the sensitive equipment and the large volume of the concrete foundation in the integrated vibration control system

***** PARTICIPATION FACTOR CALCULATION ***** Z DIRECTION

MODE	FREQUENCY	PERIOD	PARTIC. FACTOR	RATIO	EFFECTIVE MASS	CUMULATIVE MASS FRACTION	RATIO EFF. MASS TO TOTAL MASS
1	0.836217	1.1959	-0.10152E-04	0.000000	0.103060E-09	0.103621E-14	0.103621E-14
2	0.836225	1.1959	-0.13664E-05	0.000000	0.186713E-11	0.105498E-14	0.187730E-16
3	0.999446	1.0006	44.390	0.142171	1970.49	0.198123E-01	0.198123E-01
4	1.17053	0.85431	0.25194E-05	0.000000	0.634742E-11	0.198123E-01	0.638200E-16
5	1.39208	0.71835	0.13534E-04	0.000000	0.183163E-09	0.198123E-01	0.184161E-14
6	1.39209	0.71834	0.61381E-05	0.000000	0.376764E-10	0.198123E-01	0.378816E-15
7	15.4829	0.64588E-01	-0.58517E-03	0.000002	0.342419E-06	0.198123E-01	0.344285E-11
8	16.4392	0.60830E-01	0.22814E-02	0.000007	0.520460E-05	0.198123E-01	0.523295E-10
9	17.8533	0.56012E-01	312.23	1.000000	97487.6	1.00000	0.980187
10	26.0681	0.38361E-01	-0.16649E-02	0.000005	0.277181E-05	1.00000	0.278691E-10
11	27.0316	0.36994E-01	0.54793E-03	0.000002	0.300231E-06	1.00000	0.301866E-11
12	29.1708	0.34281E-01	0.10873E-03	0.000000	0.118223E-07	1.00000	0.118867E-12
13	115.289	0.86739E-02	0.15973E-02	0.000005	0.255138E-05	1.00000	0.256528E-10
14	141.899	0.70473E-02	0.93254E-01	0.000299	0.869635E-02	1.00000	0.874373E-07
15	151.632	0.65949E-02	-0.17501E-06	0.000000	0.306278E-13	1.00000	0.307947E-18
16	212.682	0.47018E-02	-0.31142E-06	0.000000	0.969820E-13	1.00000	0.975104E-18
17	215.597	0.46383E-02	-0.32564E-04	0.000000	0.106039E-08	1.00000	0.106617E-13
18	238.969	0.41846E-02	-0.74448E-05	0.000000	0.554253E-10	1.00000	0.557272E-15
19	242.566	0.41226E-02	-0.43568E-02	0.000014	0.189816E-04	1.00000	0.190850E-09
20	250.661	0.39894E-02	0.99866E-04	0.000000	0.997313E-08	1.00000	0.100275E-12
sum					99458.1		1.00000

Fig. 8.21 Modal calculation results of the integrated vibration control

where $$A = \begin{bmatrix} 0 & 0 & 1 & 0 \\ 0 & 0 & 0 & 1 \\ -\frac{k_1+k_2}{m_1} & \frac{k_2}{m_1} & -\frac{c_1+c_2}{m_1} & \frac{c_2}{m_1} \\ \frac{k_2}{m_2} & -\frac{k_2}{m_2} & \frac{c_2}{m_2} & -\frac{c_2}{m_2} \end{bmatrix},$$

$$\mathbf{b_1} = \begin{bmatrix} 0 \\ 0 \\ 1 \\ 1 \end{bmatrix}, \mathbf{b_2} = \begin{bmatrix} 0 \\ 0 \\ -\frac{1}{m_1} \\ \frac{1}{m_2} \end{bmatrix},$$

$$\mathbf{C} = \begin{bmatrix} 1\,0\,0\,0 \\ 0\,1\,0\,0 \\ 0\,0\,1\,0 \\ 0\,0\,0\,1 \end{bmatrix},$$

$$\mathrm{d}_1 = \begin{bmatrix} 0 \\ 0 \\ 0 \\ 0 \end{bmatrix}, \text{ and } \mathbf{d_2} = \begin{bmatrix} 0 \\ 0 \\ 0 \\ 0 \end{bmatrix}.$$

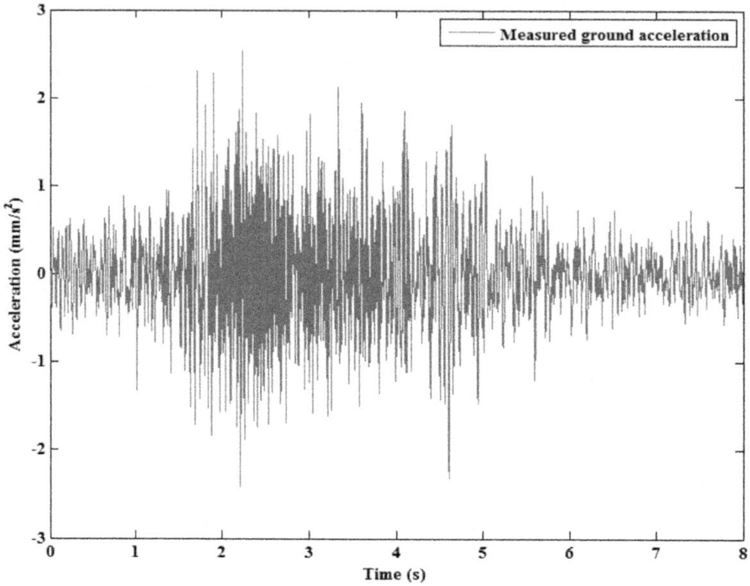

(a) Time domain

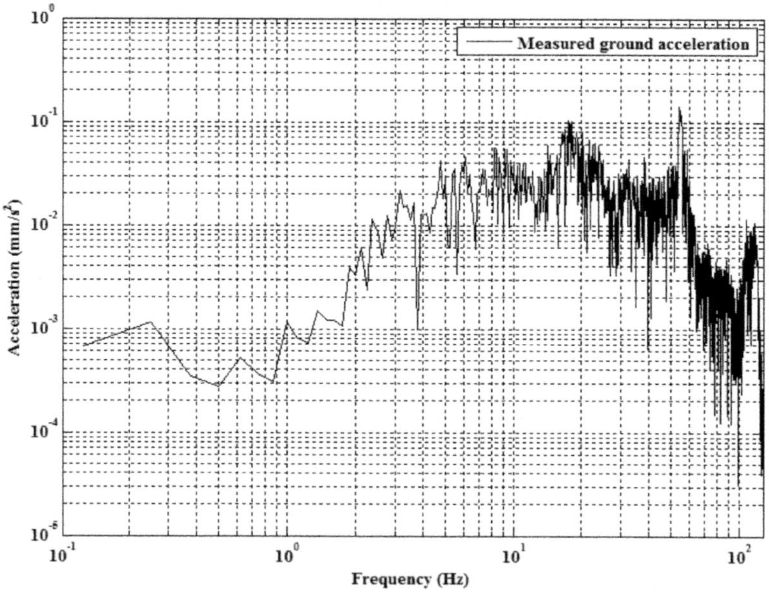

(b) Frequency domain

Fig. 8.22 Measured ground acceleration

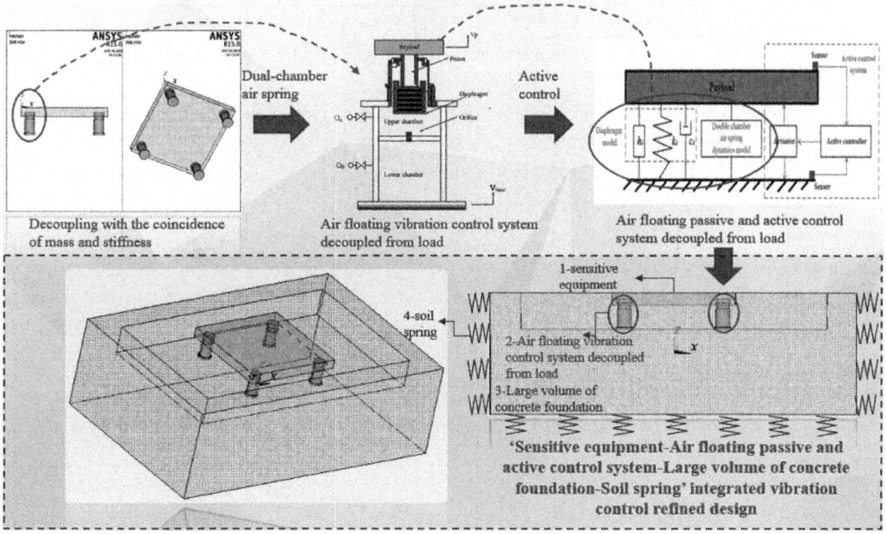

Fig. 8.22 Refined design for sensitive equipment-air floating passive and active control system-large volume of concrete foundation-soil spring integrated vibration control

In order to reasonably configure the weight matrix $\boldsymbol{Q}$ and $\boldsymbol{R}$ in the LQR control, parameters optimization was performed based on the PSO algorithm [9]. The parameters configuration of the PSO algorithm includes 100 particle populations, 100 iterations, $\lambda_1 = 2$ and $\lambda_2 = 1$ acceleration factors, $w = 0.99^t$ inertia weight coefficient (t is the t-th iteration), and Fitness $= \min\left(\left\|\dot{x}_2\right\|_\infty\right)$ fitness objective function, which aims to optimize and reduce the velocity of the sensitive equipment.

The search range of $\boldsymbol{Q}$ was set to $\left[-1\times10^6, -1\times10^6, -1\times10^6, -1\times10^6\right] \sim \left[1\times10^6, 1\times10^6, 1\times10^6, 1\times10^6\right]$, and the search range for $\boldsymbol{R}$ was defined as $1\times10^{-4} \sim 1$. Figure 8.24 shows the convergence curve of the fitness function. After calculation, the optimal gain of the LQR controller is as follows: $[-1.0763, 0.4430, 0.00393, 0.8259]\times10^5$.

Figure 8.25 compares the analytical calculation results of the system with and without active control. It shows that the optimized LQR controller can effectively achieve the active control effect and improve the lack of low-frequency control capability of passive control and the disadvantages of resonance zone amplification. In fact, the primary stage vibration isolation system has a 15% damping ratio, but there is still obvious amplification where the resonance peak cannot be completely suppressed. On this basis, the analysis of active control based on the finite element method was carried out. The calculation results indicated in Fig. 8.26 show that the strategy proposed herein can effectively achieve active control while significantly improving the passive control system.

Figure 8.27 compares the active control force and the control response between the sensitive equipment's analytical and finite element methods. It can be seen that

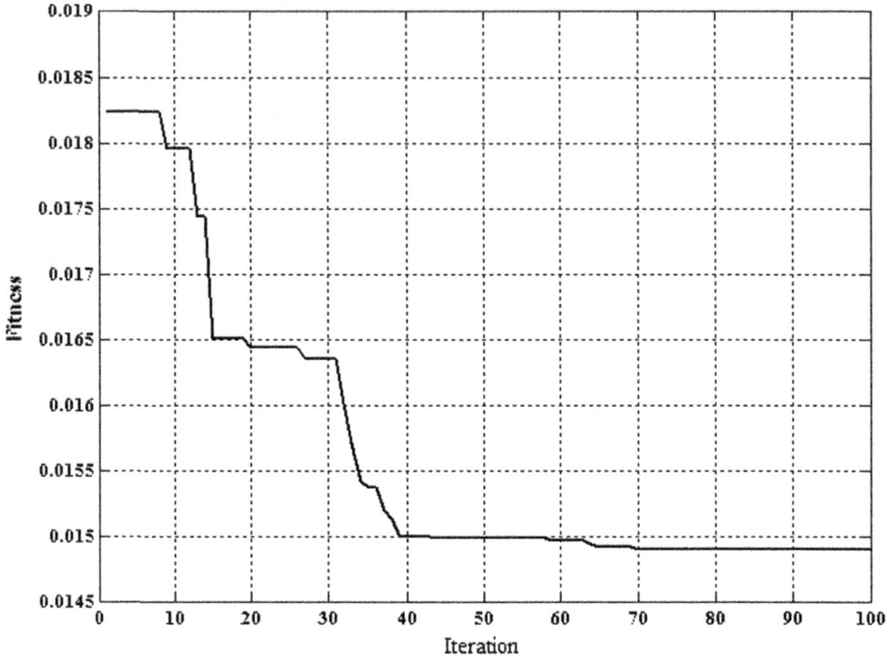

Fig. 8.24 Fitness convergence curve

the fitting effect between the two objects is good, further confirming the rationality of this study's method. Figure 8.27b shows that the response of the two objects is slightly different in the high-frequency domain because the system has presented the control object's and base's local mode (See Fig. 8.28 for the local high-order mode shapes). This phenomenon is different from the analytical calculation of the mass point system. Accordingly, the results of active control using the finite element method are more convincing.

In this study, the integrated design and calculation analysis of the sensitive equipment-air floating passive and active control system-large volume of concrete foundation-soil spring refined vibration control was carried out for micro-vibration control of an actual ultra-precision equipment project. In general, the developed techniques in this book, such as refined analysis of passive control, refined analysis of active control, system decoupling, and passive and active control decoupled from load and soil spring simulation, were utilized as a systematic application case of the key technologies developed by the authors.

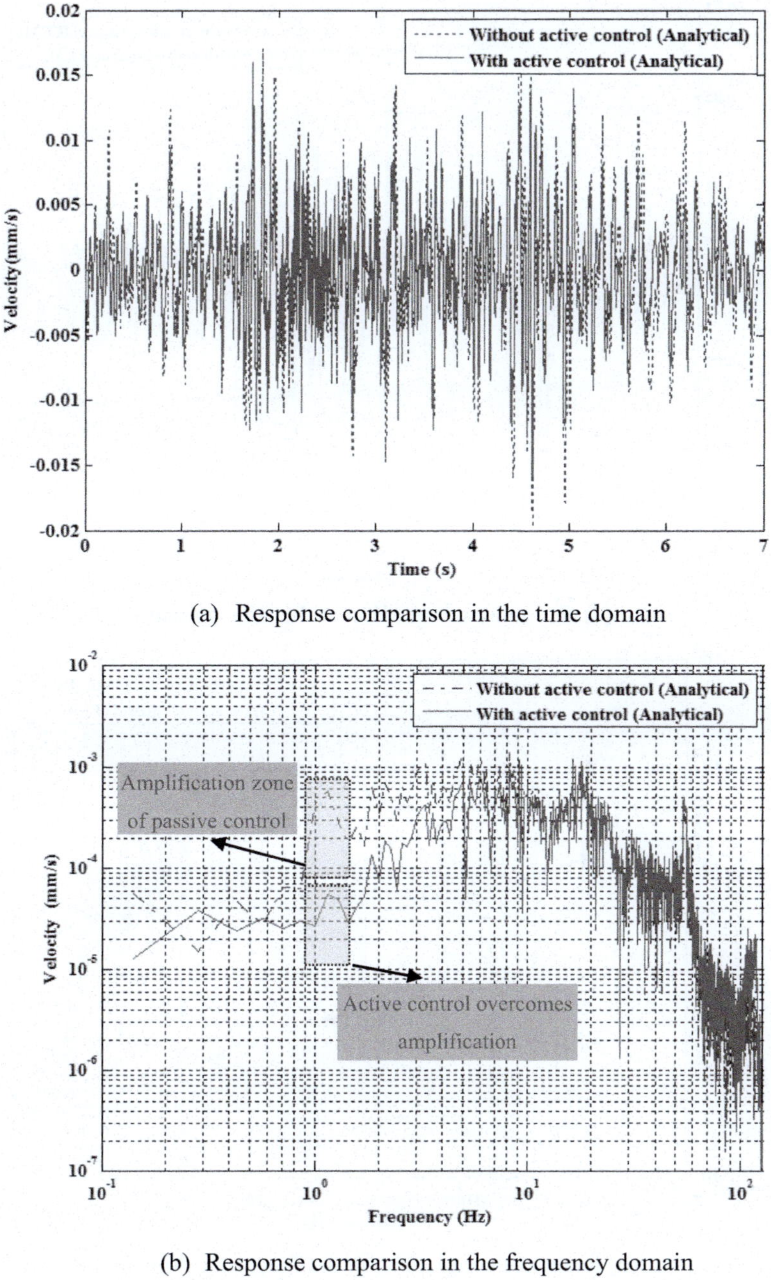

Fig. 8.25 Comparison between the analytical calculation results with and without active control

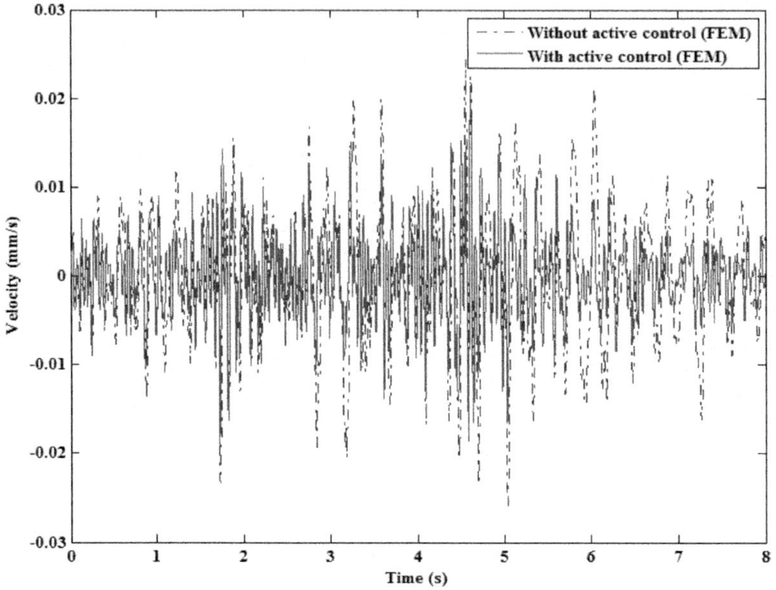

(a) Response comparison in the time domain

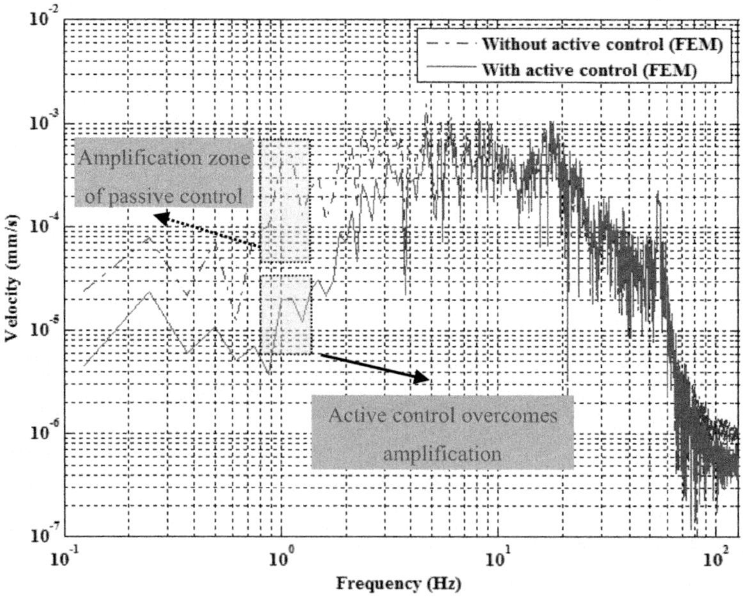

(b) Response comparison in the frequency domain

Fig. 8.26 Comparison of finite element calculation results with and without active control

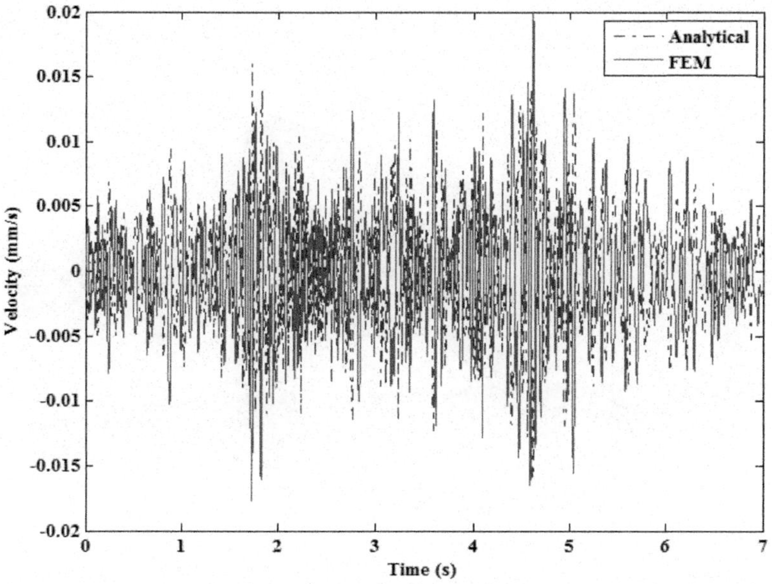

(a) Response comparison of control objects in the time domain

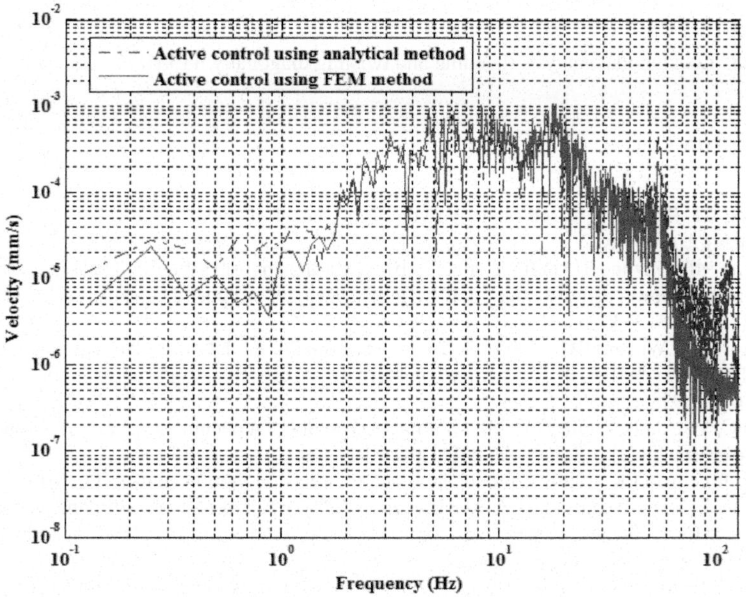

(b) Response comparison of control objects in the frequency domain

Fig. 8.27 Comparison between the analytical and the finite element calculation results under active control

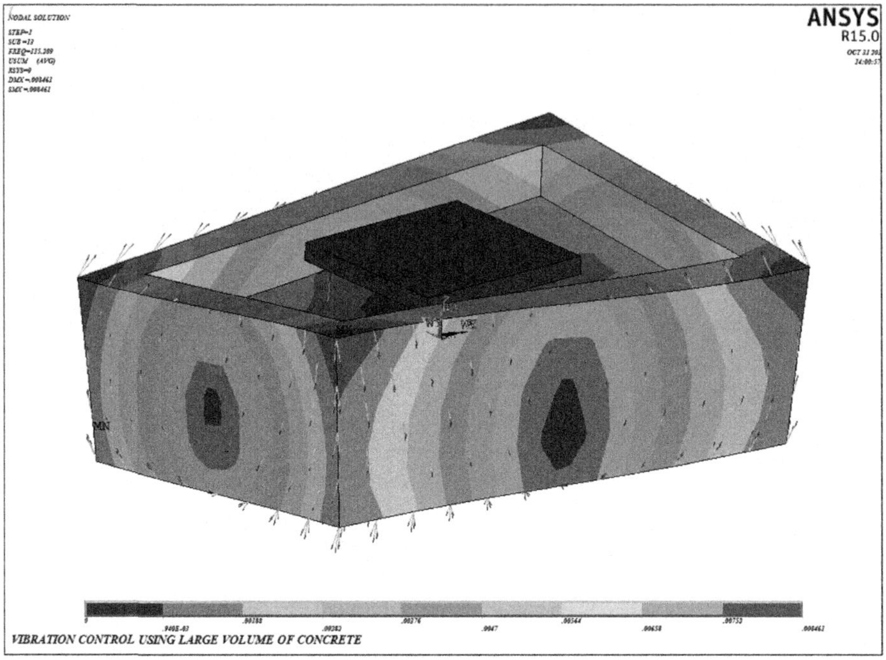

Fig. 8.28 Local modal analysis (bending and twisting diagonally, 115 Hz)

References

1. Zhang CL (2003) Research on active vibration control of micro manufacturing platform [D]. Zhejiang University
2. Hoque ME, Mizuno T, Ishino Y et al (2010) A six-axis hybrid vibration isolation system using active zero-power control supported by passive weight support mechanism [J]. J Sound Vib 329(17):3417–3430
3. Taylor JH, Wilson BH (1995) A frequency-domain model-order-deduction algorithm for nonlinear systems [C]. In: Proceedings of international conference on control applications. IEEE, pp 1053–1058
4. Erin C, Wilson B, Zapfe J (1998) An improved model of a pneumatic vibration isolator: theory and experiment [J]. J Sound Vib 218(1):81–101
5. Subrahmanyan PK (1999) A model approach to precision motion control [D]. MIT
6. Nasri M, Nezamabadi-pour H, Maghfoori M (2007) A PSO-based optimum design of PID controller for a linear brushless DC motor [J]. World Acad Sci Eng Technol Int J Electr Inf Eng 1(2):171–175
7. Chen C-K, Kuo H-H, Yan J-J et al (2009) GA-based PID active queue management control design for a class of TCP communication networks [J]. Expert Syst Appl 36(2 (Part 1)):1903–1913
8. GB50040: Standard for design of dynamic machine foundation [S]
9. Eberhart R, Kennedy J (1995) A new optimizer using particle swarm theory [C]. In: Proceedings of the sixth international symposium on micro machine and human science. IEEE, pp 39–43

Chapter 9
Low Frequency Passive and Active Control Using Quasi-zero Stiffness

Abstract In this chapter, low frequency passive and active control using quasi-zero stiffness is performed out. The principle of negative stiffness and quasi-zero stiffness approached by parallel positive and negative stiffnesses are investigated. Based on this, passive and active control based on quasi-zero stiffness are carried out and a low-frequency vibration control is achieved for micro-vibration control.

Low-frequency micro-vibration passive and active control system based on a dual-chamber air spring using a refined analysis technique was performed in Chap. 8. In general, an air spring floating vibration control system is cost-effective for high-load precision equipment, but it is not recommended for desktop precision equipment due to the need to provide uninterrupted gas sources and supporting gas circuit control systems. Therefore, this chapter focuses on low-frequency control systems based on quasi-zero stiffness (QZS).

In recent years, research on quasi-zero stiffness based on the parallel connection of positive and negative stiffnesses has become a hot spot, with many scholars carrying out designs and analyses of quasi-zero stiffness systems. Reference [1] started with the new configuration mechanism and active control and studied the critical vibration isolation mechanism unit of ultra-low frequency realized by positive and negative stiffnesses in parallel. Reference [2] used the positive stiffness of a straight beam and the buckling behavior of a bistable beam to produce negative stiffness that achieves its zero stiffness characteristics. Besides, they designed a constant force micro-motion platform based on a compliant mechanism. Reference [3] proposed a new method to perform ultra-low frequency vibration isolation through the parallel connection of positive and negative stiffnesses springs. Reference [4] developed a negative stiffness vibration isolation device, where the dynamic stiffness is much lower than the static stiffness. Reference [5] designed an ultra-low frequency system consisting of a vertical spring and two tilt springs based on quasi-zero stiffness.

W. Huang and J. Xu, *Optimized Engineering Vibration Isolation, Absorption and Control*, Springer Tracts in Civil Engineering, https://doi.org/10.1007/978-981-99-2213-0_9

9.1 Principle of Negative Stiffness

The negative stiffness system is critical to achieving quasi-zero stiffness. Additionally, the negative stiffness can be formed using the compression link in an unstable equilibrium state [1, 6, 7], as shown in Fig. 9.1. It can be seen that two connecting rods of equal length L are hinged in the middle, while the other end is supported by a horizontal preload F_p, and the connecting rod's central hinge is subjected to a concentrated external force F_v. Under certain conditions, the system can reach equilibrium. Δ is the vertical deformation of the hinge without load.

Calculation equation of presented system's vertical stiffness:

$$K_v = \frac{df}{dz} = -2F_v\left(\frac{1}{\sqrt{L^2 - \Delta^2}} + \frac{\Delta^2}{\sqrt{\left(L^2 - \Delta^2\right)^3}}\right) \tag{9.1}$$

where df/dz is the stiffness operator, and the denominator and numerator reflect the changes in force and deformation, respectively.

In actual engineering, the preload F_p can be provided by the steel spring. Thus, a force is applied on the steel spring to produce initial deformation, then pre-tightened by nut jacking, as shown in Fig. 9.2. Suppose that the stiffness of the horizontal spring is k_h, and the initial deformation of the spring after preloading is Δ_h, then the external force F_v that brings the system to equilibrium is:

$$F_v = -2k_h\left(\Delta_h - L + \sqrt{L^2 - \Delta^2}\right)\frac{\Delta}{\sqrt{L^2 - \Delta^2}} \tag{9.2}$$

Moreover, the vertical stiffness of the system is calculated as follows:

$$K_v = \frac{df}{dz} = 2k_h\left[-1 + (L - \Delta_h)\frac{L^2}{\left(L^2 - \Delta^2\right)^{3/2}}\right] \tag{9.3}$$

Rewrite Eq. (9.3) into a dimensionless stiffness expression is as follows:

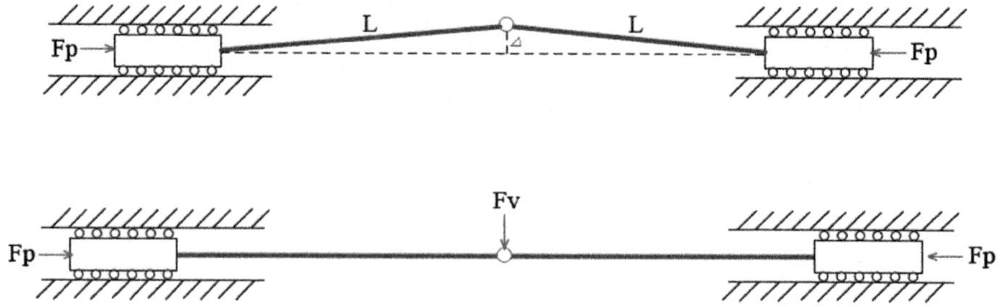

Fig. 9.1 Linkage mechanism with negative stiffness

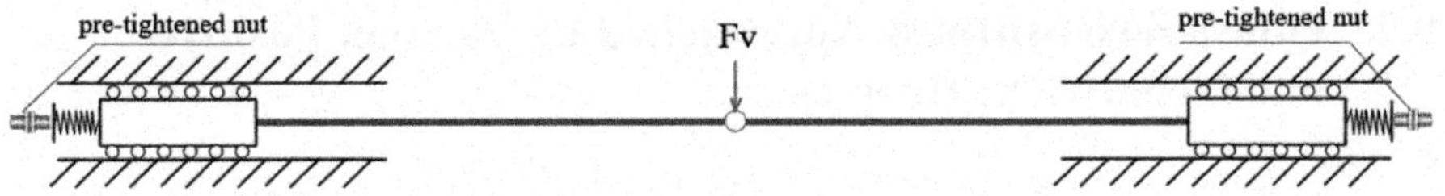

Fig. 9.2 Negative stiffness linkage mechanism with pre-tightened spring

$$\hat{k} = -1 + (1 - \alpha)\frac{1}{\left(1 - \hat{\Delta}^2\right)^{3/2}} \tag{9.4}$$

where $\hat{k} = K_v / 2k_s$, $\hat{\Delta} = \Delta / L$, and $\alpha = \Delta_h / L$.

It can be seen from Fig. 9.3 that the system exhibits negative stiffness characteristics when $\alpha > 0.8$, the system has constant negative stiffness when $\alpha = 1$, and the system consistently exhibits negative stiffness when $\alpha > 1$. The stiffness nonlinearity increases with the increase in α. Therefore, when designing a quasi-zero stiffness system, it is essential to ensure that $\alpha \geq 1$, and that is $\Delta_h \geq L$.

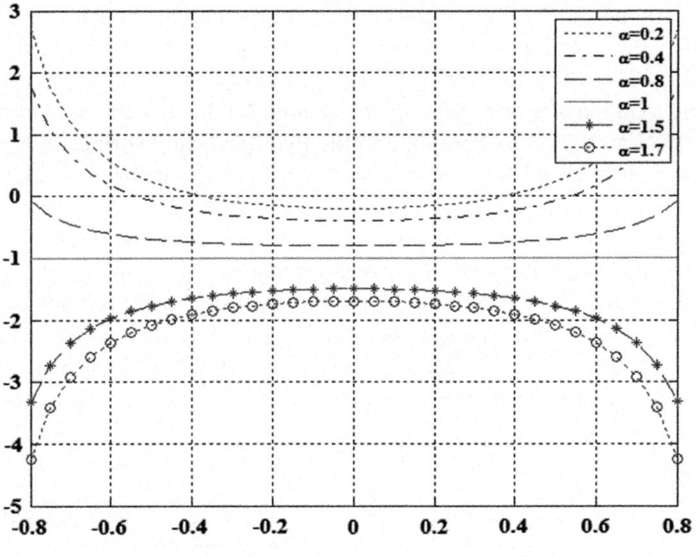

Fig. 9.3 Variation curve of negative stiffness characteristics

9.2 Quasi-zero Stiffness Approached by Parallel Positive and Negative Stiffnesses

As described in Sect. 9.1, the system behaves as a negative stiffness characteristic when α is taken as a specific value, as shown in Fig. 9.4. In theory, the system is close to zero stiffness when $k_v \approx -K_v$. On the other hand, different masses can be supported by adjusting the spring stiffness k_v and the initial deformation given by the nut's vertical pre-tightening.

9.3 Passive Control Based on Quasi-zero Stiffness

The finite element analysis was performed in ANSYS 15.0. The beam system was simulated using the *beam188* element, the spring was simulated using the *combin14* element, and the C_p was used to simulate the boundary conditions for the two-bar hinge. The calculation process starts by defining the simulation parameters according to Sect. 9.2, performing finite element modeling, applying preload, conducting static analysis (with the status, activate the prestress), and modal analysis (with the status, activate the prestress). The parallel system of positive and negative stiffnesses is shown in Fig. 9.4, where the round rod diameter is $D = 0.04$ m, the rod length is $L = 0.25$ m, the load mass is $m = 8.36$ kg ($F_v = 83.6$ N), the horizontal spring stiffness is $k_h = 200$ N/m, the preload applied to the two horizontal springs is $F_p = 100$ N, and the vertical spring stiffness is $k_v = 9.19$ N/m. The finite element model is shown in Fig. 9.5. The natural vertical frequency of the system is calculated as 1.08 Hz. The first mode of vibration is shown in Fig. 9.6. This vibration mode is the vertical overall vibration mode upon parallel connecting the positive and negative stiffnesses, which is the expected vibration mode and natural frequency after achieving quasi-zero stiffness. White noise with an amplitude of 1 m/s^2 was applied as the global vertical acceleration of the system for a duration of 10 s. The displacement response of the node at the load position was extracted with the FFT result shown in Fig. 9.7.

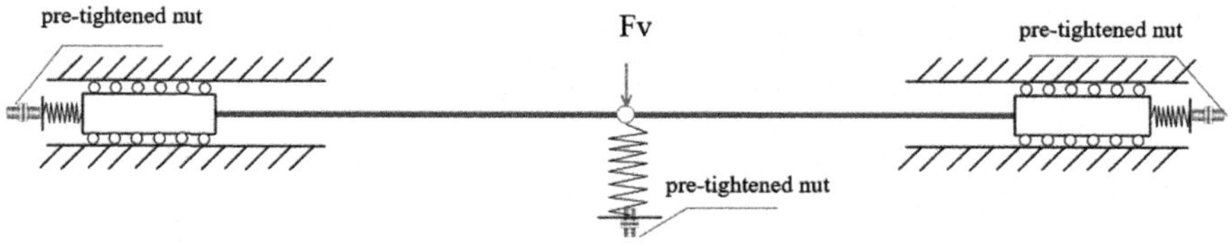

Fig. 9.4 Quasi-zero stiffness approached by parallel positive and negative stiffnesses

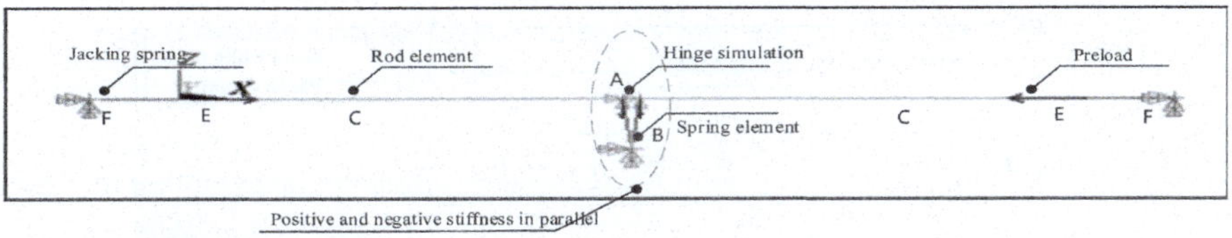

Fig. 9.5 Finite element model for quasi-zero stiffness system

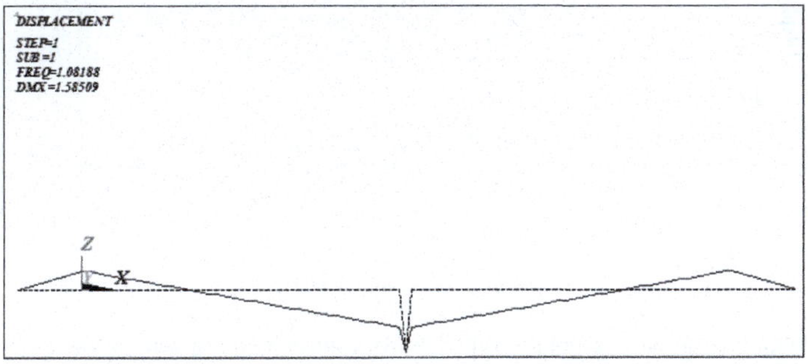

Fig. 9.6 First mode and low natural frequency (1.08 Hz) achieved by quasi-zero stiffness

9.4 Active Control Based on Quasi-zero Stiffness

The calculation results in Figs. 9.6 and 9.7 indicate that the parallel positive and negative stiffnesses can achieve a lower natural frequency. In order to further improve the vibration control effect, PID active control was adopted herein, as shown in Fig. 9.8. The proportional, integral, and differential gains of the PID controller are – 60, – 5, and – 5, respectively. The active control program was embedded in the ANSYS/APDL environment. The designed program extracts the control response of a specific node and feedbacks it to the PID controller in real-time based on the transient calculation results to achieve a closed-loop active control. The process is shown in Fig. 9.9. In this study, the sensor's pick-up point and the actuator's action point were selected at the middle hinge of the two rods. Figure 9.10 shows the active control effect. It can be seen that active control can effectively suppress resonance, which is greatly improved compared to passive control. Figure 9.11 shows the active controller-driven actuator's output force.

In this chapter, the design of low-frequency vibration control was carried out based on the principle of quasi-zero stiffness approached by parallel positive and negative stiffnesses. Additionally, the modal calculation was performed under the prestress activation state according to the spring preload principle, and a low natural frequency was achieved. Immediately afterward, a finite element-based active control analysis was conducted. The results show that active control can significantly improve the passive control effect, which verifies the designed strategy's effectiveness and opens up a new idea for a low-frequency design of micro-vibration control. The

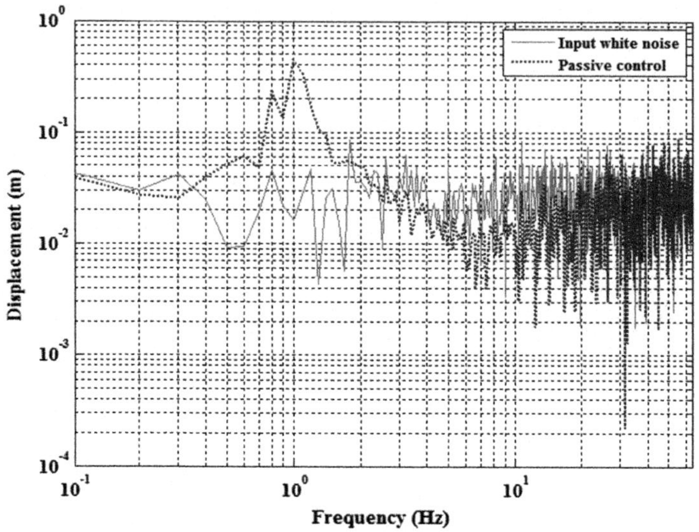

Fig. 9.7 Frequency domain response of the passive control by quasi-zero stiffness

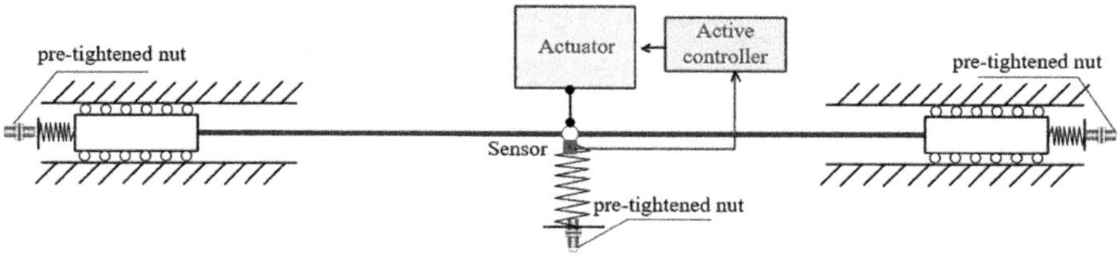

Fig. 9.8 Active control system based on quasi-zero stiffness

Fig. 9.9 Finite element analysis procedure of active control

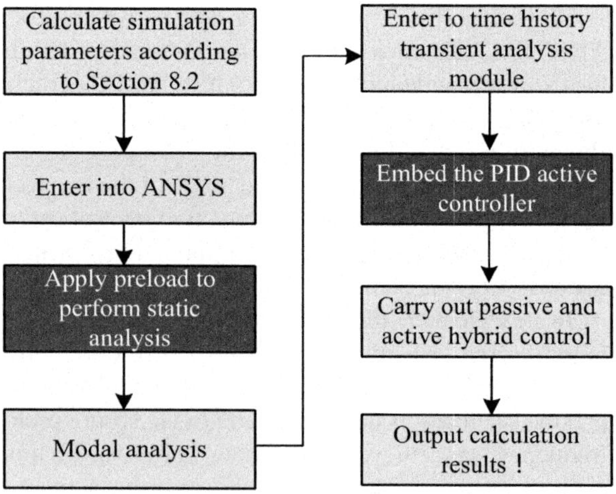

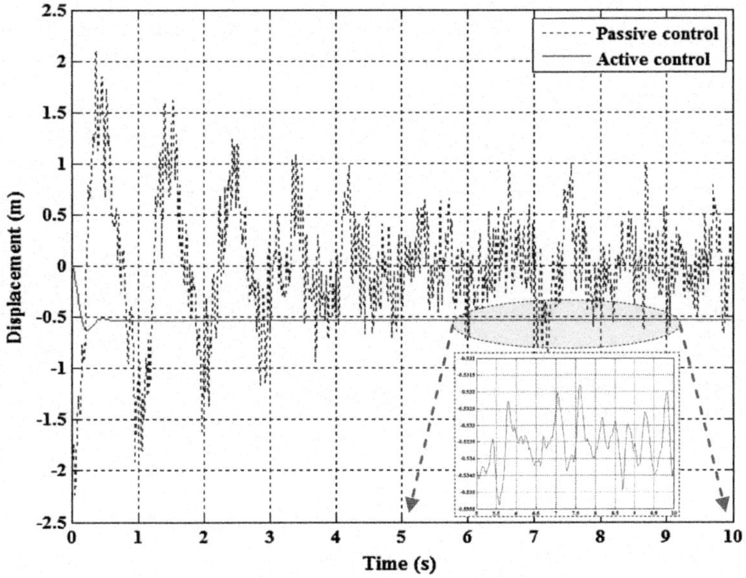

(a) Time domain response

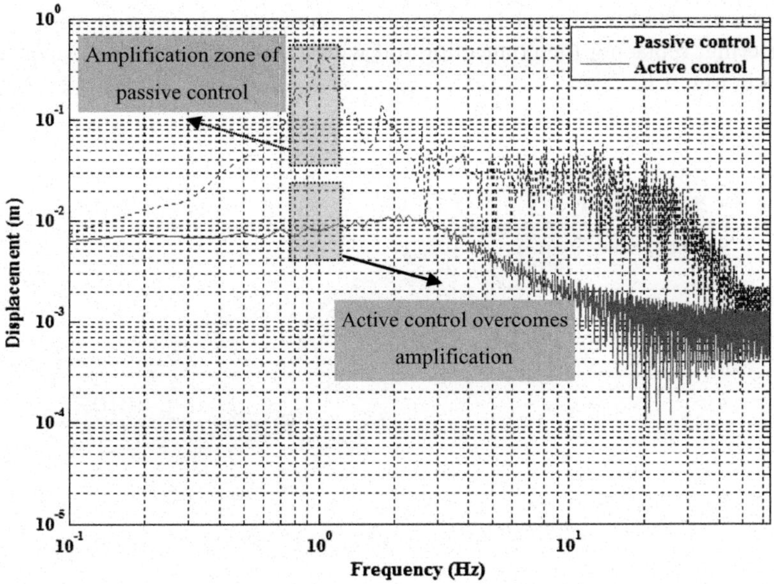

(b) Frequency domain response

Fig. 9.10 Response comparison of passive and active control

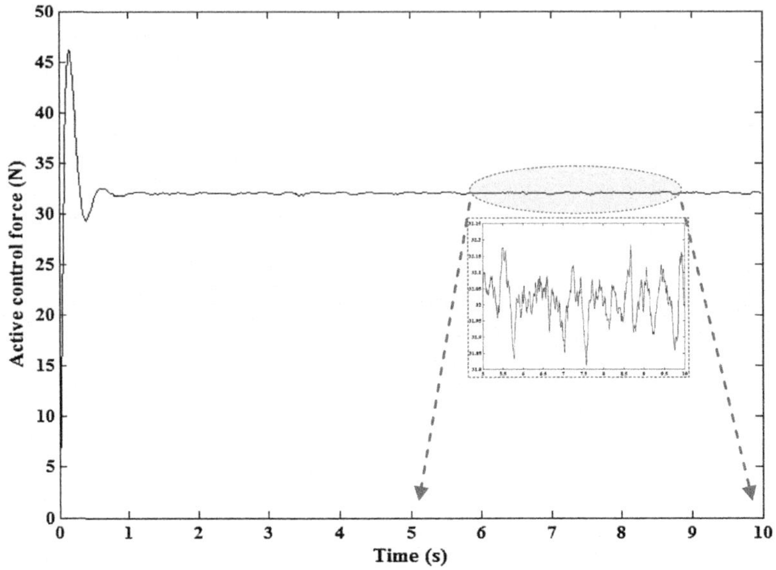

Fig. 9.11 Output force of actuator of active control

above-mentioned hinge mechanism was considered as a connecting rod. In actual engineering, replacing the connecting rod with a thin plate is more convenient, as shown in Fig. 9.12.

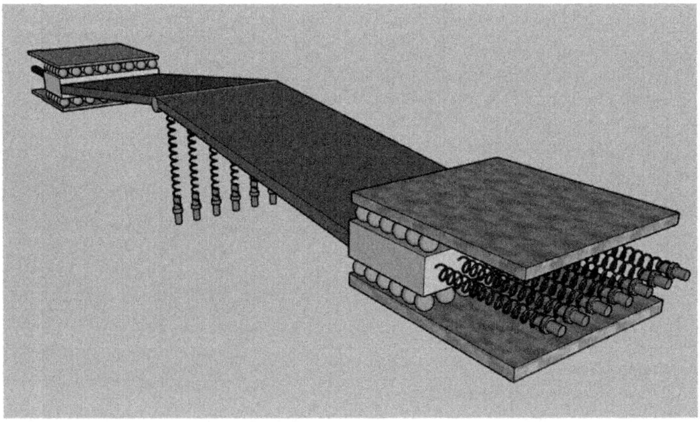

Fig. 9.12 Parallel positive and negative stiffnesses using two thin hinged plates

References

1. Wenjiang W (2014) Study on active precision vibration isolation system based on parallel of positive and negative stiffness springs [D]. Wuhan: Huazhong University of Science and Technology
2. Hu J, Chen X (2018) Optimized design of a micro-motion stage with zero stiffness [J]. Opt Precis Eng 26(6):1430–1440
3. Zhang JZ, Dong S, Li D (2004) Study on new type vibration isolation system based on combined positive and negative stiffness [J]. Nanotechnol Precis Eng 2(4):314–318
4. Carrella A, Brennan MJ, Waters TP (2007) Static analysis of a passive vibration isolator with quasi-zero-stiffness characteristic [J]. J Sound Vib 301(3–5):678–689
5. Carrella A, Brennan MJ, Kovacic I et al (2009) On the force transmissibility of a vibration isolator with quasi-zero-stiffness [J]. J Sound Vib 322(4–5):707–717
6. Platus DL (1992) Negative-stiffness-mechanism vibration isolation systems. In: Proceedings of the SPIE, vibration control in microelectronics, optics, and metrology, vol 1619, pp 44–54
7. Carrella A (2008) Passive vibration isolators with high-static-low-dynamic-stiffness [D]. University of Southampton, Institute of Sound and Vibration Research

Chapter 10
Dynamic Vibration Absorption and Performance Optimization for Equipment, Floor and High-rise Building Structure

Abstract In this chapter, the main system with damping and without damping of passive dynamic vibration absorption are studied firstly, and based on this, active dynamic vibration absorption using LQR control and PSO technique and semi-active dynamic vibration absorption using a MRD is proposed. Followed by, dynamic vibration absorption for floor structure using TMD and ATMD is proposed, and active control for ATMD is achieved in FEM. Subsequently, dynamic vibration absorption for 76-story benchmark structure using TMD and ATMD is also studied, and fluctuating wind speed field simulation using the DIT-FFT-WAWS method is adopted and the PSO is used for optimizing the control parameters.

Severe vibrations generated by power equipment affect the surrounding environment, building structure, personnel comfort, and equipment service life. As a result, effective vibration control measures must be taken [1, 2]. These measures are typically taken at the bottom of large and medium-sized equipment, which must be removed, hoisted, and restored, with certain construction difficulties and risks. It is possible to install a dynamic vibration absorber (DVA) on the top of the equipment [3, 4]. Vibration absorption is a way of adding a subsystem to the primary system to absorb and reduce its vibration. Due to their simple structure and ability to effectively reduce the vibration of mechanisms and equipment with little frequency change, dynamic vibration absorbers have become an essential means of controlling vibration in engineering.

Ormondroyd and Hartog [5] established a single-degree-of-freedom mechanical model of an undamped vibration absorber in a paper published in 1928 and found that a damped dynamic vibration absorber has a better damping effect and optimal damping. Hahnkamn [6] mentioned that the two amplitude amplification curve intersections in the main system with a dynamic absorber are fixed points that do not change with the absorber's damping. Hence, these two fixed points (P and Q) and this feature help introduce the optimal coherent condition. Nishihara and Matsuhisa [7] analytically divided the optimal parameters of a damped system and applied the simplest mathematical model to design the damped dynamic vibration absorber. Warburto and Ayorinde [8] highlighted that when a damper with a single degree

W. Huang and J. Xu, *Optimized Engineering Vibration Isolation, Absorption and Control*, Springer Tracts in Civil Engineering, https://doi.org/10.1007/978-981-99-2213-0_10

of freedom system is equipped with a DVA, the amplification factor curves corresponding to different damping ratios do not conform to the P and Q fixed-point theories. In contrast, when the damping ratio of the main system is sufficient, the amplification factor curves will be consistent with the fixed-point theory. This proves that when the main system is damped, the analytical solution for dynamic absorber parameter optimization and design is a complex nonlinear problem. As a result, numerical algorithms are used to optimize the parameters.

In recent years, the development of dynamic vibration absorption technology has become increasingly mature and widely used in vehicles [9], building and bridge engineering (also known as tuned mass dampers (TMD)) [10, 11], and other fields. With the rapid development of high-rise buildings, more attention has been paid to wind-induced vibration, and effective control strategies have been conducted on comfort and safety. Generally, these can be classified as active, semi-active, passive, and hybrid control methods. A tuned mass damper (TMD) is one of the oldest passive control devices. This device was first introduced by Frahm [12] and then used in various research and practical works. Rana and Soong [13] performed a parametric study to enhance the understanding of some essential TMD characteristics. A standard SDOF system was used to develop two different TMD optimization criteria [14]. Bekdaş and Nigdeli [15] used a metaheuristic optimization method-harmony search (HS) to find the optimal TMD parameters.

On the other hand, active tuned mass dampers (ATMD) have emerged in the research due to various restrictions in the TMD, such as the unideal control effect. Abdel-Rohman [16] presented a process for designing optimal ATMD. A control scheme consisted of an ATMD in which the control action is achieved by a fuzzy logic controller (FLC) and the ability to handle nonlinear behavior is studied in Samali et al. [17].

10.1 Passive Dynamic Vibration Absorption

10.1.1 Main System Without Damping

Consider the passive dynamic vibration absorption system in which the main system does not have damping, as shown in Fig. 10.1. The mass and stiffness of the main system are m_1 and k_1, respectively, the mass, stiffness, and damping of the dynamic vibration absorption system are m_2, k_2, and c_2, respectively, the vibration load generated by the power equipment is $F(t)$, the power equipment vibration responses are $x_1, \dot{x}_1, \ddot{x}_1$ respectively, the vibration response of the dynamic absorber are $x_2, \dot{x}_2, \ddot{x}_2$, respectively.

The dynamic equation is as follows:

$$m_1 \ddot{x}_1 + c_2\left(\dot{x}_1 - \dot{x}_2\right) + k_2(x_1 - x_2) + k_1 x_1 = F(t) \tag{10.1a}$$

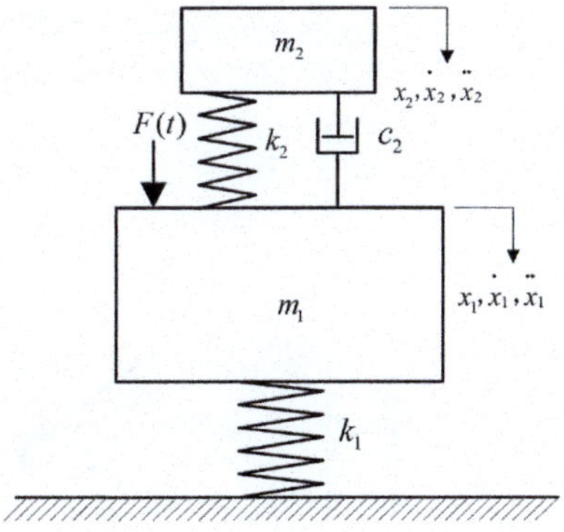

Fig. 10.1 Passive dynamic vibration absorption system with the undamped main system

$$m_2 \ddot{x}_2 + c_2\left(\dot{x}_2 - \dot{x}_1\right) + k_2(x_2 - x_1) = 0 \tag{10.1b}$$

If the vibration load is a simple harmonic excitation $F(t) = F_0 \sin\omega t$, then the system has the following amplitude ratio relationship [18–20]:

$$\left|\frac{X_1}{X_{\mathrm{st}}}\right| = \sqrt{\frac{\left(\gamma^2 - \lambda^2\right)^2 + (2\gamma\lambda\zeta_2)^2}{\left[(1-\lambda^2)(\gamma^2-\lambda^2) - \mu\gamma^2\lambda^2\right]^2 + \left[1 - (1+\mu)\lambda^2\right]^2 (2\gamma\lambda\zeta_2)^2}} \tag{10.2}$$

The static deformation of the main system is $X_{st} = F_0/k_1$, the mass ratio is $\mu = m_2/m_1$, the excitation vibration frequency ratio is $\lambda = \omega/\omega_{n1}$, the natural frequency ratio is $\gamma = \omega_{n2}/\omega_{n1}$, the natural frequency of the main system is $\omega_{n1} = \sqrt{k_1/m_1}$, the dynamic vibration absorber's natural frequency is $\omega_{n2} = \sqrt{k_2/m_2}$, and the dynamic vibration absorber's damping ratio is $\zeta_2 = c_2/(2m_2\omega_{n2})$. Figure 10.2 indicates that the amplitude ratio curves at different damping ratios ζ_2 all intersect at points P and Q ($\mu = 0.25$ and $\gamma = 0.7$). Besides, it shows that the results are consistent with the P and Q fixed points theory. Based on this theory, the parameters of dynamic vibration absorption can be optimized. The goal herein is to equalize the two points of P and Q to minimize the amplitude of the main system and maximize these two points in the curve. The optimal parameters are $\gamma = 1/(1+\mu)$ and $\zeta_{2\mathrm{opt}} = \sqrt{3\mu/[8(1+\mu)]}$. Figure 10.3 shows the same height amplitude ratio curve after optimization based on the P and Q fixed point theory.

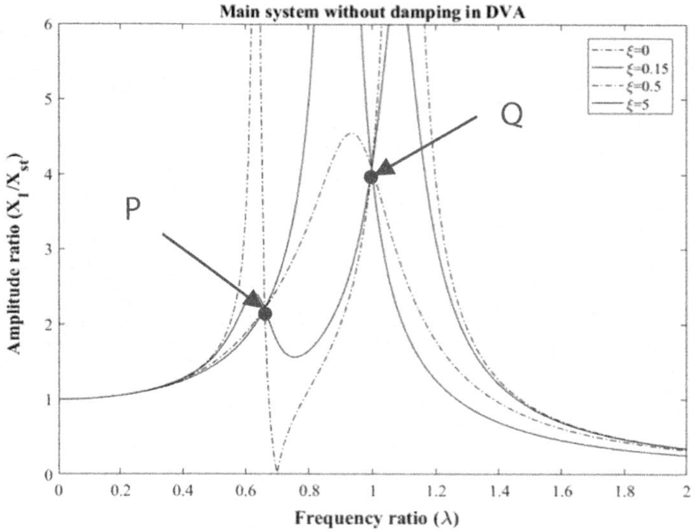

Fig. 10.2 P and Q are fixed points of the main system without dynamic vibration absorption damping

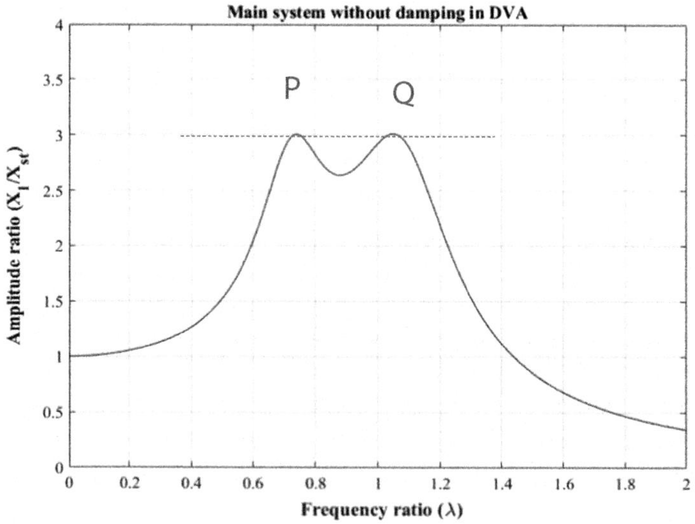

Fig. 10.3 Curves of the same height amplitude ratio after optimization of P and Q fixed points

10.1.2 Main System with Damping

Consider the passive dynamic vibration absorption system in which the main system is damped, as shown in Fig. 10.4, where the damping is c_1, and the main system's damping ratio is $\zeta = c_1 / (2m_1\omega_{n1})$.

The equation of motion is defined as follows:

$$m_1 \ddot{x}_1 + c_2\left(\dot{x}_1 - \dot{x}_2\right) + k_2(x_1 - x_2) + c_1 \dot{x}_1 + k_1 x_1 = F(t) \tag{10.3a}$$

$$m_2 \ddot{x}_2 + c_2\left(\dot{x}_2 - \dot{x}_1\right) + k_2(x_2 - x_1) = 0 \tag{10.3b}$$

The system has the following amplitude ratio [18, 19]:

$$\left|\frac{X_1}{X_{st}}\right| = \sqrt{\frac{\left(\gamma^2 - \lambda^2\right)^2 +}{\left\{\gamma^2 - \left[1 + 4\gamma\zeta_1\zeta_2 + (1+\mu)\gamma^2\right]\lambda^2 + \lambda^4\right\}^2 +}} \sqrt{\frac{(2\gamma\lambda\zeta_2)^2}{\left\{2\lambda\left\{\left(\gamma^2 - \lambda^2\right)\zeta_1 + \gamma\left[1 - (1+\mu)\lambda^2\right]\zeta_2\right\}\right\}^2}} \tag{10.4}$$

The amplitude ratio curve of the passive dynamic vibration absorption system with damping in the main system is shown in Fig. 10.5 ($\mu = 0.3$, $\gamma = 0.8$, $\zeta_1 = 0.17$, and $\zeta_2 = 0.03$).

The design of passive dynamic vibration absorption systems with damping in the main system is not similar to the undamped one since it is based on the P and Q fixed-point theory. In general, its optimal configuration must be debugged according to the mutual influence of μ, γ, ζ_1, and ζ_2 to achieve an ideal transmission efficiency,

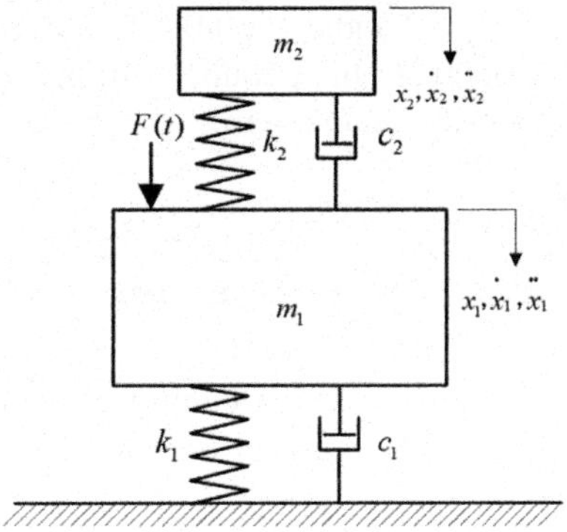

Fig. 10.4 Passive dynamic vibration absorption system with the damped main system

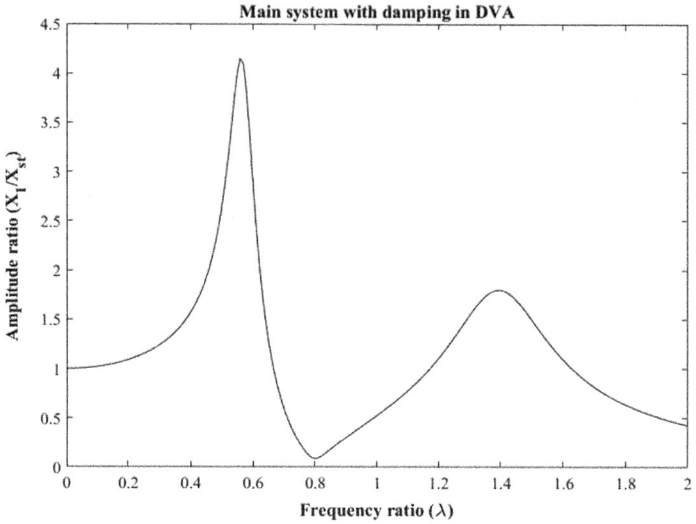

Fig. 10.5 The amplitude ratio curve of the passive dynamic vibration absorption system with a damped main system

and its process is more cumbersome. In order to solve this problem, particle swarm optimization (PSO) [21, 22] is introduced for optimization design. If m_1, k_1, and c_1 of the main system are given ($\zeta_1 = 0.17$), and the dynamic absorber's mass ratio is given ($\mu = 0.3$), and then the dynamic absorber's stiffness coefficient k_2 and damping coefficient c_2 can be optimized. In this case, there are two scenarios: the vibration load frequency is fixed to $\lambda = 0.6$; the vibration load frequency is not fixed to $\lambda \in [0.01, 2]$. The parameter optimization range of γ and ζ_2 are defined as [0.1, 0.05] ~ [2, 0.45]. Equation (10.4) is defined as the optimized fitness function. The acceleration factors in the PSO algorithm are 1 and 2, the inertia weight factor is 0.99^t, and t is the t-th iteration. Figure 10.6 shows the convergence curves of the fitness in both cases. The optimized parameters obtained in the first working condition are $\gamma = 0.595$ and $\zeta_2 = 0.05$, and the optimized parameters obtained in the second working condition are $\gamma = 2$ and $\zeta_2 = 0.05$. Therefore, this strategy can be used in actual engineering to optimize the parameter design of dynamic vibration absorption systems with a damped main system.

10.2 Active Dynamic Vibration Absorption

The sensing active control actuation system is typically added to form an active dynamic vibration absorption system based on the passive dynamic vibration absorption research, as shown in Fig. 10.7. Besides, the control can be adjusted in real-time

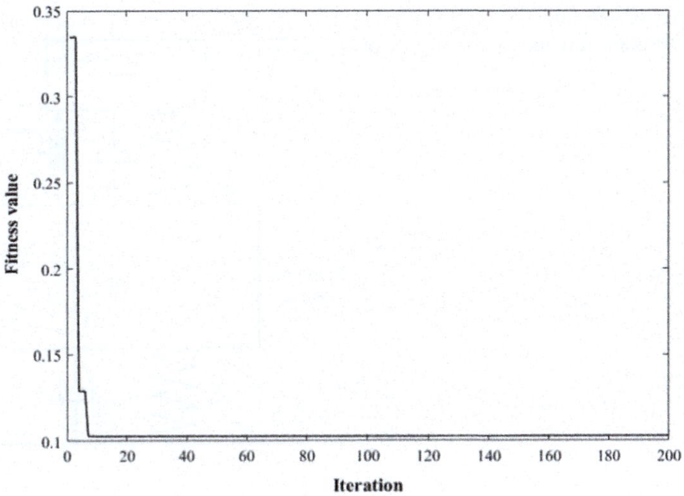

(a) The excitation frequency is fixed

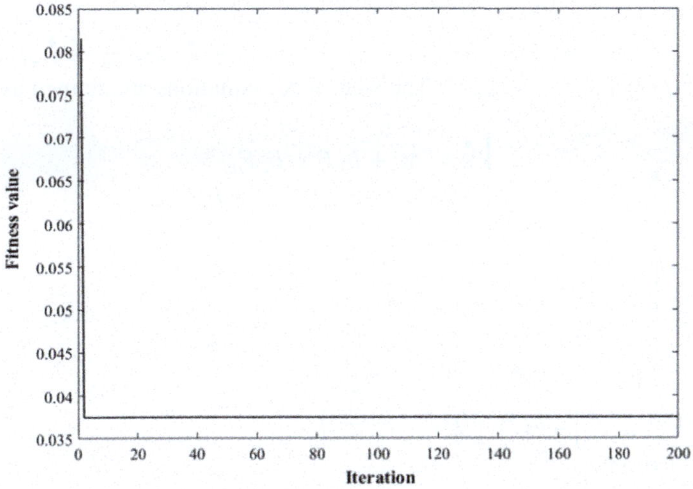

(b) The excitation frequency is not fixed

Fig. 10.6 The fitness value convergence of parameter optimization for passive dynamic vibration absorption with a damped main system

according to the controlled object and excitation characteristics to derive the actuator to output active control force $F_a(t)$.

The dynamic equation is defined as follows:

$$m_1 \ddot{x}_1 + c_2\left(\dot{x}_1 - \dot{x}_2\right) + k_2(x_1 - x_2) + c_1 \dot{x}_1 + k_1 x_1 = F(t) - F_a(t) \tag{10.5a}$$

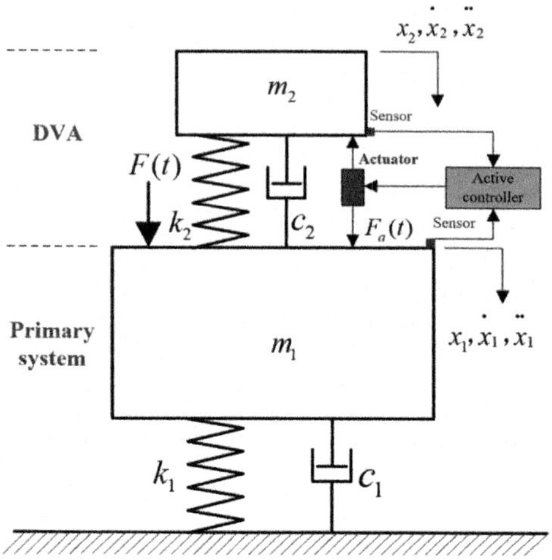

Fig. 10.7 Active dynamic vibration absorption system

$$m_2 \ddot{x}_2 + c_2\left(\dot{x}_2 - \dot{x}_1\right) + k_2(x_2 - x_1) = F_a(t) \tag{10.5b}$$

The state space variables are $z_1 = x_1, z_2 = x_2, z_3 = \dot{x}_1$, and $z_4 = \dot{x}_2$, and the state vector is $z(t) = [z_1, z_2, z_3, z_4,]^{\mathrm{T}}$. The state space equations are defined as follows:

$$\begin{cases} \dot{z} = Az + b_1 F + b_2 F_a \\ y(t) = Cz + d_1 F + d_2 F_a \end{cases} \tag{10.6}$$

where $A = \begin{bmatrix} 0 & 0 & 1 & 0 \\ 0 & 0 & 0 & 1 \\ -\frac{k_1+k_2}{m_1} & \frac{k_2}{m_1} & -\frac{c_1+c_2}{m_1} & \frac{c_2}{m_1} \\ \frac{k_2}{m_2} & -\frac{k_2}{m_2} & \frac{c_2}{m_2} & -\frac{c_2}{m_2} \end{bmatrix}$,

$b_1 = \left[0\, 0\, \frac{1}{m_1}\, 0\right]^{\mathrm{T}}, b_1 = \left[0\, 0 -\frac{1}{m_1}\, \frac{1}{m_2}\right]^{\mathrm{T}}$,

$C = \begin{bmatrix} 1\,0\,0\,0 \\ 0\,1\,0\,0 \\ 0\,0\,1\,0 \\ 0\,0\,0\,1 \end{bmatrix}$,

$d_1 = \begin{bmatrix} 0 \\ 0 \\ 0 \\ 0 \end{bmatrix}$, and $d_2 = \begin{bmatrix} 0 \\ 0 \\ 0 \\ 0 \end{bmatrix}$.

Figure 10.8 depicts the control flow for a DVA that considers linear quadratic regulator (LQR) active control [23, 24], where G is the feedback gain generated by the LQR control algorithm. Figure 10.9 shows the control flow for a DVA that considers proportional-integral-derivative (PID) active control [25, 26]. The following transfer functions can be derived as

$G_1(s) = \frac{\Delta_3}{\Delta_1\Delta_3-\Delta_2^2}$ and $G_2(s) = \frac{\Delta_2-\Delta_3}{\Delta_1\Delta_3-\Delta_2^2}$, where $\Delta_1{=}m_1s^2+c_1s+k_1+c_2s+k_2$, $\Delta_2{=}c_2s+k_2$, $\Delta_3{=}m_2s^2+c_2s+k_2$, and $s=j\omega$.

When LQR active control is performed, the weight matrix Q and R selection directly impact the control gain (G), control force, and control effect. The proportional, integral, and derivative control parameters in PID active control directly influence the control effect. Most control parameters are generally set and selected based on the engineer's experience, where the debugging process is complicated. For this reason, the above parameters are optimized based on PSO. The fitness function is defined as $\left\|\dot{x}_1\right\|_\infty$. The parameter search range of the weight matrix Q in LQR control is $1\times10^{-6}\sim1\times10^7$, and the parameter search range of the weight matrix R is $1\times10^{-6}\sim1$. In the PID control, the search range of the proportional factor k_p, integral factor k_i, and differential factor k_d is $-1\times10^5\sim1\times10^5$. The parameter settings of PSO are the same as before. $m_1=1200\,\text{kg}$, $k_1=1\times10^6\,\text{N/m}$, $c_1=1.6\times10^4\,\text{Ns/m}$, $m_2=60\,\text{kg}$, $k_2=1.5\times10^4\,\text{N/m}$, $c_2=1\times10^3\,\text{Ns/m}$. The excitation amplitude is $F_0=1000$ N, and the frequency is $f=25$ Hz. The feedback control gain obtained by optimizing LQR control is $G=[-2.413\times10^7, 3.333\times10^{-5}, -0.302\times10^7, 5.666\times10^3]$. The parameters obtained by optimizing PID control are $k_p=2.773\times10^4$, $k_i=0.0621\times10^4$, and $k_d=-2.031\times10^4$.

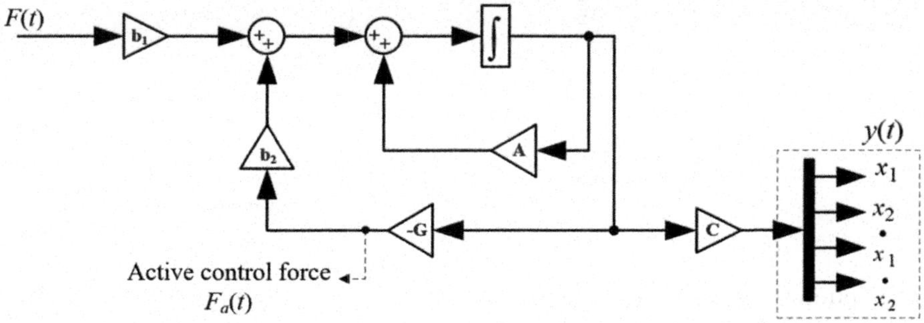

Fig. 10.8 LQR active control

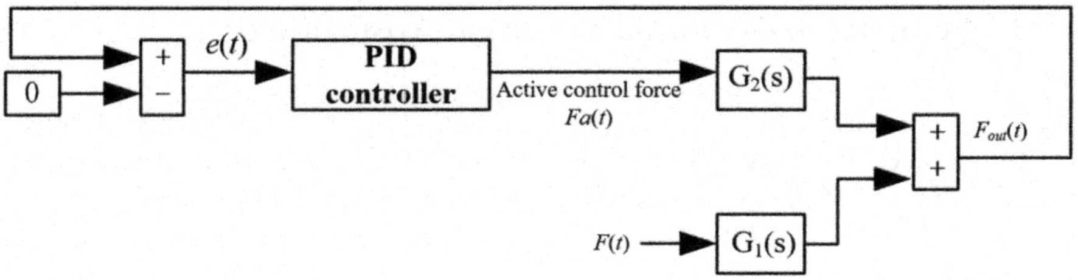

Fig. 10.9 PID active control

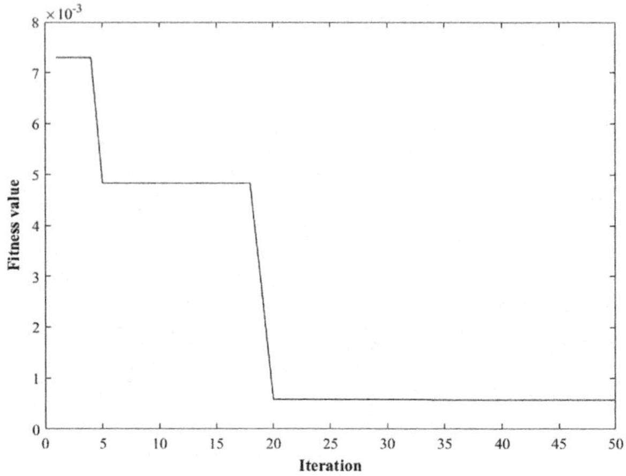

Fig. 10.10 PSO optimized LQR fitness value convergence curve

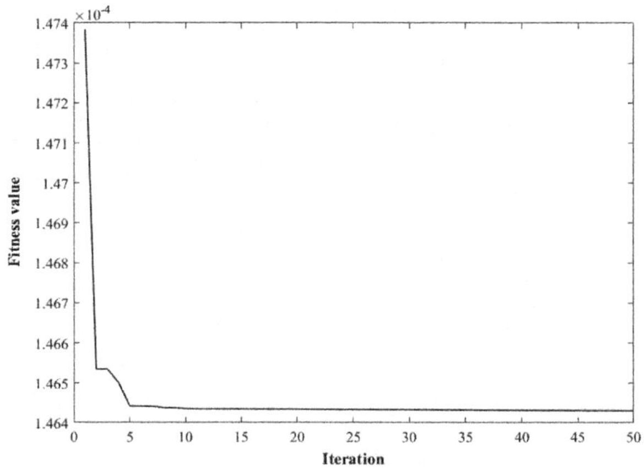

Fig. 10.11 PSO optimized PID fitness value convergence curve

The optimized fitness value convergence curve is shown in Figs. 10.10 and 10.11. Figure 10.12 compares the passive and active dynamic vibration absorption effects.

10.3 Semi-active Dynamic Vibration Absorption

Based on the previous active dynamic vibration absorption research, it is considered to replace the sensing-active control actuation system with a magnetorheological damper to form a semi-active dynamic vibration absorption system, as shown in Fig. 10.13. Semi-active systems can overcome the problems of complex sensing/actuation systems, large energy consumption, and time lag and robustness in active systems. The outside supply needs only a small amount of energy supply to make the magnetorheological damper work normally.

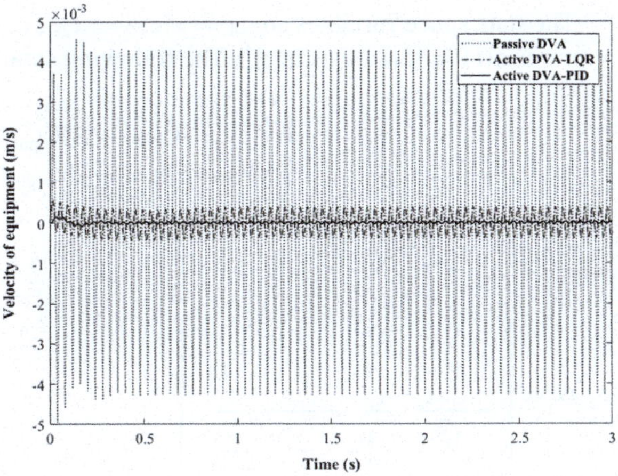

Fig. 10.12 Comparison between passive and active dynamic vibration absorption effects

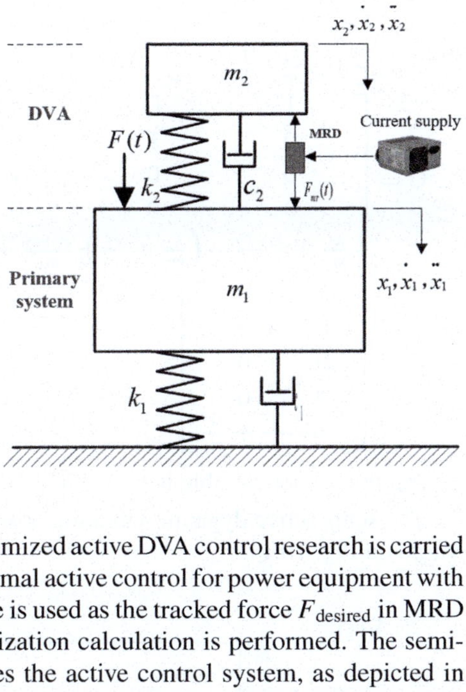

Fig. 10.13 Semi-active dynamic vibration absorption system

Based on the research in Sect. 5.5, optimized active DVA control research is carried out to equivalently achieve PSO-H$_\infty$ optimal active control for power equipment with the active control force F_a, and this force is used as the tracked force F_{desired} in MRD semi-active control, and equivalent realization calculation is performed. The semi-active DVA control using MRD replaces the active control system, as depicted in Fig. 10.13, where the damping force generated by MRD is F_{mr}(t), and the dynamic equation at this time is as follows:

$$m_1 \ddot{x}_1 + c_2\left(\dot{x}_1 - \dot{x}_2\right) + k_2(x_1 - x_2) + c_1 \dot{x}_1 + k_1 x_1 = F(t) - F_{\text{mr}}(t) \quad (10.7\text{a})$$

$$m_2 \ddot{x}_2 + c_2\left(\dot{x}_2 - \dot{x}_1\right) + k_2(x_2 - x_1) = F_{\text{mr}}(t) \quad (10.7\text{b})$$

The state space variables are $z_1 = x_1, z_2 = x_2, z_3 = \dot{x}_1$, and $z_4 = \dot{x}_2$, and the state vector is $z(t) = [z_1, z_2, z_3, z_4,]^{\mathrm{T}}$. The following state space equations are obtained:

$$\begin{cases} \dot{z}(t) = \mathrm{A}z(t) + \mathrm{BF}_{\mathrm{input}}(t) \\ y(t) = \mathrm{C}z(t) + \mathrm{DF}_{\mathrm{input}}(t) \end{cases} \tag{10.8}$$

where $F_{\mathrm{input}} = \begin{bmatrix} F(t) \\ F_{\mathrm{mr}}(t) \end{bmatrix}$,

$$\mathrm{A} = \begin{bmatrix} 0 & 0 & 1 & 0 \\ 0 & 0 & 0 & 1 \\ -\frac{k_1+k_2}{m_1} & \frac{k_2}{m_1} & -\frac{c_1+c_2}{m_1} & \frac{c_2}{m_1} \\ \frac{k_2}{m_2} & -\frac{k_2}{m_2} & \frac{c_2}{m_2} & -\frac{c_2}{m_2} \end{bmatrix},$$

$$\mathrm{B} = \begin{bmatrix} 0 & 0 \\ 0 & 0 \\ \frac{1}{m_1} & -\frac{1}{m_1} \\ 0 & \frac{1}{m_2} \end{bmatrix},$$

$$\mathrm{C} = \begin{bmatrix} 1 & 0 & 0 & 0 \\ 0 & 1 & 0 & 0 \\ 0 & 0 & 1 & 0 \\ 0 & 0 & 0 & 1 \\ \frac{k_2}{m_2} + \frac{k_1+k_2}{m_1} & -\left(\frac{k_2}{m_2} + \frac{k_2}{m_1}\right) & \frac{c_1+c_2}{m_1} + \frac{c_2}{m_2} & -\left(\frac{c_2}{m_1} + \frac{c_2}{m_2}\right) \end{bmatrix},$$

$$\mathrm{D} = \begin{bmatrix} 0 & 0 \\ 0 & 0 \\ 0 & 0 \\ 0 & 0 \\ -\frac{1}{m_1} & \frac{1}{m_1} + \frac{1}{m_2} \end{bmatrix},$$

and the output variable is $y(t) = \begin{bmatrix} x_1 & x_2 & \dot{x}_1 & \dot{x}_2 & \ddot{x}_2 - \ddot{x}_1 \end{bmatrix}^{\mathrm{T}}$.

The semi-active dynamic vibration absorption based on MRD can be calculated based on Fig. 10.14.

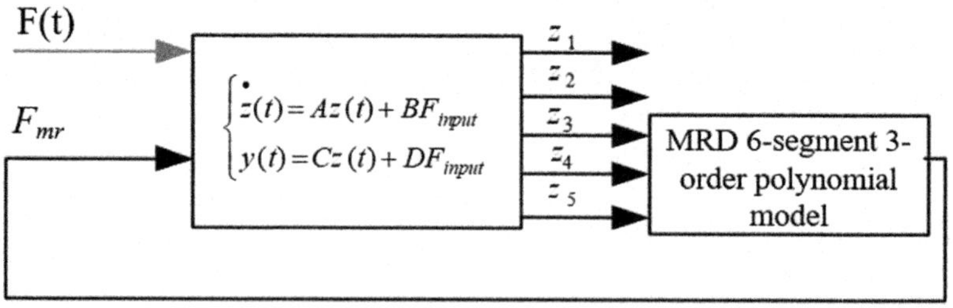

Fig. 10.14 Schematic diagram of semi-active control with an MRD

10.4 Dynamic Vibration Absorption for Floor Structure

Pedestrian loads and equipment vibrations on large space building structure floors, such as lecture halls, dance rooms, and concert halls, affect comfort and require vibration control measures.

The engineering floor structure's length, width, and thickness in Fig. 10.15 are 9, 8.1, and 0.3 m, respectively. The cross-sectional dimensions of the four-sided beam are 0.2 × 0.35 m, the multi-ribbed beam section size is 0.4 × 0.85 m, and the structural material is C35 concrete. The four corners of the floor cover are constrained by three-direction displacement, and the rotation is released without restriction. A vertical vibration load acts on the floor span, as shown in Fig. 10.16. In order to control the vibration response of the floor, dynamic vibration absorption measures were taken into account, and the tuned mass dampers (TMD-1 to TMD-9) were arranged according to Fig. 10.15.

A finite element analysis was performed using ANSYS 19.2, where the beam189 element simulates the beam, and the shell181 element simulates the floor structure. The tuned mass damper was simulated by combining mass21 and combin14. The vertically dominant mode of the floor structure without tuned mass dampers occurs in the first order (warping), the natural frequency is $f_0 = 4.12$ Hz, the mode participation mass is 71048.9 kg, and the mode shapes are shown in Fig. 10.17. The mass ratio of TMD-1 to TMD-9 is $\mu = 0.01$, and its natural frequency and damping ratio were designed as $f_{\text{TMD}} = \sqrt{f_0/(1+\mu)}$ and $\zeta_{\text{TMD}} = \sqrt{3\mu/[8(1+\mu)]}$ [18], respectively.

Figure 10.18 compares the vibration control response of the floor with and without tuned mass dampers. The vibration pick-up point was optionally selected as the installation location of TMD-4. It can be seen that installing a tuned mass damper

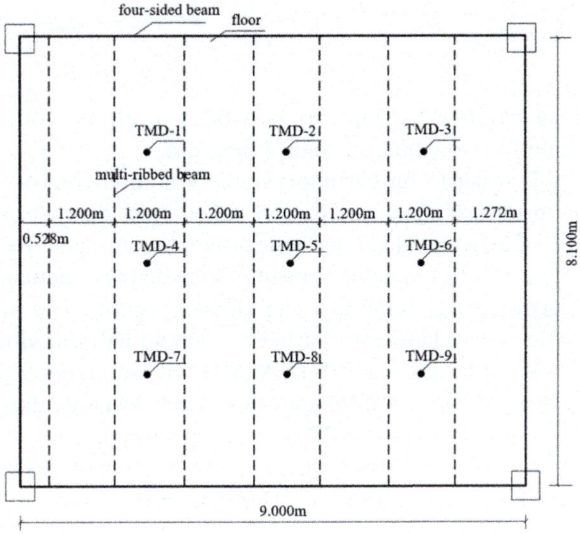

Fig. 10.15 Floor structure and dynamic vibration absorber layout

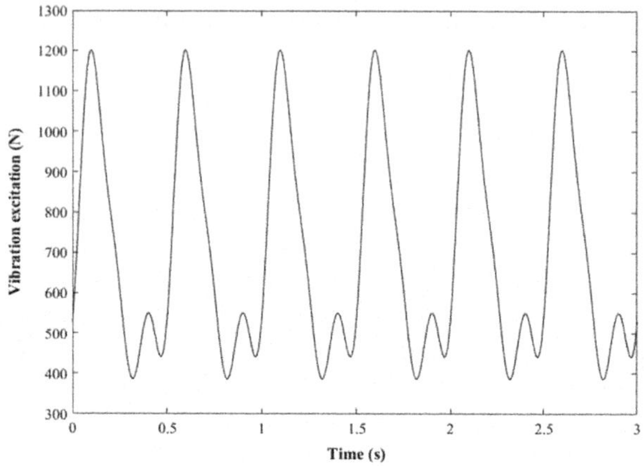

Fig. 10.16 Vertical vibration load acting on the floor

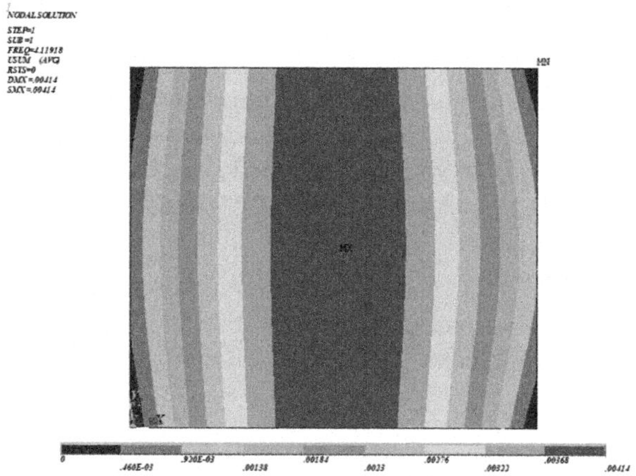

Fig. 10.17 The first-order natural mode shape cloud of the floor mainly vertical-based vibration

can significantly improve the vibration response of the floor structure and which is an effective vibration control measure.

In order to further improve the vibration control response of the floor structure, consider changing the passive-tuned mass damper to an active-tuned mass damper [27, 28], where it has the same principle same as active dynamic vibration absorption. Hence, it has an added sensing-active control actuation system, and it provides an active control force that can change according to the external environment in the tuned mass damper to further counteract harmful vibrations.

In order to carry out ATMD vibration control calculation in the ANSYS finite element environment, this study develops the calculation program shown in Fig. 10.19 based on APDL and adopts the PID active control. The vibration pick-up point was selected as the installation location of TMD-4, where TMD-4 was changed to ATMD, and the other TMD characteristics were kept unchanged. The proportional,

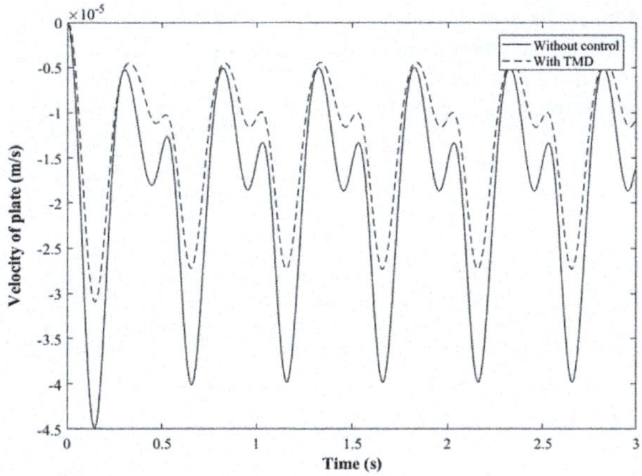

Fig. 10.18 Comparison between floor vibration response with and without tuned mass dampers

integral, and differential parameters in the PID controller are -3×10^6, -2.5×10^6, and 0, respectively.

Figure 10.20 compares the vibration control response of the floor structure of TMD, ATMD, and non-tuned mass damper. It can be seen that the active tuning mass damper can significantly improve the passive control effect. Figure 10.21 shows the actuator output in ATMD. Figures 10.20 and 10.21 verify the feasibility and effectiveness of the calculation strategy proposed in Fig. 10.19.

This section systematically investigated the passive and active control of dynamic vibration absorption (DVA) for power equipment and building structure floor. The particle swarm optimization (PSO) algorithm was introduced to optimize the design of the parameters of passive DVA with damping in the main system, the weight matrices of linear quadratic regulator (LQR) active control, and the three control factors of proportional-integral-derivative (PID) active control. In the vibration control of the building floor structure, based on the utilization of a passive tuned mass damper (TMD), a novel design and calculation strategy for active tuned mass

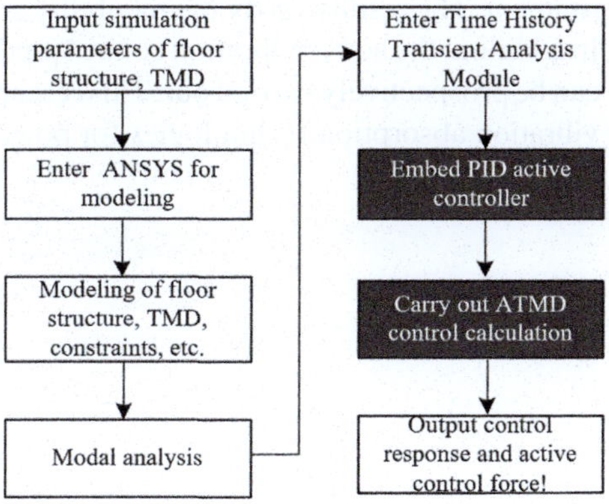

Fig. 10.19 ATMD control flow chart based on ANSYS/APDL

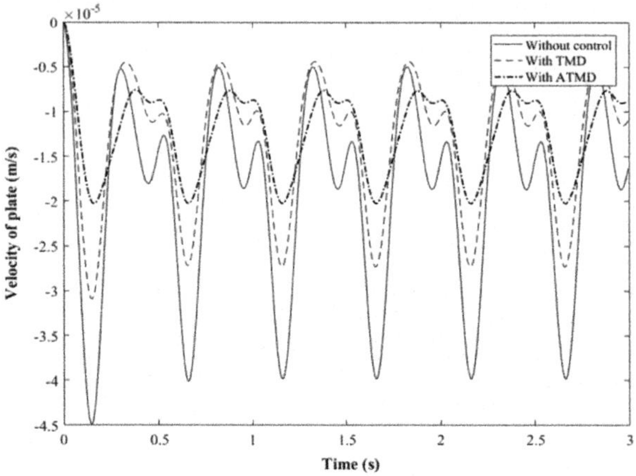

Fig. 10.20 Comparison between the vibration control response of passive and active tuned mass damper

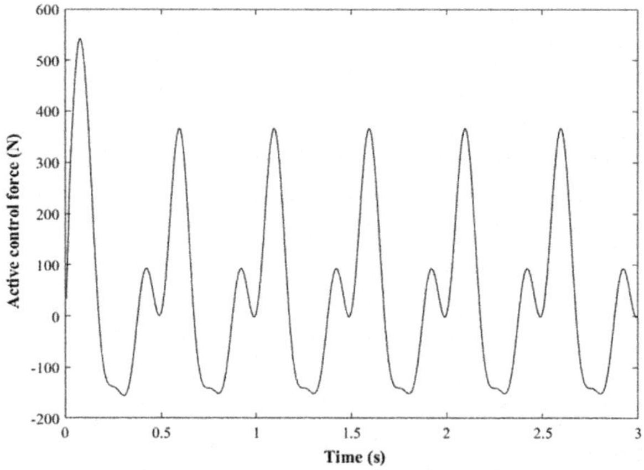

Fig. 10.21 Active control force in ATMD

damper (ATMD) vibration control was proposed in the ANSYS finite element environment. The results verify the reliability and effectiveness of the developed strategy. In addition, the active vibration control prediction for practical engineering structures can be prospectively carried out. This research can further develop advanced dynamic vibration absorption technologies for power equipment and building structures.

10.5 Dynamic Vibration Absorption for High-rise Building Structure

10.5.1 Fluctuating Wind Speed Field Simulation Using the DIT-FFT-WAWS Method

The simulation of random fluctuating wind strategies can be generally divided into two approaches: linear regression filtering techniques, such as the autoregressive (AR) [29] and autoregressive moving average (ARMA) [30–32], harmonic synthesis method based on trigonometric series superposition. The former requires leaser computational efforts but provides poor accuracy. In contrast, the latter can possess higher precision and be unconditionally stable. Shinozuka and Jan [33] simulated the random wind field by WAWS and computed the structural responses. Recently, extensive research on further improving WAWS has been conducted constantly to improve its efficiency. Yang [34] applied the fast Fourier transform (FFT) to WAWS for the first time, significantly improving harmonic terms superposition efficiency. In order to further improve the FFT efficiency, Sun [35] has used the decimation in time (DIT) strategy (a technique in the field of signal processing) to accelerate the superposition.

An M-dimensional, zero means, stationary, and random process is assumed as:

$$\mathrm{v}(t) = \{\mathrm{v}_1(t), \mathrm{v}_2(t), \cdots, \mathrm{v}_M(t)\}^T \tag{10.9}$$

The cross-spectral density function matrix is:

$$S(f) = \begin{bmatrix} s_{11}(f) & s_{12}(f) & \cdots & s_{1M}(f) \\ s_{21}(f) & s_{22}(f) & \cdots & s_{2M}(f) \\ \cdots & \cdots & \cdots & \cdots \\ s_{M1}(f) & s_{M2}(f) & \cdots & s_{MM}(f) \end{bmatrix} \tag{10.10}$$

where $S_{ii}(f)$ and $S_{ij}(f)$ are the self-spectral density function matrix and cross-spectral density function matrix, respectively, and i and $j = 1, 2, \ldots, M$.

The Cholesky decomposition of Eq. (10.10) can be performed as follows:

$$S(f) = H(f)H^*(f)^T \tag{10.11}$$

$$H(f) = \begin{bmatrix} H_{11}(f) & 0 & \cdots & 0 \\ H_{21}(f) & H_{22}(f) & \cdots & 0 \\ \cdots & \cdots & \cdots & \cdots \\ H_{M1}(f) & H_{M2}(f) & \cdots & H_{MM}(f) \end{bmatrix} \tag{10.12}$$

where $H(f)$ and $H^*(f)$ are the conjugate transpose matrices whose elements have the following properties:

$$H_{jj}(f) = H_{jj}(-f),\ j = 1, 2, \cdots, M \tag{10.13}$$

$$H_{ij}(f) = \left|H_{ij}^{*}(f)\right| e^{i\theta_{ij}(f)} \tag{10.14}$$

$$\theta_{ij}(f) = \tan^{-1}\left(\frac{\mathrm{Im}\left(H_{ij}(f)\right)}{\mathrm{Re}\left(H_{ij}(f)\right)}\right) \tag{10.15}$$

where $i = 1, 2, \cdots, M$, $j = 1, 2, \cdots, i-1$, and $i > j$.

A stochastic process vector $\mathrm{v}(t)$ can be defined as:

$$\begin{aligned} &\mathrm{v}_i(t) \\ &= \sum_{j=1}^{i} \sum_{l=1}^{N} \left|H_{ij}\left(f_{jl}\right)\right| \cdot \sqrt{2\Delta\omega} \cos\left[2\pi f_{jl} t + \theta_{ij}\left(f_{jl}\right) + \Phi_{jl}\right], \\ &i = 1, 2, \cdots, M \end{aligned} \tag{10.16}$$

where $f_{jl} = \frac{\omega_{jl}}{2\pi} = \frac{(l-1/2)\Delta\omega}{2\pi}$, $\Delta\omega = \omega_{\mathrm{up}}/N$ is frequency increment, N is a positive and large enough integer, ω_{up} is the cutoff frequency that must be chosen according to the actual frequency characteristics of the target structure, and Φ_{jl} is a random number between 0 and 2π.

The time step Δt should satisfy Eq. (10.17) in order to prevent aliasing between mutually adjacent frequencies when wind speed data is generated by Eq. (10.16).

$$\Delta t \leq \pi / \omega_{\mathrm{up}} \tag{10.17}$$

Equation (10.11) shows that there is a huge amount of computations since Cholesky decomposition must be performed for each frequency f.

Divide the $N = 2^{\beta}$ sequences $x(i)(i = 0, 1, 2, \cdots, N-1)$ into two parts.

$$\begin{cases} x(2j) = x_1(j) \\ x(2j+1) = x_2(j) \end{cases} \quad \left(j = 0, 1, 2, \cdots, \tfrac{N}{2} - 1\right) \tag{10.18}$$

Discrete Fourier transform (DFT) of the front and rear halves of the presented sequences can be achieved as follows:

$$X(k) = X_1(k) + W^k X_2(k)\ \left(k = 0, 1, 2, \cdots, \tfrac{N}{2} - 1\right) \tag{10.19}$$

$$X\left(k + \frac{N}{2}\right) = X_1(k) - W^k X_2(k)\ \left(k = 0, 1, 2, \cdots, \tfrac{N}{2} - 1\right) \tag{10.20}$$

where $W = \exp(-j4\pi/N)$, and j is the imaginary unit.

The $x(i)$ FFT form can be obtained by repeating the above process. The analysis shows that the computation amount is proportional to N^2 when the direct operation is performed, yet, it is proportional to $N \log_2 N$ if DIT-FFT is used.

In order to introduce DIT-FFT, Eq. (10.16) should be rewritten as follow:

$$y_i(p\Delta t) = \sqrt{2\Delta\omega}\sum_{i=1}^{j} M_{ij}(p\Delta t)\exp\left[i\left(\frac{j\Delta\omega}{2}p\Delta\right)t\right] \tag{10.21}$$

$$M_{xj}(p\Delta t) = \sum_{l=1}^{2N-1} B_{xj}(l\Delta\omega)\exp\left[i\left(\frac{2\pi lp}{2N}\right)\right] \tag{10.22}$$

$$B_{ij}(l\Delta\omega) = \begin{cases} H_{ij}\left(l\Delta\omega + \frac{1}{2}\Delta\omega\right)\exp\left(i\Phi_{ij}\right), & 0 \le l \le N \\ 0, & N \le l \le 2N \end{cases} \tag{10.23}$$

Then Eq. (10.22) is accelerated by using the DIT-FFT-WAWS technique,

$$M_{ij}(p\Delta t) = \mathbf{DIT} - \mathbf{FFT}\left[B_{ij}(l\Delta\omega)\right] \tag{10.24}$$

where $\mathbf{DIT} - \mathbf{FFT}[\cdot]$ denotes the FFT based on DIT.

10.5.2 TMD and ATMD Control for 76-story Benchmark Structure

The tall building considered herein is a 76-story 306.1 m height office tower proposed for the city of Melbourne, Australia [36]. The total mass of the building, including heavy machinery in the plant rooms, is 153000 t. The total volume of the building is 510000 m^3, resulting in a mass density of 300 kg/m^3, which is a typical concrete structure. The building is slender, with a height-to-width ratio of 7.3. Therefore, it is a wind sensitive. Meantime, $\overline{v}_{10} = 15\,\text{m/s}$, the ground roughness coefficient is $\alpha = 0.03$, the atmospheric density is 1.25 kg/m^3, and the geomorphologic factor is 0.4. The shape coefficients can be obtained from the load code for the design of building structures [37].

Order reduction operation must be performed for the original model to avoid overly complex analysis. In addition, the condensation model can well preserve the original structure's dynamic characteristics. Besides, sensors were disposed of on floors 1, 30, 50, 55, 60, 65, 70, 75, and 76 (Fig. 10.22).

In this paper, Davenport fluctuating wind speed field [38] was simulated, where the self-spectral density function is denoted as:

■

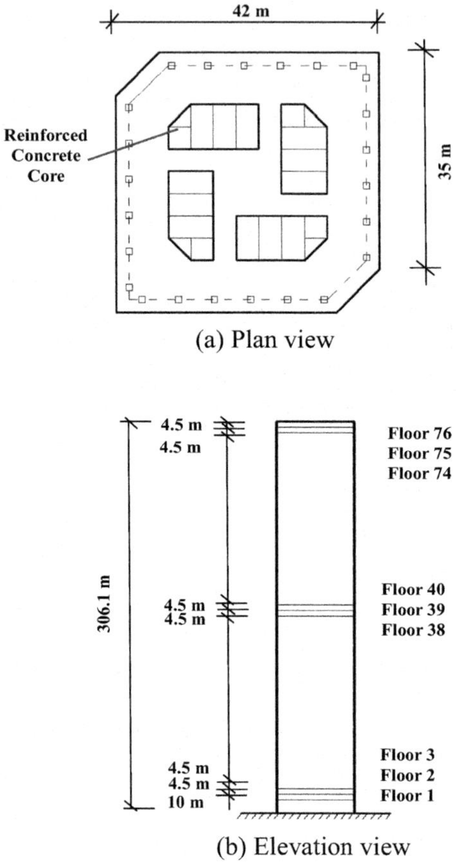

Fig. 10.22 Schematic diagram of the Benchmark structure

where $x = \frac{1200f}{\overline{v}_{10}}$ is the turbulence integral scale factor, f is the frequency of the fluctuating wind, α is the ground roughness coefficient, and $\overline{v}_{10}$ is the average wind speed at 10 m height.

The cross-spectral density function of two random floors $\langle \mathbf{m}, \mathbf{n} \rangle$ of the benchmark structure can be written as:

$$S_{ii}(f) = 4\alpha v_{10^{-2}} x^2/[f(1+x^2)^{4/3}] \tag{10.25}$$

where ch(f) is the coherence function.

In this study, only the planes downwind were considered. Thus, ch(f) can be expressed as:

$$\mathrm{ch}(f) = \exp\left[\frac{-2f\cdot\sqrt{c_x^2(x_i-x_j)^2 + c_z^2(z_i-z_j)^2}}{\overline{v}(z_i)+\overline{v}(z_j)}\right] \tag{10.26}$$

where (x_i, z_i) and (x_j, z_j) are the coordinates of the considered planes, c_x and c_z are the attenuation coefficients, and $c_x = 6, c_z = 10$.

The phase angle $\Psi(f)$ can be determined as follows:

$$\Psi(f) = \begin{cases} \frac{\pi f \Delta z}{4V(z)} & \frac{f\Delta z}{V(z)} \leq 0.1 \\ 1.25 - \frac{10\pi \Delta z}{V(z)} & 0.1 < \frac{f\Delta z}{V(z)} \leq 0.125 \\ \text{rand} & \frac{f\Delta z}{V(z)} > 0.125 \end{cases} \tag{10.27}$$

Figures 10.23 and 10.24 show the speed process and spectral density function curves of the 60th and 76th floors in the benchmark structure. Compared to the Davenport objective spectrum, the simulated spectrum is almost the same as the trend and peak value. Therefore, using DIT-FFT-WAWS to simulate the fluctuating wind for the benchmark structure is feasible. On this basis, fluctuating wind load processes are listed in Fig. 10.25.

A high-rise building that has n stories is considered here, and TMD is disposed on the k-th floor. The motion equations are formulated as follows:

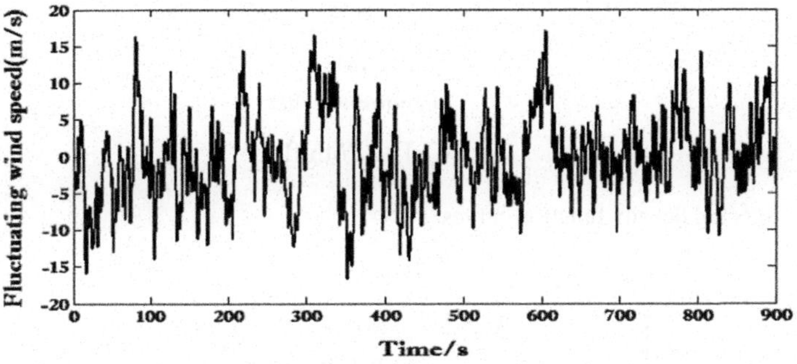

(a) The 60th floor

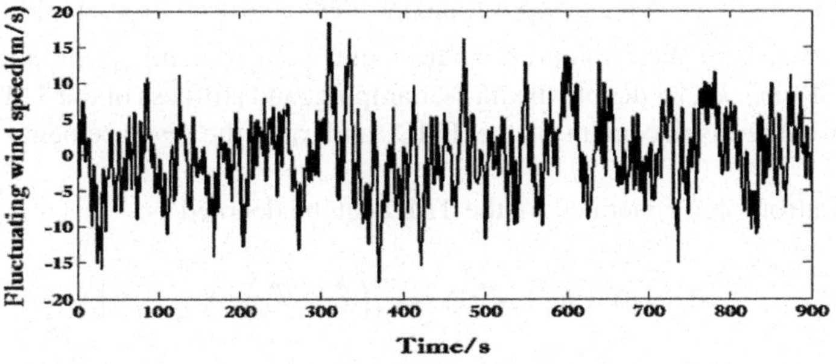

(b) The 76th floor

Fig. 10.23 Wind speed field

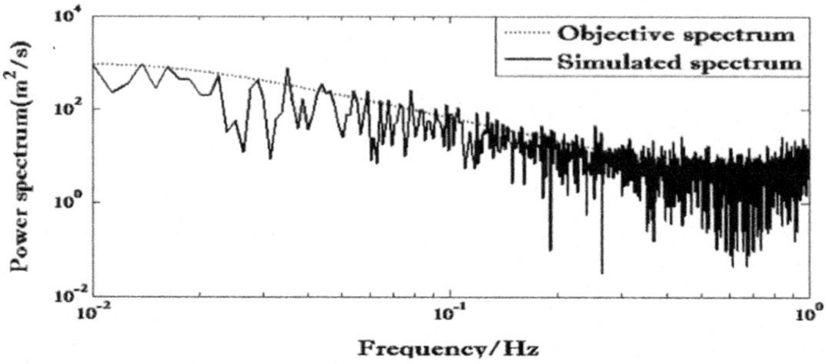

(a) The 60th floor

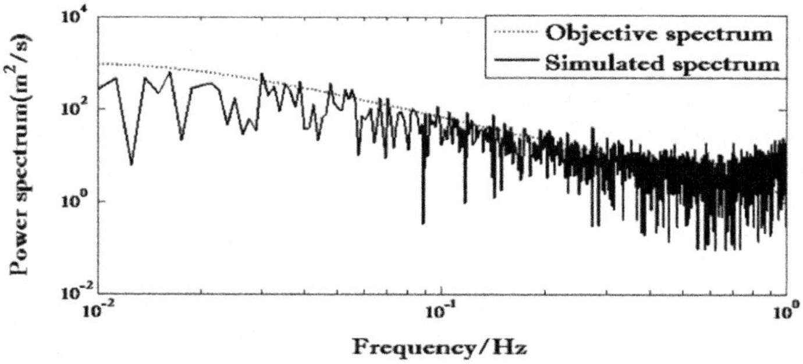

(b) The 76th floor

Fig. 10.24 Spectral density function curves

$$\begin{cases} M\ddot{X} + C\dot{X} + KX = F_{\text{TMD}} + F_{\text{wind}} \\ m_d\ddot{x}_d + c_d\left(\dot{x}_d - \dot{x}_k\right) + k_d(x_d - x_k) = 0 \end{cases} \tag{10.28}$$

where M, C, K are the mass, damping, and stiffness matrix of the tall building, respectively, X, $\dot{X}$, $\ddot{X}$ are the structure displacement, velocity, and acceleration matrix, respectively, m_d, k_d, c_d denote the mass, damping, and stiffness of the TMD, respectively, x_d is the displacement of the TMD, and x_k is the displacement of the k-th floor.

The control force generated by the TMD can be denoted as:

$$F_{\text{TMD}} = \left[0, \cdots 0, k_d(x_d - x_k) + c_d\left(\dot{x}_d - \dot{x}_k\right), 0, \cdots, 0\right]^T \tag{10.29}$$

where k represent the rows and n represents the total number of rows.

The dynamic equations can be derived as follows:

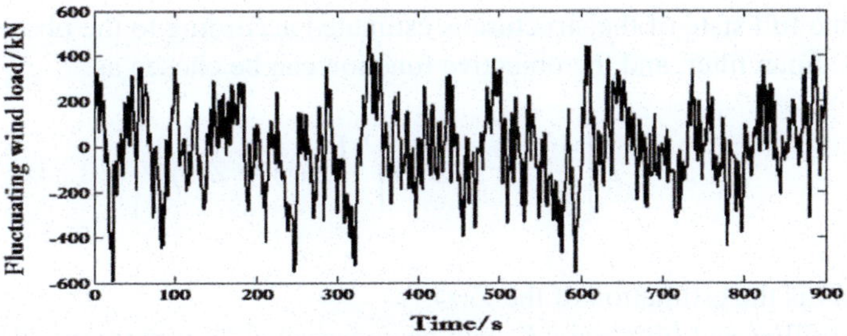

(a) The 60th floor

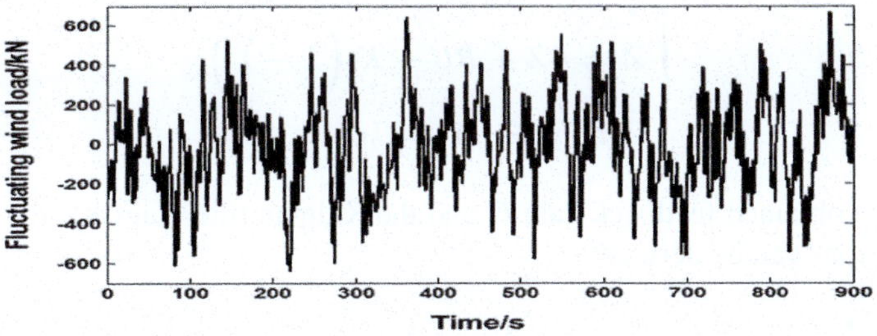

(b) The 76th floor

Fig. 10.25 Fluctuating wind load

$$\begin{cases} M\ddot{X} + C\dot{X} + KX = F_{\text{wind}} + \text{EU} \\ m_d\ddot{x}_d + c_d\left(\dot{x}_d - \dot{x}_k\right) + k_d(x_d - x_k) = F_{\text{actuator}} \end{cases} \tag{10.30}$$

where F_{wind} is the wind load vector, U is the control force vector, E is the location matrix of the control force, and F_{actuator} is the control force exerted on the structure and the ATMD system.

Equation (10.30) can be written in the state space form as follows:

$$\begin{cases} \dot{Z} = AZ + BU + DF_{\text{wind}}(t) + \varepsilon_1(t),\ Z(0) = 0 \\ Y = C_0AZ + C_0BU + \varepsilon_2(t) \end{cases} \tag{10.31}$$

where $\varepsilon_1(t)$ and $\varepsilon_2(t)$ are the input noise and measurement noise, respectively, and they are both Gauss zero mean white noise processes.

Firstly, the control force $U(t)$ can be designed using the LQR algorithm as:

$$U(t) = -\text{GZ}(t) \tag{10.32}$$

where G is the gain matrix of the continuous state space equations.

Then the full state of the structure is estimated according to the observed output and the Kalman filter, and the objective function can be chosen as:

$$J_e = E\left[\left\{Z(t) - \widehat{Z}(t)\right\}^T \left\{Z(t) - \widehat{Z}(t)\right\}\right] \tag{10.33}$$

where $\widehat{Z}(t)$ is the estimation of the $Z(t)$.

Kalman filter and filter gain K_e can be constructed by minimizing the objective function as follows:

$$\begin{cases} \dot{\widehat{Z}} = A\widehat{Z} + BU + K_e\left(Y - \widehat{Y}\right) \\ \widehat{Z}(t_0) = \widehat{Z}_0, \widehat{Y} = C_0\widehat{Z} \end{cases} \tag{10.34}$$

With the obtained feedback gain G and the Kalman filter gain K_e, the following state equations can be derived:

$$\begin{cases} \dot{\widehat{Z}} = (A - \mathrm{BG} - K_eC_0A + K_eC_0\mathrm{BG}) + K_eY + \mathrm{DF}_{\mathrm{wind}} \\ Y = C_0\left(\dot{Z} - \mathrm{DF}\right), Z(0) = 0 \end{cases} \tag{10.35}$$

In the process of TMD or ATMD control, the whole time T_f can be discrete with an equal step Δt, and the performance index, namely integral of the absolute value of error (IAE), can be defined as:

$$\mathrm{IAE}_i = \sum_{t=1\Delta t}^{T_f} |x_i(t)| \tag{10.36}$$

where $t = k\Delta t$ is an arbitrary discrete time, x_i is the displacement of the i-th floor with respect to the ground level, in which it denotes the error in the proposed control, $T_f = 900\,s$, and $\Delta t = 0.1\,s$.

The fitness function is defined as:

$$J = \sum_{i=1}^{76} \mathrm{IAE}_i \tag{10.37}$$

By performing this index, the overall displacement of the tall building (summation of every floor displacement) in the whole time process can be reduced, and it is consistent with the final purpose of wind-induced control for high-rise structures.

In order to investigate the performance of the proposed control strategy, the PSO-based TMD/ATMD was applied to the benchmark structure, and the main idea has been presented above.

The frequency ratio of TMD (ATMD) to the first-order frequency of the main structure is denoted as β_{ratio}, the mass ratio to the total mass of the building is denoted as m_{ratio}, and the damping ratio is denoted as ζ_{ratio}.

For the PSO-based TMD, m_{ratio}, β_{ratio}, and ζ_{ratio} were optimized by evaluating the cost function as shown in Eq. (10.38). Additionally, the number of particles, particle size, number of iterations, and c_1 and c_2 were chosen as 100, 3, 50, 2.0, and 2.0, respectively.

$$\begin{aligned}
&\text{Particle} = [\dim(1), \dim(2), \dim(3)](i.e.\, m_{\text{ratio}}, \beta_{\text{ratio}}, \zeta_{\text{ratio}}) \\
&\text{Constraints:} \\
&\begin{cases} 0.001 \le m_{\text{ratio}} \le 0.009 \\ 0.001 \le \zeta_{\text{ratio}} \le 0.900 \\ 0.100 \le \beta_{\text{ratio}} \le 10.000 \end{cases}
\end{aligned} \tag{10.38}$$

For the PSO-based ATMD, m_{ratio}, β_{ratio}, ζ_{ratio}, and weight matrix Q and R were optimized by evaluating the cost function as given in Eq. (10.39). Moreover, the number of particles, particle size, number of iterations, and c_1 and c_2 were set to 100, 34, 50, 2.0 and 2.0, respectively (Fig. 10.26).

$$\begin{aligned}
&\text{Particle} = \left[\frac{\dim(1)}{m_{\text{ratio}}}, \frac{\dim(2)}{\beta_{\text{ratio}}}, \frac{\dim(3)}{\zeta_{\text{ratio}}}, \right. \\
&\qquad\qquad \left. \frac{\dim(4) \quad \ldots \quad \dim(33)}{q}, \frac{\dim(34)}{R} \right] \\
&\text{Constraints:} \\
&\begin{cases} 0.001 \le m_{\text{ratio}} \le 0.009 \\ 0.001 \le \zeta_{\text{ratio}} \le 0.900 \\ 0.100 \le \beta_{\text{ratio}} \le 10.000 \\ 1 \le q_i \le 1 \times 10^5, \ (i = 1 \sim 30) \\ 0.01 \le R \le 1 \end{cases}
\end{aligned} \tag{10.39}$$

where $q = [\dim(4), \ldots, \dim(33)]$, and $Q = \text{diag}(q)$.

As seen in Table 10.1, m_{ratio}, β_{ratio}, and ζ_{ratio} are almost the same for two different strategies. This can indicate that the optimal parameters of the mass damper system are consistent with each other. Thus, the active algorithm may play a critical role, as shown in Figs. 10.27, 10.28 and 10.29.

In order to compare the results with the original benchmark structure study, the parameters were set the same as the literature [36], namely TMD mass = 500 t, the natural frequency = 0.16 Hz, and the damping ratio = 0.2. The weighting matrix is defined as follows:

■

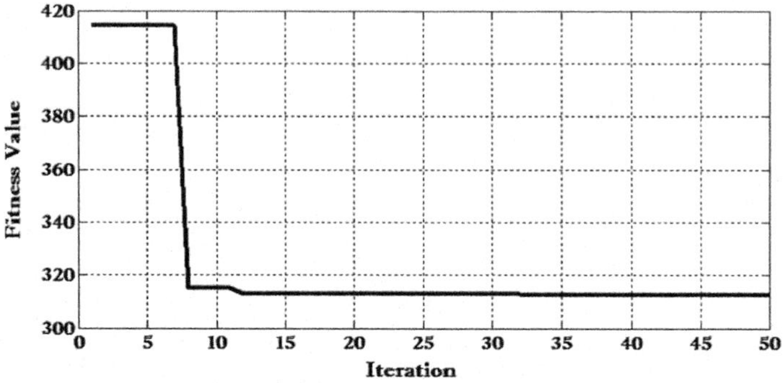

(a) PSO-based TMD

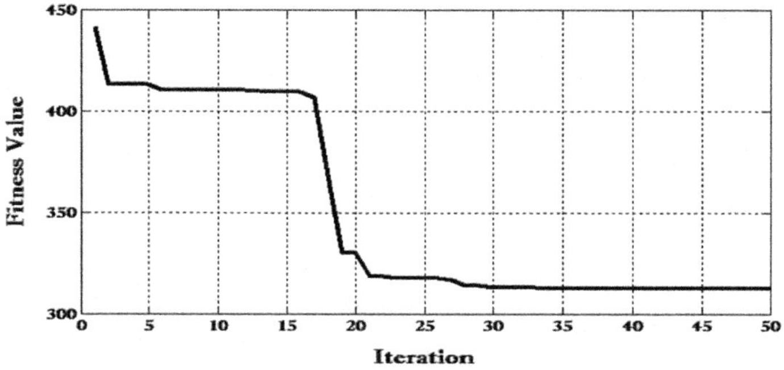

(b) PSO-based ATMD

Fig. 10.26 Fitness convergence

Table 10.1 Optimal parameters

PSO based TMD
$m_{\text{ratio}} = 0.009$; $\zeta_{\text{ratio}} = 0.0946$; $\beta_{\text{ratio}} = 1.031$.
PSO based ATMD
$m_{\text{ratio}} = 0.009$; $\zeta_{\text{ratio}} = 0.0946$; $\beta_{\text{ratio}} = 1.031$;
$Q = \text{diag}[0.907, 0.00001, 0.939, 0.344, 0.963, 0.0555, 1.000, 0.00001, 1.000, 0.0354, 0.951, 1.000, 0.0292, 0.0325, 0.929, 0.427, 0.376, 0.954, 0.922, 0.0171, 0.0736, 1.000, 0.197, 0.511, 0.00001, 0.465, 0.248, 0.850, 0.856, 0.0292] \times 10^5$; $R = 0.901$.

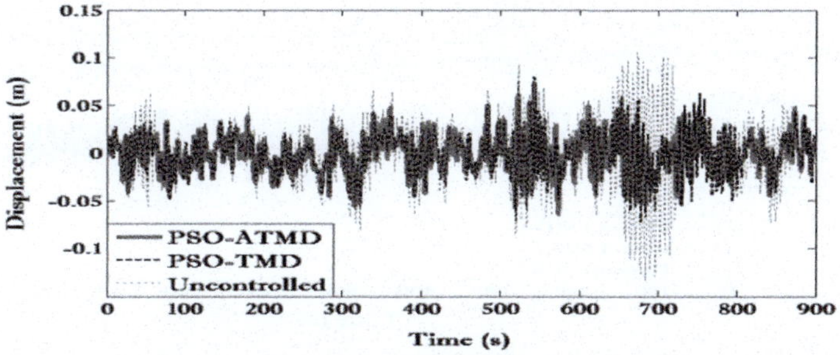

(a) The 55th floor

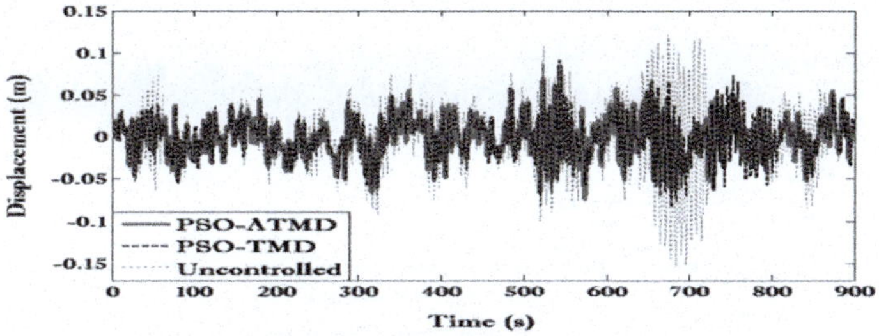

(b) The 60th floor

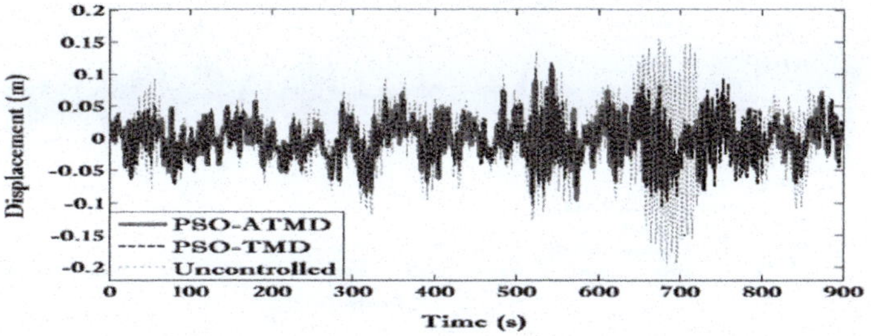

(c) The 70th floor

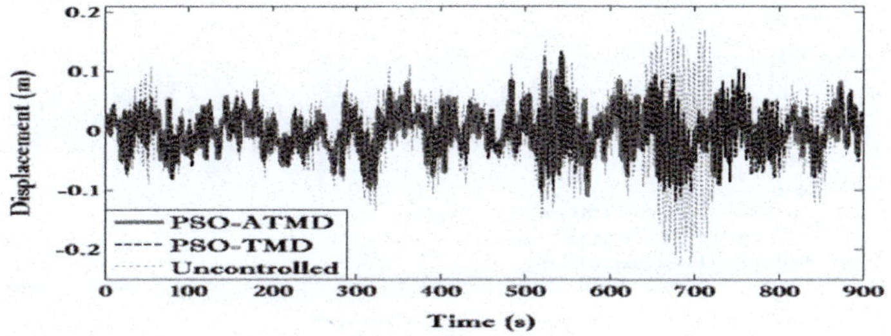

(d) The 76th floor

Fig. 10.27 Structure displacement for comparison

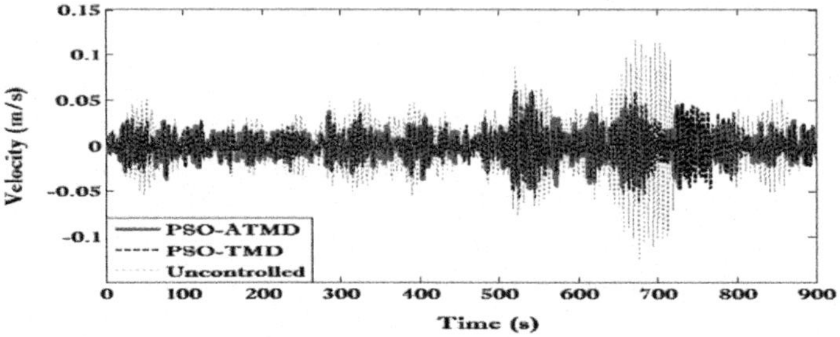

(a) The 55th floor

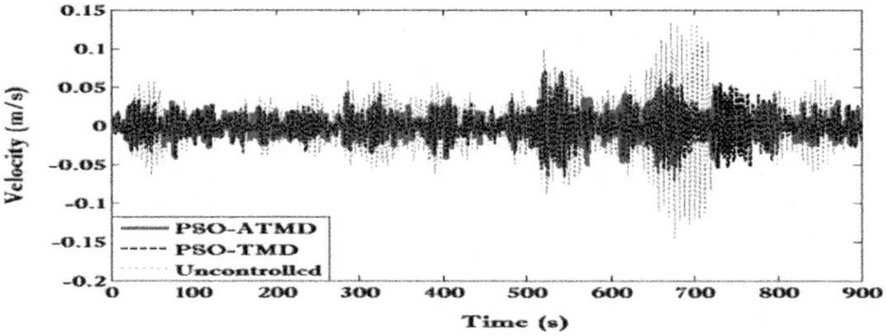

(b) The 60th floor

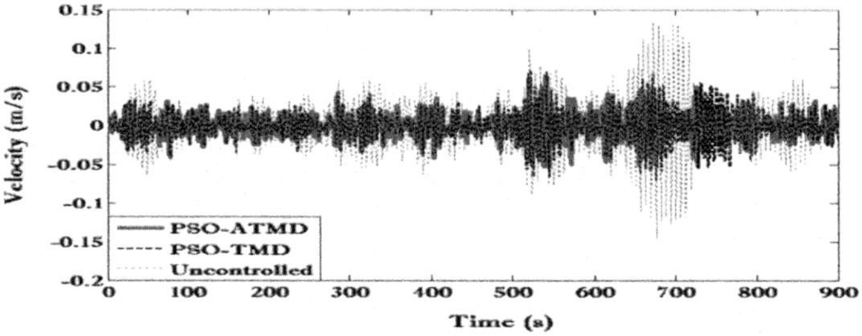

(c) The 70th floor

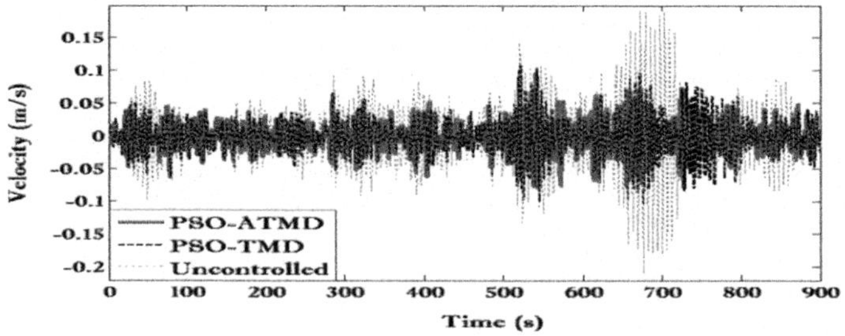

(d) The 76th floor

Fig. 10.28 Structure velocity for comparison

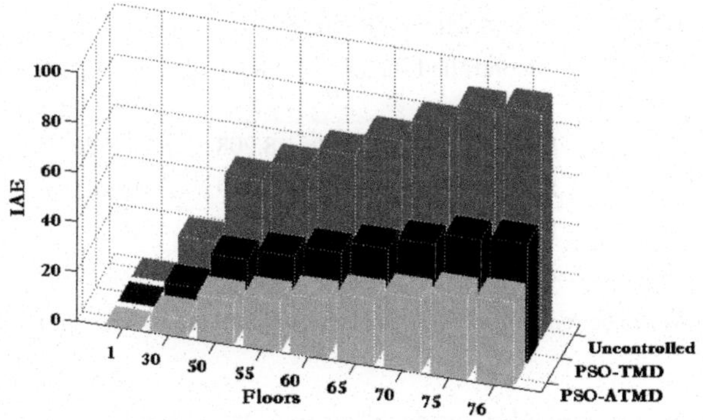

Fig. 10.29 IAE index in comparison

$Q = \mathrm{diag}[1, 1, 1, 1, 1, 1, 1, 1, 1, 0, 1, 1, 1, 1, 1, 10^5, 10^5, 10^5, 10^5, 0, 1, 1, 1, 1, 10^5, 10^5, 10^5, 10^5, 10^5, 0]$, $R = 15 \times 10^{-2}$, and the peak responses of different working conditions are all listed in Table 10.2.

Figures 10.27 and 10.28 and Tables 10.2, 10.3, and 10.4 show the reduction ratios with respect to the uncontrolled responses. It can be seen that PSO-based ATMD controls the wind-induced vibration of the high-rise structure in a better way compared to the PSO-based TMD. Particularly, it can be clearly seen by comparing the acceleration reduction, such as the 76th floor. Additionally, the proposed strategy can effectively achieve the optimal parameters of the TMD/ATMD, and the controlled responses are better than the empirical approach.

This study investigated the application of DIT-FFT-WAWS to achieve a simulated fluctuating wind load process of a high-rise building, considering vertical correlation. Generally, the presented algorithm is simple and fast. Besides, good accuracy can

Table 10.2 Peak displacement (cm)

Building floor	Uncontrolled	TMD	PSO-TMD	Reduction (%)	ATMD	PSO-ATMD	Reduction (%)
55	13.349	8.916	7.930	40.595	7.718	7.336	45.045
60	15.395	10.257	9.141	40.624	8.905	8.467	45.002
70	19.662	13.039	11.669	40.652	11.390	10.838	44.878
76	22.396	14.817	13.290	40.659	12.987	12.363	44.798

Table 10.3 Peak velocity (cm/s)

Building floor	Uncontrolled	TMD	PSO-TMD	Reduction (%)	ATMD	PSO-ATMD	Reduction (%)
55	12.582	8.215	6.127	51.303	7.074	5.763	54.196
60	14.471	9.450	7.163	50.501	8.113	6.649	54.053
70	18.410	12.031	9.441	48.718	10.267	8.513	53.758
76	20.927	13.681	10.925	47.794	11.648	9.718	53.562

Table 10.4 Peak acceleration (cm/s^2)

Building floor	Uncontrolled	TMD	PSO-TMD	Reduction (%)	ATMD	PSO-ATMD	Reduction (%)
55	13.352	9.923	8.208	38.526	7.539	6.256	53.145
60	16.804	11.835	9.369	44.245	7.522	6.687	60.205
70	19.851	12.855	12.764	35.701	9.153	8.691	56.219
76	27.825	22.332	19.797	28.852	11.905	9.370	66.325

Note The reduction (%) is the PSO-TMD/PSO-ATMD with respect to uncontrolled.

be guaranteed. Compared to the Davenport spectrum, the simulated power spectral density function curves of fluctuating wind speed can be highly consistent with the objective trend. Therefore, it can provide a strong guarantee for accurate wind load simulation.

A novel PSO-based TMD (ATMD) controller was proposed to reduce the benchmark building subjected to the simulated fluctuating wind. This study emphasizes designing and optimizing the parameters of the TMD (ATMD) control scheme for achieving the best results in reducing the high-rise building. The design of the two controllers was formatted as an optimization problem based on the IAE performance index, which was solved by the PSO technique.

The numerical results show that both controllers can effectively reduce the wind-induced vibration of the tall building, and the innovative method is applicable. It is worth mentioning that PSO-based ATMD can perform better than the other, mainly due to the active controller in view of the nearly identical optimal parameters of the mass damper system.

Conventional control strategies of structure vibration, combined with artificial intelligence, can be further improved by using a specific performance index that can reflect the typical characteristic of the controlled object. This idea can deal with wind-induced vibration control and give inspiration for other structural vibrations, such as seismic vibration.

References

1. Farshidianfar A, Saghafi A, Kalami SM et al (2012) Active vibration isolation of machinery and sensitive equipment using H_∞ control criterion and particle swarm optimization method [J]. Meccanica 47(2):437–453
2. Chang S, Sun W, Cho SG et al (2016) (2016) Vibration control of nuclear power plant piping system using Stockbridge damper under earthquakes [J]. Sci Technol Nucl Installations 5014093:1–12
3. Brewer GA (1979) Dynamic vibration absorber [J]. J Acoust Soc Am 66(4):1233
4. Ari M, Faal RT, Zayernouri M (2020) Vibrations suppression of fractionally damped plates using multiple optimal dynamic vibration absorbers [J]. Int J Comput Math 97(4):851–874
5. Ormondroyd J, Hartog JPD (1928) The theory of the dynamic vibration absorber [J]. ASMR J Appl Mech 50(7):11–22

6. Hahnkamm E (1933) Die dampfung von fundamentschwingungen bei veranderlicher erregerfrequenz [J]. Ing Arch 4:191–201
7. Nishihara O, Matsuhisa H (1997) Design and tuning of vibration control devices via stability criterion [J]. Trans Jpn Soc Mech Eng 10(1):165–168
8. Warburton GB, Ayorinde EO (1980) Optimal absorber parameters for simples system [J]. Earthq Eng Struct Dyn 8(3):197–217
9. Du Y (2015) Research on application of dynamic vibration absorber in the automobile vibration control [D]. Hefei University of Technology
10. Farshidianfar A, Soheili S (2013) Optimization of TMD parameters for earthquake vibrations of tall buildings including soil structure interaction [J]. Int J Optim Civil Eng 3(3):409–429
11. Casciati F, Giuliano F (2009) Performance of multi-TMD in the towers of suspension bridges [J]. J Vib Control 15(6):821–847
12. Frahm H (1911) Device for damping vibrations of bodies, U.S. Patent No. 989,958. Washington, DC: U.S. Patent and Trademark Office
13. Rana R, Soong TT (1998) Parametric study and simplified design of tuned mass dampers. Eng Struct 20(3):193–204
14. Marano GC, Greco R, Bernardino C (2010) A comparison between different optimization criteria for tuned mass dampers design. J Sound Vib 329(23):4880–4890
15. Bekdaş G, Nigdeli SM (2011) Estimating optimum parameters of tuned mass dampers using harmony search. Eng Struct 33(9):2716–2723
16. Abdel-Rohman M (1984) Optimal design of active TMD for buildings control. Build Environ 19(3):191–195
17. Samali B, Al-Dawod M, Kwok KCS, Naghdy F (2004) Active control of cross wind response of 76-story tall building using a fuzzy controller. J Eng Mech 130(4):492–498
18. Ren MZ (2011) Analysis, control and calculation method of mechanical vibration [M]. Machinery Industry Press
19. Zhang D (2015) Optimizing the parameters of dynamic vibration absorber and researching on the active control [D]. Chang'an University
20. Wei W (2015) Study of control strategy of variable mass dynamic vibration absorber [D]. Chang'an University
21. Eberhart R, Kennedy J (1995) A new optimizer using particle swarm theory [C]. In: MHS'95 Proceedings of the sixth international symposium on micro machine and human science. IEEE, pp 39–43
22. Huang W, Jian X et al (2016) Active vibration control for power equipment using particle swarm optimization technique [J]. J Hefei Univ Technol 39(4):494–498
23. Zhang JQ, Peng H, Zhang J et al (2018) Weight parameters optimization for the weight matrices in a vehicle suspension LQR active control [J]. J Vib Shock 37(22):214–219
24. Schulz SL, Gomes HM, Awruch AM (2013) Optimal discrete piezoelectric patch allocation on composite structures for vibration control based on GA and modal LQR [J]. Comput Struct 128:101–115
25. Chen YL, Fu LZ, Mei DQ et al (2003) Design of PID vibration control system for ultra-precision device [J]. Chin J Eng Des 10(1):19–23
26. Yaliang C, De-qing MEI, Long-zhu FU (2003) Design of two-dimension fuzzy-PID vibration control system for ultra-precision device [J]. Chin Mech Eng 14(12):1056–1059
27. Li C, Zhang J (2007) Investigations of vibration control of tall buildings under wind loads using active tuned mass damper [J]. J Earthq Eng Eng Vibr 27(5):160–165
28. Bani-Hani KA (2007) Vibration control of wind-induced response of tall buildings with an active tuned mass damper using neural networks [J]. Struct Control Health Monit 14(1):83–108
29. Iannuzzi A, Spinelli P (1987) Artificial wind generation and structural response. J Struct Eng 113(12):2382–2398
30. Lazzari M, Saetta AV, Vitaliani RV (2001) Non-linear dynamic analysis of cable-suspended structures subjected to wind actions. Comput Struct 79(9):953–969
31. Li Y, Kareem A (1990) ARMA systems in wind engineering. Probab Eng Mech 5(2):49–59

32. Torres JL, Garcia A, De Blas M, De Francisco A (2005) Forecast of hourly average wind speed with ARMA models in Navarre (Spain). Sol Energy 79(1):65–77
33. Shinozuka M, Jan CM (1972) Digital simulation of random processes and its applications. J Sound Vib 25(1):111–128
34. Yang J-N (1973) On the normality and accuracy of simulated random processes. J Sound Vib 26(3):417–428
35. Sun Z (2007) Computer simulation and analysis of wind loading of structures. Doctoral dissertation, Nanjing University of Aeronautics and Astronautics, China
36. Yang JN, Agrawal AK, Samali B, Wu J-C (2004) Benchmark problem for response control of wind-excited tall buildings. J Eng Mech 130(4):437–446
37. Design code (2012) Load code for the design of building structures, Beijing, China
38. Davenport AG (1965) The relationship of wind structure to wind loading. In: Proceedings of symposium on wind effect on building and structures, London, pp 390–402

Chapter 11
Optimal Sensors Deployment

Abstract In this chapter, a probabilistic sensing model for sensor deployment is proposed firstly, and an improved discrete particle swarm optimization algorithm is developed for optimal sensor deployment (OSD) on planar structure and in spatial structure, and the numerical examples indicate that the gbest solution can obtain the OSD whether the sensors are uniform or combinational.

Sensors play an essential role in the field of vibration testing, data collection, and health monitoring of engineering structures. In general, appropriate sensor deployment is directly related to sensing efficiency and economic cost. This chapter proposes a discrete particle swarm optimization algorithm based on a probabilistic sensor model to solve the problem of optimal sensors deployment on two-dimensional planar structures and three-dimensional spatial structures.

11.1 Probabilistic Sensing Model

Sensors have been widely used in health monitoring, critical area monitoring, and military fields related to civil engineering structures [1–5]. Besides, optimal deployment technology has always been a key issue in using sensors. In actual engineering, the maximum detection distance of the sensor is generally determined. Moreover, it is generally assumed that the sensor's detection is evenly distributed in all directions [6]. When the target is within the maximum sensor's detection distance, it can be recorded as 1, whereas when the position of the target exceeds the maximum detection distance of the sensor, the sensor cannot detect the target at this time, and the detection result can be recorded as 0. Although this representation format of 0/1 sensor detection result is simple, it is quite different from the actual situation. Because in actual engineering, the detection ability of the sensor is often greatly affected by the environment, structural form, and other factors. Hence, some scholars consider the sensor's detection model as a form of probability distribution [7–9].

The problem of optimal sensors deployment is arranging a reasonable number of sensors, selecting their type to monitor a given structure area, and ultimately

W. Huang and J. Xu, *Optimized Engineering Vibration Isolation, Absorption and Control*, Springer Tracts in Civil Engineering, https://doi.org/10.1007/978-981-99-2213-0_11

maximizing the sensor coverage area [9–11]. The sensor's detection probability is generally considered a function of the distance between the sensor and the target, which is also affected by environmental factors and measurement noise [12]. Wu et al. [13] proposed a novel probabilistic sensing model and defined the model as a class of NP-complete problems based on combinatorial optimization [14]. The issue herein is that the NP-complete problem is a complicated optimization problem that is difficult to be solved in polynomial time, and conventional optimization methods are challenging to achieve. Therefore, intelligent algorithms are urgently needed.

In optimizing sensor deployment, a civil engineering structure's monitored area R (surveillance region) can be a two-dimensional planar or three-dimensional spatial structure. For a two-dimensional planar structure, if the case is a rectangular plane (such as a common plate structure), the distance between any target detection point on the structure and the sensor is recorded as r, and the probability that the sensor detects the point is defined as the target detected probability (P). If a set of sensors are given in advance and denoted as set (S), the sensors can be of the same type or different types, which can be distinguished according to their maximum detection distance.

In the optimal sensors deployment strategy, the probability of being detected at any point on the planar structure is defined as the probability of being detected by any (or more) sensors in the sensors' group. If the position of the sensor is recorded as r_s, the distance of any point r on the planar structure between the sensor can be written as $\|r_s - r\|$ and shown in Fig. 11.1. Assuming that the sensor coordinate is (S_x, S_y), and the coordinate of any point on the planar structure is (r_x, r_y), and thus $\|r_s - r\|$ can be written as $\sqrt{(S_x - r_x)^2 + \left(S_y - r_y\right)^2}$. If the planar structure is discretized, the discrete accuracy of the x-direction and y-direction are recorded as d_x and d_y, respectively. Besides, the coordinates of any point on the sensor and the planar structure can be written in the form of $\left(m \times d_x, n \times d_y\right)$, m, and n, representing the number of nodes for any point away from the origin.

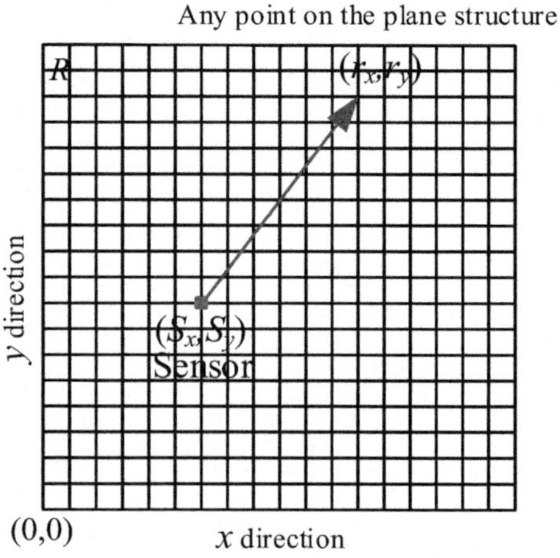

Fig. 11.1 Schematic diagram of sensors detection on plane

Wu et al. [13] proposed an imperfect sensor approximation strategy to characterize the detection characteristics of the sensor and express it with a probabilistic model. Theoretically, the detection probability $0 \leq P \leq 1$ is a function closely related to the sensor's maximum detection range, as shown in Fig. 11.2. The detection probability of a single sensor can be expressed using the following Gaussian function:

$$P\left(S_k, \tau, \alpha_{s_k}\right) = e^{-\frac{\tau^2}{2\alpha_{s_k}^2}} \tag{11.1}$$

where s_k is the k-th sensor in the sensor set; τ is the straight line distance between the detection target point and the sensor on the plane structure, is recorded as $\tau = \|r_s - r\|$, and is shown in Fig. 11.1. This value must be between 0 and the maximum detection distance to limit the target detection point to be within the maximum detection range of the sensor, as shown in Fig. 11.2; α_{s_k} is the detection quality coefficient of the k-th sensor. This parameter directly determines the shape of the probability curve, and it is determined by the type of selected sensors.

Equation (11.1) is illustrated in Fig. 11.3.

The detection probability of a single sensor is described above. In engineering practice, combining multiple sensors is very common. Previously, Wu et al. [13] introduced the concept of overlap probability to characterize the crossover and overlap detection that occurs when multiple sensors are used. The probability can be interpreted as the probability of a point on the structure being detected by multiple overlapping sensors equals the sum of the detection probabilities of a single sensor minus the joint detection probability. The overlapping detection area of multiple sensors can be illustrated in Fig. 11.4, and the core algorithm is:

$$\begin{cases} \text{overlap} = P(R) \cdot P\left(S_k, \tau, \alpha_{s_k}\right) \\ P(R) = P(R) + P\left(S_k, \tau, \alpha_{s_k}\right) - \text{overlap} \end{cases} \tag{11.2}$$

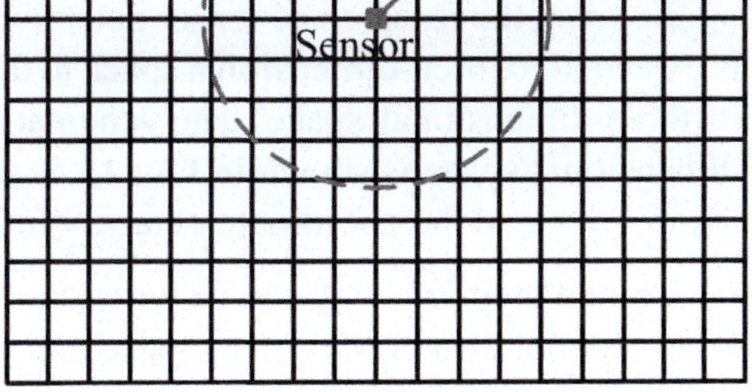

Fig. 11.2 Maximum detection range of a sensor on the plane

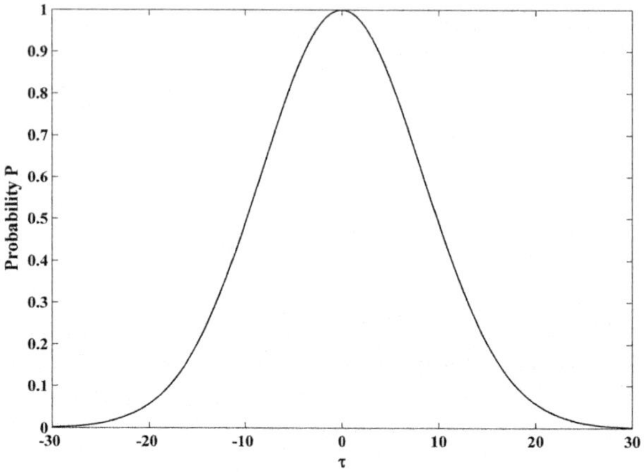

Fig. 11.3 Gaussian probability distribution

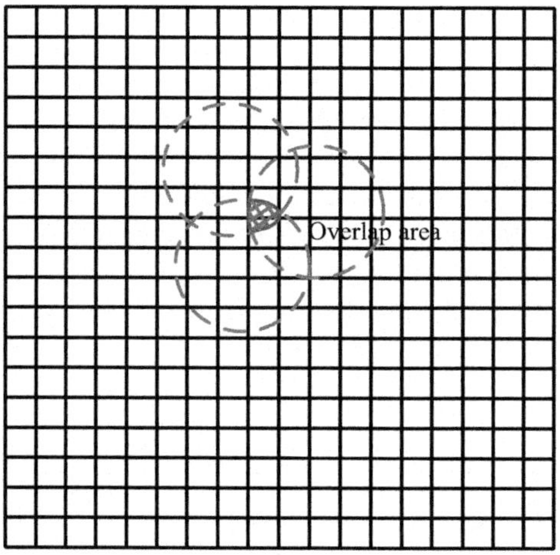

Fig. 11.4 Overlap area of multiple sensors detection on a planar structure

Equation (11.2) reflects an accumulative probability that reveals the process of multiple sensors continuously detecting target points on the structure to be arranged according to their maximum sensing distance. In general, it can be calculated using an embedded loop algorithm. The detailed algorithm description is shown in reference [13].

Based on the research of sensor detection on planar structure, the sensor detection is extended to three-dimensional space to develop the optimal deployment of sensors in three-dimensional space. The schematic diagram of sensor detection in three-dimensional space is shown in Fig. 11.5, and the sensor's coordinate is assumed as (S_x, S_y, S_z), and the coordinate of any point in the spatial structure is assumed as (r_x, r_y, r_z), and then $\|r_s - r\|$ can be written as $\sqrt{(S_x - r_x)^2 + \left(S_y - r_y\right)^2 + (S_z - r_z)^2}$. If the spatial structure is discretized, the discrete accuracy in the x-direction, y-direction, and z-direction are recorded as d_x, d_y, and d_z, respectively, and then the

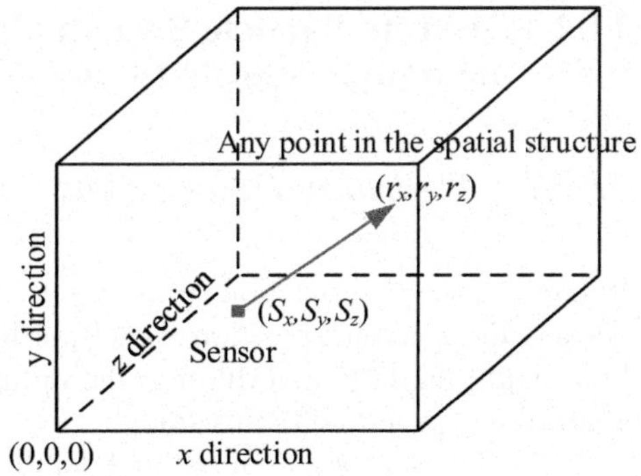

Fig. 11.5 Schematic diagram of a sensor detection in spatial structure

coordinate of the sensor and any point in the spatial structure can be written as $(m \times d_x, n \times d_y, l \times d_z)$, and m, n, and l for which they represent the number of nodes of any point away from the origin. The sensor's detection probability is still calculated based on Eqs. (11.1) and (11.2), but the distance between the target detection point and the sensor should be the straight-line distance in the spatial structure. The maximum sensor detection range in three-dimensional space is illustrated in Fig. 11.6.

Similar to the sensor optimal deployment problem, there is a traveling salesman problem [15] and a knapsack problem [16]. This optimization problem cannot be effectively solved in polynomial time, which is the non-deterministic polynomial problem (NP problem). An NP-complete problem is the most complicated type of NP problem, where obtaining the optimal solution through limited computation time and space is difficult. The optimal sensors deployment problem is a type of NP-complete problem [17], where solving it must rely on intelligent optimization methods.

The PSO algorithm is a novel swarm intelligence optimization algorithm. In the search space, particles update their speed and position through continuous learning, updating the optimal local solution, and finally, obtaining the optimal global solution. The idea of a discrete particle swarm optimization algorithm (Discrete PSO, DPSO) is proposed here to solve the problem of optimal sensors deployment.

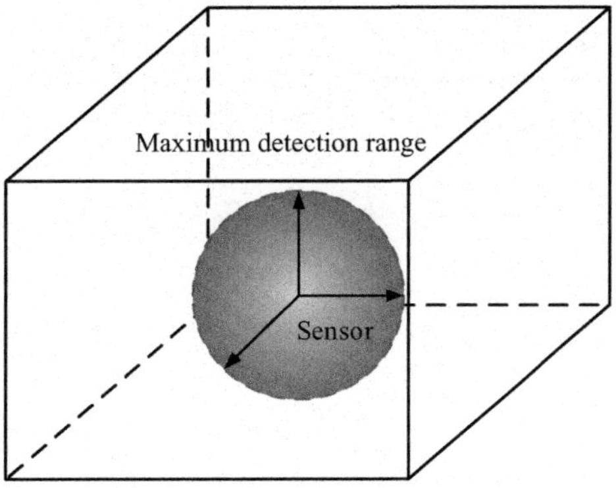

Fig. 11.6 Maximum sensor detection range in the three-dimension space

11.2 Discrete Particle Swarm Optimization Algorithm for Planar Sensor Deployment

11.2.1 Algorithm Descriptions

In the discrete optimization space, the traditional PSO algorithm is not applicable because the speed and position of each particle are discrete values. In order to successfully apply the PSO algorithm to the optimization of sensor deployment, a DPSO algorithm is proposed in this study.

Consider the problem of optimal deployment of sensors on a planar structure.

- DPSO parameter configuration

Select appropriate calculation parameters for DPSO, including the maximum number of iterations, the number of particle populations, the dimension of each particle, the learning factors c_1 and c_2, and the inertia weight factor ω

↓

- Initialization

For the discretized planar structure grid shown in Fig. 11.1, assume that the number of discrete grids in the x and y directions is 50 × 50, and the sensors to be deployed are uniform sensors with the same type and combinational sensors with different types

1. Uniform sensors: The sensor set is defined as $S = [1, 1, \ldots, 1]$, , where 1 is the first type of sensor (type 1), and k is the number

2. Combinational sensors: The sensor set is defined as $S = [1, 1, \ldots, 1, 2, 2, \ldots, 2, 3, 3, \ldots, 3, 4, 4, \ldots, 4]$, where 1, 2, 3, and 4 are the first, second, third, and fourth type of sensor (type 1, type 2, type 3, and type 4), respectively. Among them, the number of type 1 sensors is k_1, the number of type 2 sensors is k_2, the number of type 3 is k_3, the number of type 4 sensors is k_4, and the total number of all types is k

---

Next, the initialization of the particle swarm positions is explained using the following pseudocode:

swarm = zeros(50 × swarmsize,dim); % Generate a zero matrix of [50 × swarmsize, dim] to represent the particle group to be initialized, where swarmsize represents the population number, and dim represents the particle dimension

---

(continued)

(continued)

% The following for-end loop applies to uniform sensors deployment and is a binary (0–1) integer encoding problem
for i = 1 to swarmsize
    $n_1 = 50; n_2 = 50; n = n_1 \times n_2;$ % n_1 and n_2 represent the grid number of discrete grids on the plane in both the x and y direction, respectively. This study takes it as a 50 × 50 grid
    $swarm_0 = zeros(n_1, n_2);$
    //*r = randperm(n);* % This operation randomly shuffles integers from 1 to *n* to obtain a sequence of numbers
    $rn = (r \leq k);$ //% This operation can randomly select *k* positions from the above arrangement to configure *k* sensors. Note: The randomizing concept is critical to the diversity of populations, which aligns with PSO's core ideas
        $temp = true(n_1, n_2);$ % Produces an $n_1 \times n_2$ matrix with logical values 1
        $swarm_0(temp) = rn;$ % Randomly configure the given *k* sensors on a planar grid
        $swarm((((i - 1) \times 50 + 1)\ to\ (i \times 50)), 1\ to\ dim) = swarm_0;$ % Complete population initialization
end

% The following for-end loop applies to combinational sensors deployment, and it is a 0–1–2–3–4 integer coding problem
for i = 1 to swarmsize
    $swarm_0 = swarm((((i - 1) \times 50 + 1)\ to\ (i \times 50)), 1\ to\ dim);$
    $n_1 = 50; n_2 = 50; n = n_1 \times n_2;$
    r = randperm(n);
    rr = r(1 to k);% Reserve positions to be configured for four different types of sensors with a *k* total number
    for j = 1 to k
        number = rr(j);
        //*if mod(number,50) = 0*% The calculation result of the mod operation is number-a × 50, where a = floor (number/50), and the floor operation takes the largest integer that does not exceed number/50 (Gaussian rounding operation)
    h = number/50;
        else
    h = fix(number/50) + 1;% Fix operation truncates and rounds number/50
        end% h is the row position number, indicating the position of the sensor in the *x* direction of the planar structure
        l = mod(number,50);
        % *l* indicates the column position number, representing the position of the sensor in the *y* direction of the planar structure
        if l = 0
        l = 50;
        end//% This operation restores the actual configuration position of combinational sensors. Hence, it matches the corresponding position from the matrix *rr*

(continued)

(continued)

$if\, j \le k_1$
$swarm_0\,(h,l) = 1;$% Deploy type 1 sensors (k_1)
$elseif\,(j \ge (k_1 + 1))\ and\ (j \le (k_1 + k_2))$
$swarm_0\,(h,l) = 2;$ % Deploy type 2 sensors (k_2)
$elseif\,(j \ge (k_2 + 1))\ and\ (j \le (k_1 + k_2 + k_3))$
$swarm_0\,(h,l) = 3;$ % Deploy type 3 sensors (k_3)
$elseif\,(j \ge (k_3 + 1))$
$swarm_0\,(h,l) = 4;$ % Deploy type 4sensors (k_4)
end
end
$swarm((((i-1) \times 50 + 1)\ to\ (i \times 50)), 1\ to\ dim) = swarm_0;$
end

Next, update the particle swarm's speed and position, the optimal local solution (*pbest*), and the optimal global solution (*gbest*)

$$x = \begin{cases} 0.98, & v > 4 \\ \dfrac{1}{1 + e^{-v}}, & -4 \le v \le 4 \\ 0.02, & v < -4 \end{cases} \tag{11.3}$$

- Update the particle swarm's speed and position

The particle swarm speed and position are updated according to Eqs. (1.1) and (1.2). In the DPSO algorithm, the velocity value has a greater influence on the particle's position. The speed of the large value defines its position as 1 with a great probability. Otherwise, its position may be set to 0. Based on this, the position updating equation in the standard PSO algorithm is improved. This study introduces a *sigmoid* function, which can update the position of particles according to the speed value

Equation (11.3) is illustrated in Fig. 11.7

At this point, the particle swarm's updated position has been obtained, but they are still unsuitable for DPSO because the obtained position value is non-integer. In order to solve this problem, the following position value rounding strategy is proposed, which is described in detail using the following pseudocode

---

(continued)

(continued)

```
% The following nested for-end loop is used for the uniform sensors calculation
    for i = 1 to swarmsize
            x0 = x((((i − 1) × 50 + 1) to (i × 50)),1 to dim); % x is the calculated position of
the particle swarm using Eq. (11.3)
                [number1,number2] = find(x0 == 1);% This operation is to find the row and
column number of the element with x0 = 1
                n1 = length(number1 or number2);% Count the number of elements in vector
number1 or number2
            //r = randperm(n1);
            b = r((k + 1) to end);
            n2 = length(b); // % Note: This operation is based on a random selection strategy,
and this random idea is critical!
            for j = 1 to n2
            c = b(j);
            x0 (number1(c),number2(c)) = 0;
            end
            x((((i − 1) × 50 + 1) to (i × 50)),1 to dim) = x0;
    end
```

```
% The following nested for-end loop is used for to calculate the combinational sensors
for i = 1 to swarmsize
    x0 = x((((i − 1) × 50 + 1) to (i × 50)),1 to dim);
    [number1,number2] = find(x0 == 1);
    n1 = length(number1);
    r = randperm(n1);
    b = r(1 to k); n2 = length(b);
        for j = 1 to n2
            c = b(j);
            //if j ≤ k1
        x0(number1(c),number2(c)) = 1;
            elseif (j ≥ (k1 + 1)) and (j ≤ (k1 + k2))
        x0(number1(c),number2(c)) = 2;
            elseif (j ≥ (k2 + 1)) and (j ≤ (k1 + k2 + k3))
        x0(number1(c),number2(c)) = 3;
            elseif (j ≥ (k3 + 1))
        x0(number1(c),number2(c)) = 4;
            end// % Deploy k sensors with four types of combinational sensors
        end
        d = r((k + 1) to end);n3 = length(d);
        for j = 1 to n3
        e = d(j);
        x0(number1(e),number2(e)) = 0;
        end
        x((((i − 1) × 50 + 1) to (i × 50)),1 to dim) = x0;
    end
```

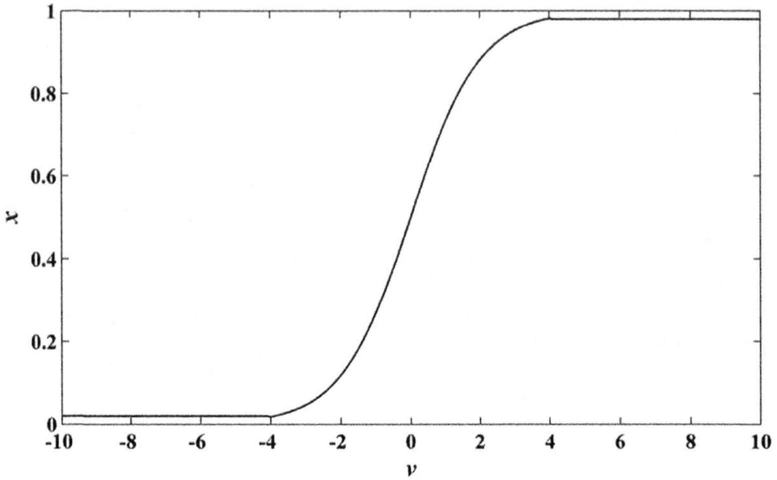

Fig. 11.7 A sigmoid function

- Calculate the objective function's fitness value

  The fitness function can be calculated according to the sensor probability detection model defined in Sect. 11.1

- Check whether the maximum number of iterations is reached

  If the maximum number of iterations is reached, the calculation is terminated, and the optimal global value *gbest* is output. Otherwise, the speed, position, and optimal local value of the particle swarm are updated until reaching the end condition

11.2.2 Optimal Sensors Deployment on a Two-dimensional Planar Structure

Based on the proposed DPSO, the deployment of the optimal sensors on a two-dimensional planar structure is studied. In this study, the discrete grids' number for the planar structure was set to 50 × 50, where the discrete accuracy is $dx = 1$ and $dy = 1$, respectively. The characteristics of the selected four types of sensors are defined in Table 11.1.

This study considers two different sensor deployments, including uniform and combinational sensors. Among them, the uniform sensor is a type 2 sensor in Table 11.1, and the sensors selected are combinational sensors that include all four types.

The parameters of DPSO were set such that the maximum number of iterations is 1000, the population is 500, the learning factors $c_1 = 2$ and $c_2 = 1$, and the inertia weight factor is $\omega = 0.99^t$ (t is the t-th iteration). The fitness function is defined as follows:

$$\text{fitness} = \frac{\text{length}\left(P(\overline{R}) > 0.95\right)}{50 \times 50}\% \tag{11.4}$$

Equation (11.4) aims to maximize the sensor's detection probability on a given planar structure to achieve maximum area coverage. In general, 0.95 was used to define desired coverage probability. This value was set according to actual needs, where $P\left(\overline{R}\right)$ sorts the detection probabilities $P(R)$ from high to low. The fitness function results are regarded as confidence in the planar structure's sensor detection result. The pseudocodes herein are as follows:

for i = 1 to 50

for j = 1 to 50

P(R)$_1$ = sort(P(R)(i,j,1 to k));% The $P(R)$ calculated by Eq. (11.2) is arranged in descending order, k is the total number of sensors

P(R)$_2$ = P(R)$_1$ (end);

$P(\overline{R})$ (i,j) = P(R)$_2$;

end

end

Table 11.1 Four different types of sensors (planar structure)

Type	Maximum detection distance	Detection quality coefficient
1	5	80
2	10	78
3	12	68
4	15	60

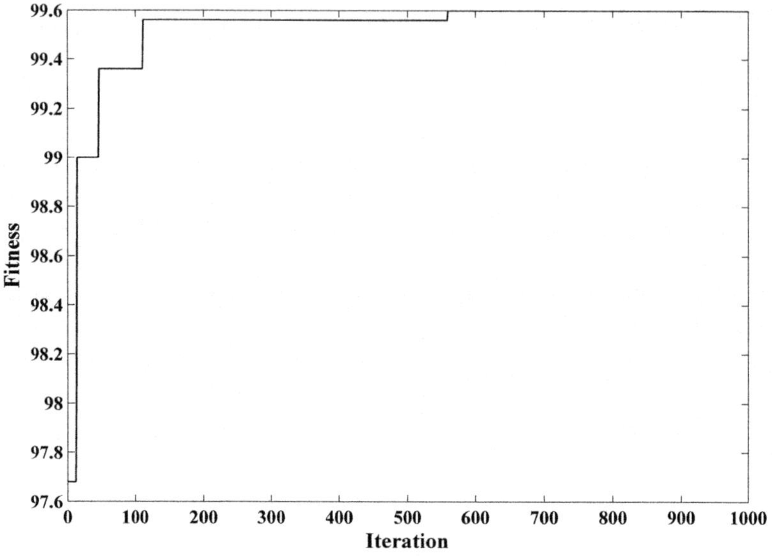

Fig. 11.8 Fitness convergence (uniform sensors/planar structure)

Example 1: Optimized Deployment of Uniform Sensors
The convergence curve of the fitness function based on DPSO optimization, assuming that in the optimization deployment study of uniform sensors, a total of 20 type 2 sensors were used, is shown in Fig. 11.8. The obtained optimal solution *gbest*, the optimal configuration scheme of the sensors, is shown in Fig. 11.9. Finally, based on this optimal sensors deployment, the detection probability coverage obtained is shown in Fig. 11.10, and the probability confidence calculated based on Eq. (11.4) is 99.94%.

Example 2: Optimized Deployment of Combinational Sensors

In the optimization deployment study of combinational sensors, three sensors of type 1 ~ type 3 and four sensors of type 4 are used. The total number of sensors is 13. At this time, the DPSO optimized fitness convergence curve is shown in Fig. 11.11, and the obtained optimal solution *gbest*, the optimal sensors deployment scheme, is shown in Fig. 11.12. Based on the optimal sensors deployment, the detection probability coverage obtained is shown in Fig. 11.13, and the probability confidence calculated based on Eq. (11.4) is 98.96%.

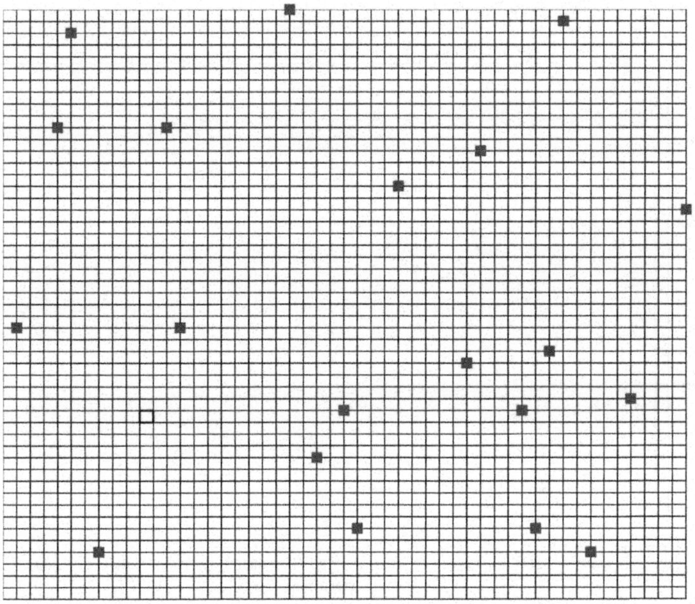

Fig. 11.9 Optimal sensors deployment (uniform sensors/planar structure)

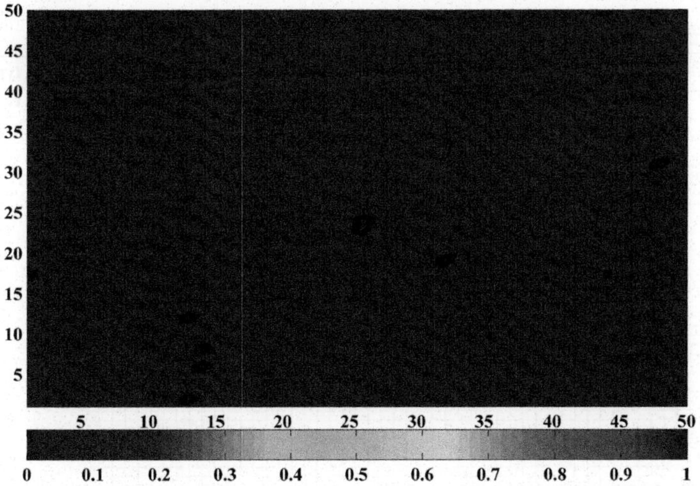

Fig. 11.10 Pseudocolor figure of the detection probability coverage (uniform sensors/planar structure)

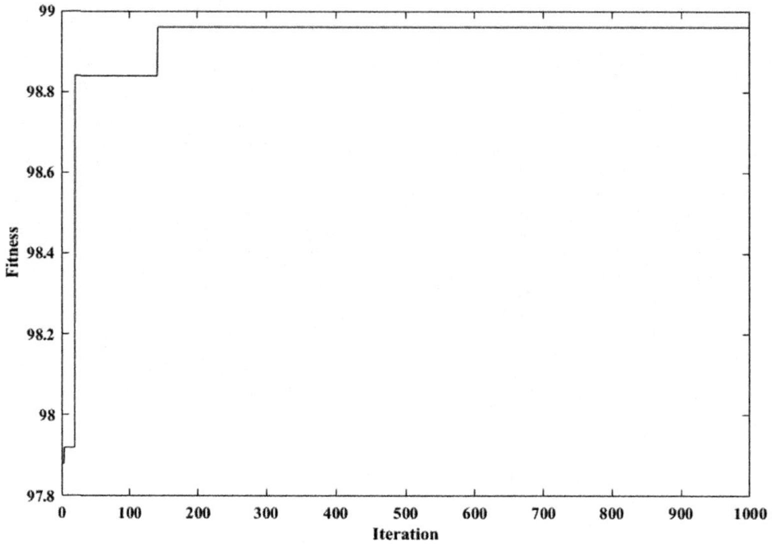

Fig. 11.11 Fitness convergence (combinational sensors/planar structure)

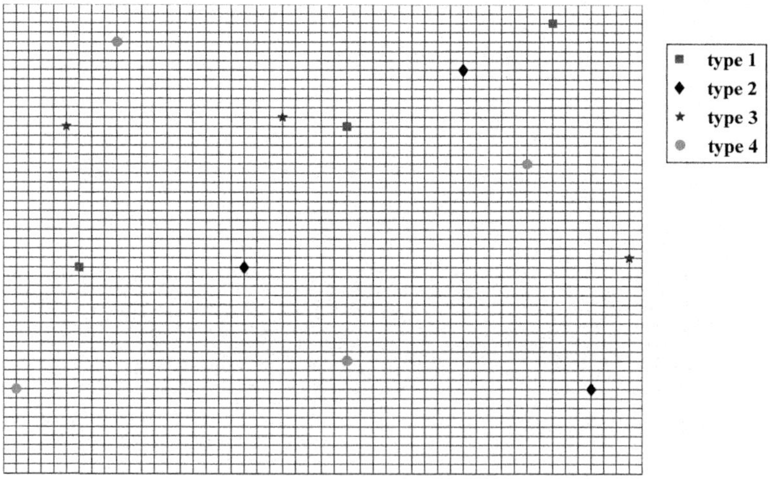

Fig. 11.12 Optimal sensors deployment (combinational sensors/planar structure)

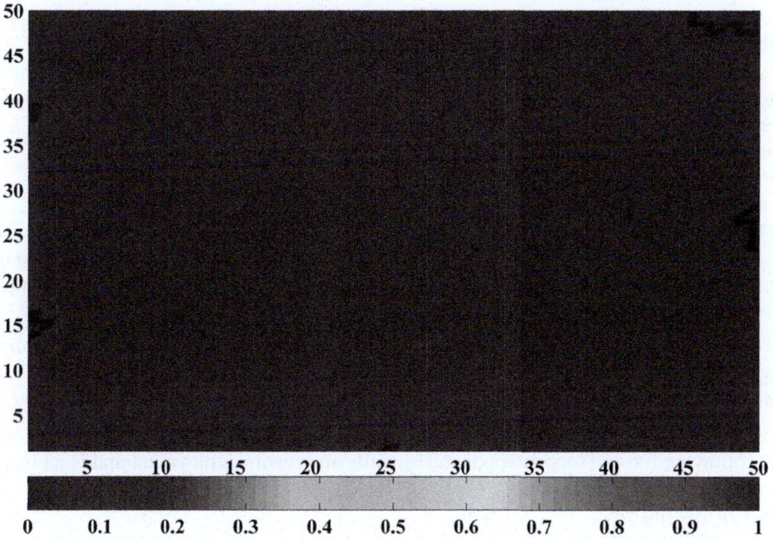

Fig. 11.13 Pseudocolor figure of the detection probability coverage (combinational sensors/planar structure)

11.3 Discrete Particle Swarm Optimization Algorithm for Spatial Sensor Deployment

11.3.1 Algorithm Descriptions

After proposing a DPSO and successfully solving the planar structure's sensors deployment optimization problem, a DPSO is proposed based on previous research to achieve the optimal deployment of sensors on a three-dimensional spatial structure. The specific steps are as follows:

- DPSO parameter configuration
  Same as 11.2.1

• Initialization

Discretize the three-dimensional spatial structure shown in Fig. 11.5, assuming that the number of discrete grids in the x-direction, y-direction, and z-direction is 50 × 50 × 50, similarly, divided the sensors to be deployed into uniform and combinational sensors

1. Uniform sensors: The sensor set is defined as $S = [1, 1, \ldots, 1]$, where 1 is the first type of sensor (type 1), and k is the number

2. Combinational sensors: The sensor set is defined as $S = [1, 1, \ldots, 1, 2, 2, \ldots, 2, 3, 3, \ldots, 3, 4, 4, \ldots, 4]$, 1, 2, 3, and 4 represent the first, second, third, and four types of sensors (type 1, type 2, type 3, and type 4), respectively. Among them, the number of type 1 sensors is k_1, the number of type 2 sensors is k_2, the number of type 3 is k_3, the number of type 4 sensors is k_4, and the total number of the four types of sensors is k

The specific operation for particle swarm position initialization is given in the form of the following pseudocodes:

swarm = zeros(50 × swarmsize,50,dim); % Generate a [50 × swarmsize, 50, dim] three-dimensional zero matrix representing the particle swarm to be initialized

% The following for-end loop is applicable to uniform sensors deployment and is a binary (0–1) integer encoding problem

for i = 1 to swarmsize

n_1 = 50;n_2 = 50;n_3 = 50;n = n_1 × n_2 × n_3; % n_1, n_2, and n_3 represent the grid number of discrete grids in the spatial space in both x, y, and z directions, respectively. This study assumes it as 50; hence, it is a 50 × 50 × 50 grid

$swarm_0$ = zeros(n_1,n_2,n_3);

//r = randperm(n);

rn = (r ≤ k); //

temp = true(n_1,n_2,n_3);

$warm_0$(temp) = rn; % The k sensors are randomly deployed in the three-dimensional spatial grid

swarm((((i − 1) × 50 + 1) to (i × 50)), 1 to 50, 1 to dim) = $swarm_0$; % Complete population initialization

end

% The following for-end loop applies to combinational sensors deployment, and it is a 0–1–2–3–4 integer coding problem

for i = 1 to swarmsize

$swarm_0$ = swarm((((i − 1) × 50 + 1) to (i × 50)), 1 to 50,1 to dim);

n_1 = 50;n_2 = 50; n_3 = 50;n = n_1 × n_2 × n_3;

r = randperm(n);

rr = r(1 to k);% Reserve deployment positions for four different types of sensors with a total number of k

(continued)

(continued)

```
        for j = 1 to k
                number1 = rr(j);
                lyaer = fix(number1/2500) + 1;% Determine which layer the sensor is on in the z direction and locate it, as shown in Fig. 11.14
                number = number1_2500 × (layer–1); % Restore its plane projection position
        if mod(number,50) = 0
    h = number/50;
        else
    h = fix(number/50) + 1;
        end
        l = mod(number,50);
        if l = 0
                l = 50;
        end
                if j ≤ k1
        swarm0 (h,l) = 1;% Deploy type 1 sensors (k1)
        elseif (j ≥ (k1 + 1)) and (j ≤ (k1 + k2))
        swarm0 (h,l) = 2; % Deploy type 2 sensors (k2)
        elseif (j ≥ (k2 + 1)) and (j ≤ (k1 + k2 + k3))
        swarm0 (h,l) = 3; % Deploy type 3 sensors (k3)
        elseif (j ≥ (k3 + 1))
        swarm0 (h,l) = 4; % Deploy type 4 sensors (k4)
                end
    end
        swarm((((i – 1) × 50 + 1) to (i × 50)), 1 to 50,1 to dim) = swarm0;
end
```

Next, the speed, position, optimal local solution *pbest*, and optimal global solution *gbest* of the particle swarm are updated

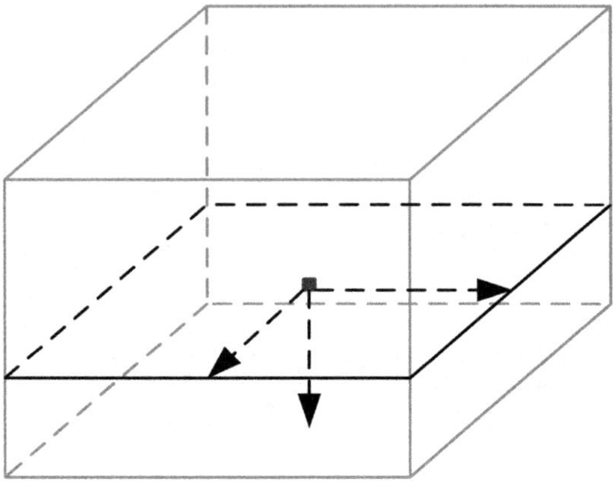

Fig. 11.14 Locate the sensor in the spatial structure

• Update the particle swarm's speed and position

Consider the *sigmoid* function shown in Eq. (11.3), and update its position according to the speed of each particle

Once the particle swarm obtains the updated position, it still requires rounding the position value to adapt it to the DPSO calculation. The pseudocodes are as follows:

% The following for-end loop is used to calculate the uniform sensors

for i = *1 to swarmsize*

$x_0 = x(((i-1)\times 50+1)$ *to* $(i\times 50))$, *1 to 50,1 to dim); %* x is the calculated particle swarm position using Eq. (11.3)

number = *find*(x_0 = = *1*);

[$number_1$,$number_2$,$number_3$] = *ind2sub(size*(x_0)*,number);%* This operation is to convert the linear index of the matrix into the corresponding subscript

n_1 = *length*($number_1$ *or* $number_2$ *or* $number_3$)*;%* Count the number of elements in vector $number_1$, $number_2$, or $number_3$

//r = *randperm*(n_1);

b = *r((k* + *1) to end);*

n_2 = *length(b);* //

for j = *1 to* n_2

c = *b(j);*

x_0 ($number_1(c)$,$number_2(c)$, $number_3(c)$) = *0;*

end

$x(((i-1)\times 50+1)$ *to* $(i\times 50))$, *1 to 50,1 to dim)* = x_0;

end

(continued)

(continued)

```
% The following for-end loop is used to calculate the combinational sensors
for i = 1 to swarmsize
        x0 = x((((i − 1) × 50 + 1) to (i × 50)), 1 to 50,1 to dim);
        number = find(x0 = 1);
        [number1,number2,number3] = ind2sub(size(x0),number);
        n1 = length(number1);
        r = randperm(n1);
        b = r(1 to k); n2 = length(b);
                for j = 1 to n2
                        c = b(j);
                        //if j ≤ k1
                x0(number1(c),number2(c), number3(c)) = 1;
                        elseif (j ≥ (k1 + 1)) and ( j ≤ (k1 + k2))
                x0(number1(c),number2(c), number3(c)) = 2;
                        elseif (j ≥ (k2 + 1)) and ( j ≤ (k1 + k2 + k3))
                x0(number1(c),number2(c), number3(c)) = 3;
                        elseif (j ≥ (k3 + 1))
                x0(number1(c),number2(c), number3(c)) = 4;
                end// % Deploy k four types of combinational sensors
        end
        d = r((k + 1) to end);n3 = length(d);
        for j = 1 to n3
        e = d(j);
        x0(number1(e),number2(e), number2(e)) = 0;
        end
        x((((i − 1) × 50 + 1) to (i × 50)), 1 to 50, 1 to dim) = x0;
        end
```

- Calculate the fitness value of the objective function
  Same as Sect. 11.2.1

- Check whether the maximum number of iterations is reached
  Same as Sect. 11.2.1

11.3.2 Optimal Sensors Deployment in Three-dimensional Spatial Structure

The optimal sensors deployment in the three-dimensional spatial structure is investigated herein based on the DPSO proposed above. The discrete grids' number for the three-dimensional space structure is $50 \times 50 \times 50$, and the discrete accuracy is $dx = 1, dy = 1$, and $dz = 1$. The characteristics of the selected four types of sensors are defined and shown in Table 11.2.

Two different sensor deployment types were considered in this study, namely: uniform and combinational sensors. Uniform sensors were selected as the type 2 sensor in Table 11.2, and the combinational sensors selected include all four types. The DPSO parameter setting is the same as that in Sect. 11.2.2.

The fitness function used in the DPSO calculation was adjusted to:

$$\text{fitness} = \frac{\text{length}\left(P\left(\overline{R}\right) > 0.95\right)}{50 \times 50 \times 50}\% \tag{11.5}$$

Similarly, the Eq. (11.5) definition aims to maximize the probability coverage of three-dimensional spatial structure detection. The calculation of the fitness function must be adjusted. The following pseudocodes for details are shown:

for i = 1 to 50
    for j = 1 to 50
        for m = 1 to 50
        $P(R)_1$ = sort(P(R)(i,j,m,1 to k));% The $P(R)$ calculated by Eq. (11.2) was arranged in descending order, and k is the total number of sensors
        $P(R)_2 = P(R)_1(end)$;
        $P\left(\overline{R}\right)(i,j,m) = P(R)_2$;
        end
    end
end

Example 3: Optimized Deployment of Uniform Sensors
Assuming that in the optimization deployment study of uniform sensors, a total of 20 type 2 sensors were used, and the convergence curve of the fitness function

Table 11.2 Four different types of sensors (spatial structure)

Type	Maximum detection distance	Detection quality coefficient
1	10	80
2	15	78
3	20	68
4	25	60

based on DPSO optimization is shown in Fig. 11.15. The obtained optimal solution gbest, the optimal sensors deployment location, is shown in Fig. 11.16. Based on the optimal sensors deployment algorithm, the obtained detection probability coverage is shown in Fig. 11.17. Moreover, based on the probability calculated by Eq. (11.4), the confidence is 99.90%.

Example 4: Optimized Deployment of Combinational Sensors

Assume that in the optimization deployment study of combinational sensors, the type 1, type 2, and type 3 sensors are used, and four sensors of type 4 are used. Accordingly, the total number of sensors is 13. At this time, the DPSO-optimized fitness convergence curve is shown in Fig. 11.18. Besides, the obtained optimal solution gbest, the optimal sensors deployment scheme, is shown in Fig. 11.19. Finally, the detection probability coverage obtained, based on the optimal sensors deployment, is shown in Fig. 11.20, and the probability confidence calculated based on Eq. (11.4) is 97.42%.

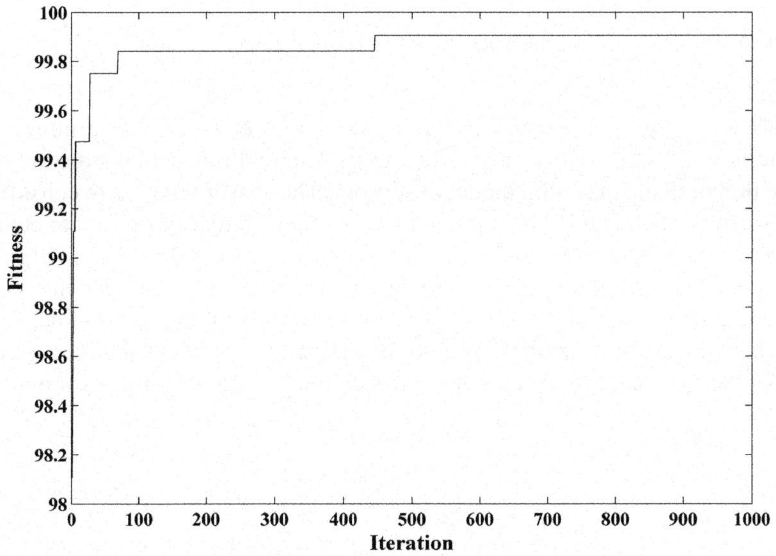

Fig. 11.15 Fitness convergence (uniform sensors/spatial structure)

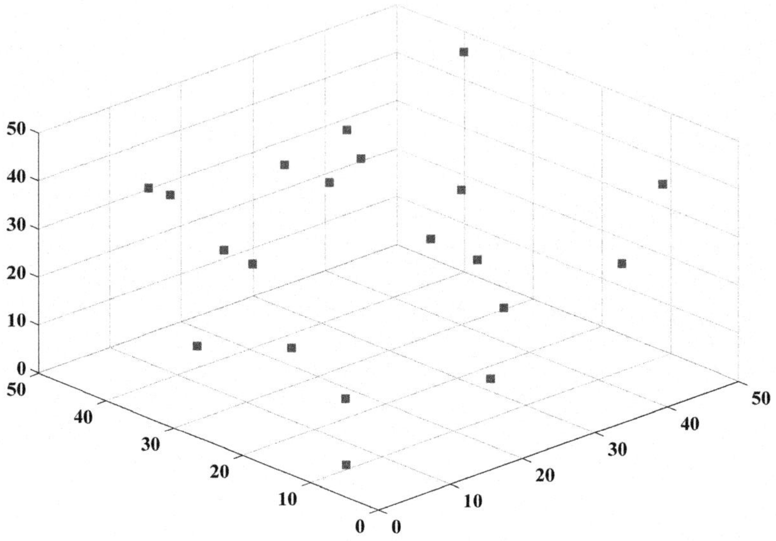

Fig. 11.16 Optimal sensors deployment (uniform sensors/spatial structure)

This chapter proposes a discrete particle swarm optimization algorithm to solve the problem of optimal sensor placement on planar and in spatial structures based on the probabilistic detection model of sensors. The sensor types were considered to be the same type for the uniform sensors case and different types for the combined sensors case. Additionally, this study proposes the targeted deployment algorithms for each of the cases. The results of the calculation examples verify that the proposed DPSO algorithm can effectively solve the problem of optimal sensors deployment on planar and spatial structures. This study has a certain guiding significance for the optimal sensors deployment on other types and geometric engineering structures.

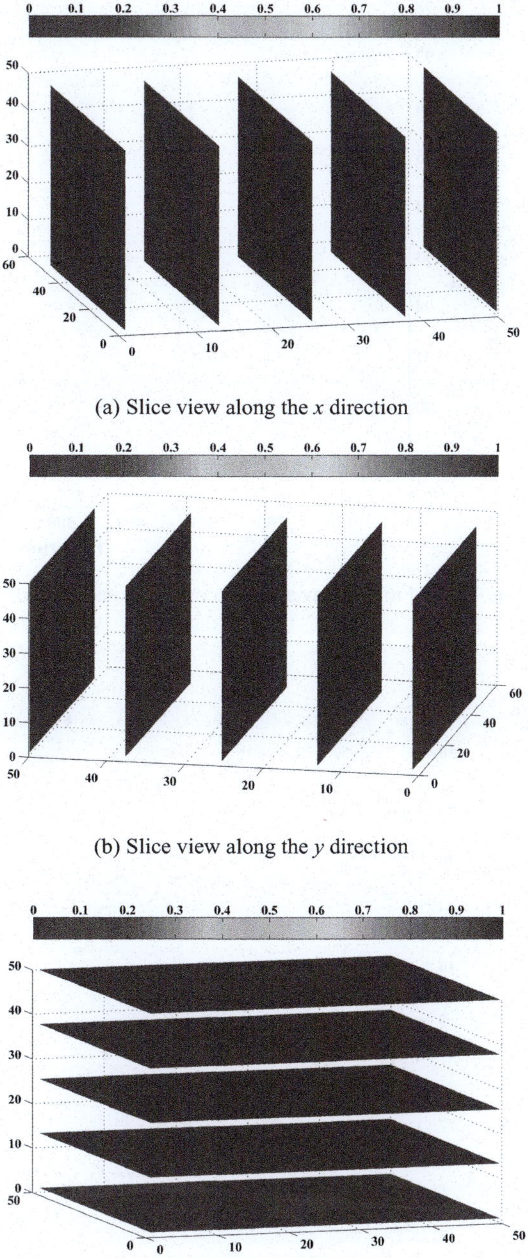

(a) Slice view along the x direction

(b) Slice view along the y direction

(c) Slice view along the z direction

Fig. 11.17 Pseudocolor figure of the detection probability coverage (uniform sensors/spatial structure)

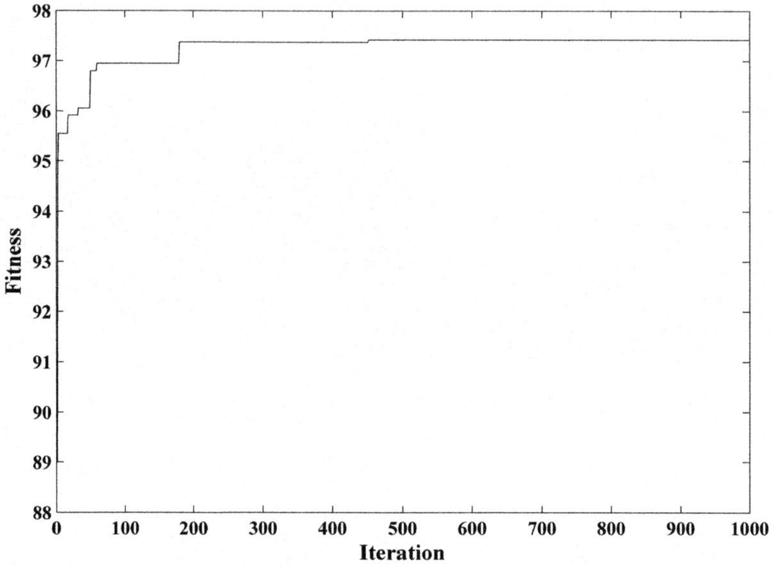

Fig. 11.18 Fitness convergence (combinational sensors/spatial structure)

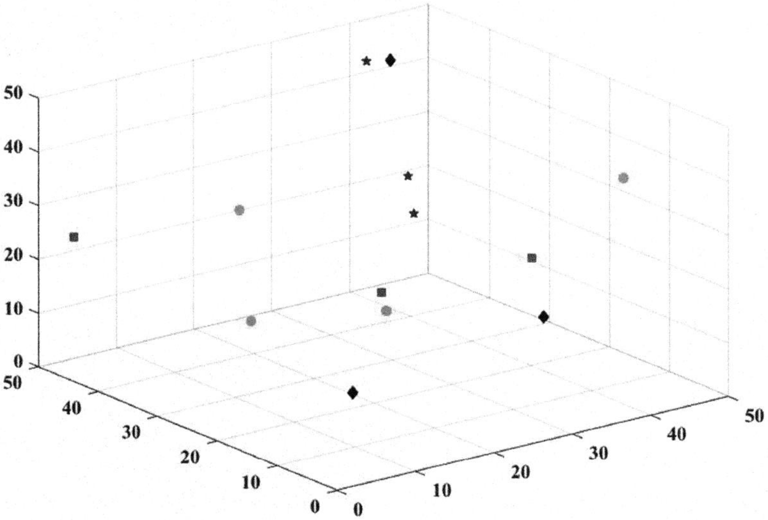

Fig. 11.19 Optimal sensors deployment (combinational sensors/spatial structure)

Fig. 11.20 Pseudocolor figure of the detection probability coverage (combinational sensors/spatial structure)

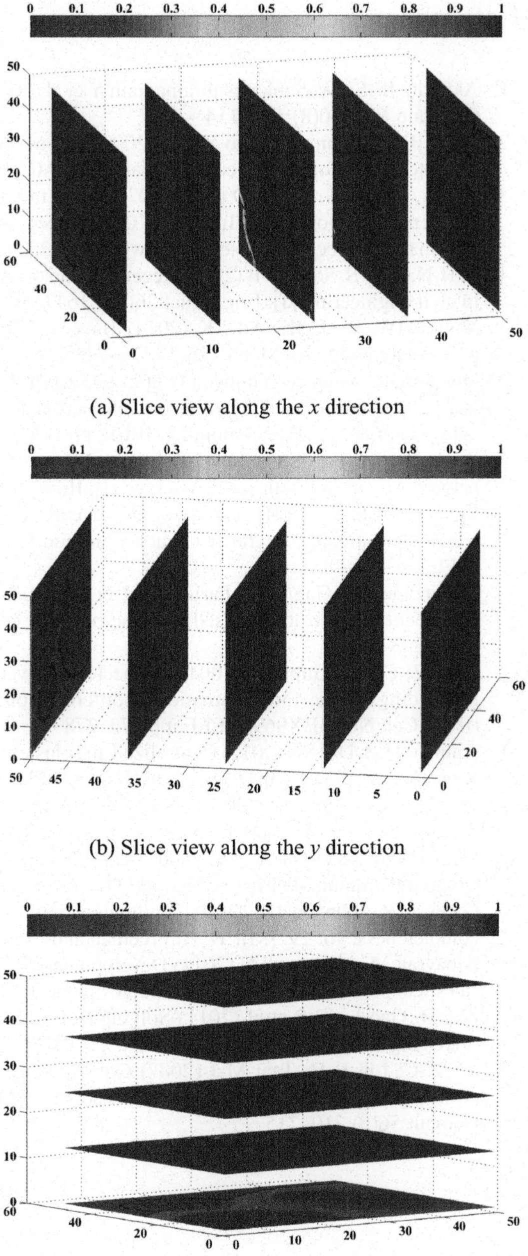

(a) Slice view along the x direction

(b) Slice view along the y direction

(c) Slice view along the z direction

References

1. Akyildiz IF, Su W, Sankarasubramaniam Y et al (2002) A survey on sensor networks [J]. IEEE Commun Mag 40(8):102–114
2. Abidi BR, Aragam NR, Yao Y et al (2009) Survey and analysis of multimodal sensor planning and integration for wide area surveillance [J]. ACM Comput Surv 41(1):7.1–7:36
3. Kim S, Pakzad S, Culler D et al (2007) Health monitoring of civil infrastructures using wireless sensor networks [C]. In: IPSN '07: Proceedings of the 6th international conference on information processing in sensor networks, pp 254–263
4. Li H-N, Li D-S, Song G-B (2004) Recent applications of fiber optic sensors to health monitoring in civil engineering [J]. Eng Struct 26(11):1647–1657
5. Vanik MW, Beck JL, Au SK (2014) Bayesian probabilistic approach to structural health monitoring [J]. J Eng Mech 126(7):738–745
6. Dingxing Z, Ming X, Yingwen C et al (2006) Probabilistic coverage configuration for wireless sensor networks [C]. In: 2006 International conference on wireless communications, networking and mobile computing. IEEE, pp 1–4
7. Wang Y-C, Tseng Y-C (2008) Distributed deployment schemes for mobile wireless sensor networks to ensure multilevel coverage [J]. IEEE Trans Parallel Distrib Syst 19(9):1280–1294
8. Liu B, TowsleyD (2004) A study of the coverage of large-scale sensor networks [C]. In: 2004 IEEE International conference on mobile ad-hoc and sensor systems (IEEE Cat. No.04EX975). IEEE, pp 475–483
9. Akbarzadeh V, Gagne C, Parizeau M et al (2013) Probabilistic sensing model for sensor placement optimization based on line-of-sight coverage [J]. IEEE Trans Instrum Meas 62(2):293–303
10. Zhou Z, Das S, Gupta H (2004) Connected k-coverage problem in sensor networks [C]. In: Proceedings of the 13th international conference on computer communications and networks (IEEE Cat. No.04EX969). IEEE, pp 373–378
11. Ammari HM, Das SK (2012) Centralized and clustered k-coverage protocols for wireless sensor networks [J]. IEEE Trans Comput 61(1):118–133
12. Rapaić MR, Kanović Ž, Jeličić ZD (2008) Discrete particle swarm optimization algorithm for solving optimal sensor deployment problem [J]. J Autom Control 18(1):9–14
13. Wu Q, Rao NSV, Du X et al (2007) On efficient deployment of sensors on planar grid [J]. Comput Commun 30(14–15):2721–2734
14. Garey MR, Johnson DS (2002) Computers and intractability: a guide to the theory of NP-completeness, vol 29. [M]. W. H. Freeman and Company
15. Albayrak M, Allahverdi N (2011) Development a new mutation operator to solve the traveling salesman problem by aid of genetic algorithms [J]. Expert Syst Appl 38(3):1313–1320
16. Zou D, Gao L, Li S et al (2011) Solving 0–1 knapsack problem by a novel global harmony search algorithm [J]. Appl Soft Comput 11(2):1556–1564
17. Ke W-C, Liu B-H, Tsai M-J (2007) Constructing a wireless sensor network to fully cover critical grids by deploying minimum sensors on grid points is NP-complete [J]. IEEE Trans Comput 56(5):710–715

The manufacturer's authorised representative in the EU is Springer Nature Customer Service Centre GmbH, Europaplatz 3, 69115 Heidelberg, Germany. If you have any concerns regarding our products, please contact ProductSafety@springernature.com

Printed and bound by CPI Group (UK) Ltd, Croydon, CR0 4YY
01/10/2026
02225853-0007